"十三五"
国家重点图书

Series on Advanced Electronic Packaging Technology and Key Materials

先进电子封装技术与关键材料丛书

汪正平（C.P. Wong） 刘胜（Sheng Liu） 朱文辉（Wenhui Zhu） 主编

Modeling, Analysis, Design and Tests for Electronics Packaging beyond Moore

超摩尔时代电子封装建模、分析、设计与测试

张恒运（Hengyun Zhang） 车法星（Faxing Che）
林挺宇（Tingyu Lin） 赵文生（Wensheng Zhao） 著

· 北京 ·

内 容 简 介

本书系统介绍了超摩尔时代先进封装理论模型、分析与新的模拟结果。内容涉及 2.5D、3D、晶圆级封装的电性能、热性能、热机械性能、散热问题、可靠性问题、电气串扰等问题，提出了基于多孔介质体积平均理论的建模方法并应用于日渐复杂的先进封装结构，以及模型验证、设计和测试，并从原理到应用对封装热传输进行了很好的介绍。同时，引入并分析了碳纳米管、金刚石镀层、石墨烯等新材料的应用性能。本书针对产品开发阶段封装的热管理和应力管理方面进行了详细阐述，在封装性能测试、蒸汽层散热器、微槽道冷却、热电制冷等方面也有涉及。相应的试验测试和案例分析也便于读者提高对封装性能表征和评估方法的理解。

本书可作为从事微电子封装行业人员的参考资料，也可供高等院校相关专业研究生和高年级本科生学习参考。

图书在版编目（CIP）数据

超摩尔时代电子封装建模、分析、设计与测试＝Modeling, Analysis, Design and Tests for Electronics Packaging beyond Moore：英文/张恒运等著. —北京：化学工业出版社，2020.12

（先进电子封装技术与关键材料丛书/汪正平，刘胜，朱文辉主编）

ISBN 978-7-122-37952-8

Ⅰ.①超… Ⅱ.①张… Ⅲ.①电子技术-封装工艺-英文 Ⅳ.①TN05

中国版本图书馆 CIP 数据核字（2020）第 214407 号

本书由化学工业出版社与 Elsevier 出版公司合作出版。版权由化学工业出版社所有。本版本仅限在中国内地（大陆）销售，不得销往中国台湾地区和中国香港、澳门特别行政区。

责任编辑：吴　刚　李玉峰　　　　　　　封面设计：关　飞
责任校对：边　涛

出版发行：化学工业出版社(北京市东城区青年湖南街 13 号　邮政编码 100011)
印　　装：北京虎彩文化传播有限公司
710mm×1000mm　1/16　印张27¼　字数752千字　2021年3月北京第1版第1次印刷

购书咨询：010-64518888　　　　　　　　售后服务：010-64518899
网　　址：http://www.cip.com.cn
凡购买本书，如有缺损质量问题，本社销售中心负责调换。

定　　价：398.00 元　　　　　　　　　　版权所有　违者必究

Preface to the Series

The technical level and development scale of the integrated circuit (IC) industry is one of the important indicators to measure a country's industrial competitiveness and comprehensive national strength, and is the source of modern economic development. The application of IC has already become routine in various industries, such as military satellites, radar, civilian automotive electronics, smart equipment, and consumer electronics, etc. At present, the IC industry has formed three major industrial chains of design, manufacturing and packaging testing, which have become the indispensable pillar in the IC industry.

IC packaging is an indispensable process in the IC industry, which is the bridge from chip to device and device to system. It is a key fundamental manufacturing part of the IC industry and a competitive commanding height for the core device manufacturing of the IC industry.

With the rapid development of IC technology, higher and higher requirements for miniaturization, multi-function, high reliability and low cost of electronic products are put forward. Facing this situation, the electronic packaging materials and technologies are undergoing rapid development, promoting lots of advanced packaging materials. Advanced electronic packaging materials and technologies are the core of IC packaging.

In order to promote the development of China's advanced electronic packaging industry and meet the urgent needs of researchers ranged from teaching and scientific study to engineering developing in the field of electronic packaging, the editorial committee has invited famous specialists to write the *Series on Advanced Electronic Packaging Technology and Key Materials* in recent years (English version). The series includes: "*Advanced Polyimide Materials*", "*From LED to Solid State Lighting*", "*Freeform Optics for LED Packages and Applications*", "*Modeling, Analysis, Design and Tests for Electronics Packaging beyond Moore*", "*TSV(through-silicon via technology) Package*" etc.

This series of books systematically describes the advanced electronic packaging from three aspects: advanced packaging materials, advanced packaging technologies and advanced packaging simulation design methods. This series covers the most advanced packaging materials such as polyimide materials and packaging technologies such as freeform optical technology, TSV(through-silicon via technology) packaging, and advanced packaging simulation design methods such as multi-physics analysis and applications. In addition, this series also makes a planning outlook and forecast for the development trend of advanced electronic packaging.

This series of books is of great worth for workers engaged in scientific research, production and application in electronic packaging and related industries, and also has great reference significance to teachers and students of related majors in higher education institutions.

We believe that the publication of this series of books will play a positive role in promoting the development of China's IC industry and advanced electronic packaging industry.

Finally, we would like to express our sincere gratitude to our colleagues who have worked hard in the preparation of this series. We also express our heartfelt thanks to those who participated in organizing the publication of this series!

C.P. Wong
IEEE Fellow
Member of Academy of Engineering of the USA
Member of Chinese Academy of Engineering
Former Bell Labs Fellow
Dean of Engineering, The Chinese University of Hong Kong
Regents' Professor, Georgia Institute of Technology, Atlanta, GA 30332, USA

Sheng Liu, Ph.D.
IEEE Fellow, ASME Fellow
Chang Jiang Scholar Professor
Dean, School of Power and Mechanical Engineering,
Executive Director, Institute of Technological Sciences,
Wuhan University,
Wuhan, Hubei, China

Wenhui Zhu, Ph.D.
National Invited Professor
College of Mechanical and Electrical Engineering,
Central South University
Changsha, Hunan, China

Contents

Preface		vii
Acknowledgments		xi
About the authors		xiii

1 Introduction — 1
- 1.1 Evolution of integrated circuit packaging — 1
- 1.2 Performance and design methodology for integrated circuit packaging — 6
- References — 11
- Further reading — 12

2 Electrical modeling and design — 13
- 2.1 Fundamental theory — 13
- 2.2 Modeling, characterization, and design of through-silicon via packages — 28
- References — 53

3 Thermal modeling, analysis, and design — 59
- 3.1 Principles of thermal analysis and design — 59
- 3.2 Package-level thermal analysis and design — 74
- 3.3 Numerical modeling — 89
- 3.4 Package-level thermal enhancement — 97
- 3.5 Air cooling for electronic devices with vapor chamber configurations — 101
- 3.6 Liquid cooling for electronic devices — 106
- 3.7 Thermoelectric cooling — 112
- References — 126

4 Stress and reliability analysis for interconnects — 131
- 4.1 Fundamental of mechanical properties — 131
- 4.2 Reliability test and analysis methods — 149
- 4.3 Case study of design-for-reliability — 227
- References — 239

5	**Reliability and failure analysis of encapsulated packages**	**245**
	5.1 Typical integrated circuit packaging failure modes	245
	5.2 Heat transfer and moisture diffusion in plastics integrated circuit packages	245
	5.3 Thermal- and moisture-induced stress analysis	263
	5.4 Fracture mechanics analysis in integrated circuits package	270
	5.5 Reliability enhancement in PBGA package	282
	References	290
	Further reading	292
6	**Thermal and mechanical tests for packages and materials**	**293**
	6.1 Package thermal tests	293
	6.2 Material mechanical test and characterization	306
	References	322
7	**System-level modeling, analysis, and design**	**325**
	7.1 System-level thermal modeling and design	325
	7.2 Mechanical modeling and design for microcooler system	348
	7.3 Codesign modeling and analysis for advanced packages	370
	References	389

Appendix 1 Nomenclature	393
Appendix 2 Conversion factors	403
Index	405

Preface

In more than 50 years, the miraculous development of semiconductor devices has been witnessed by the increasing number and shrinking gate size of transistors in the integrated circuit (IC) chip and packaging. From 2 transistors in the first-generation IC by Jack Kilby in 1959 to around 5 billion transistors in the latest processor for a mobile phone, the scaling of Moore's law has been relentlessly followed. Because of the physical limitations, the device scaling would stop at the 7−5 nm technology node. Instead, more and more attention is being paid to the heterogeneous and hybrid integration of multiple chips in a package. Dissimilar chips such as central processing units, memory, graphic processing units, RF chips, MEMS sensors, and optical devices are to be integrated into one package, which is often called as more than Moore (MtM) or beyond-Moore technologies. New applications such as mobile communication, cloud computing, big data, Internet of things, energy storage and conversion, automatic driving, and, more recently, artificial intelligence are becoming the driving forces for the innovation of electronic devices and systems.

Many advanced electronic packages are being conceptualized and manufactured to meet the various application scenarios. Among them, the 2.5D and 3D IC packages and their hybrid combinations are viewed as the most promising technologies. Fan-out wafer-level packaging provides a low-cost heterogeneous integration solution for multiple chips.

The combination of these packaging technologies leads to increased design complexity, decreased design margins, and increased possibility of failure in signal integrity, process, thermal performance and reliability, which may hamper the technical progress. Many efforts have been made on the development of 2.5D and 3D packages based on through-silicon via (TSV) technologies to accumulate experience and knowledge in the new areas. With the increasing development in packaging architecture and concepts, it is important to understand the new phenomena and mechanism associated with the packages. The modeling, simulation, and analysis tools with validated accuracy and efficiency are to be developed, though not yet, at the same pace to address the new issues and to predict and guide the packaging design and process. A proper use of the modeling and analysis tool can greatly save design time, minimize the defects, and shorten the time to market.

Nonetheless, fundamental understandings with effective analysis tools are still lacking in the prediction of electrical and thermal performance and thermomechanical reliability. New challenges and issues relate to various aspects: the cross talk between TSVs, the thermal performance with insulated TSVs, thermal management for 3D IC packaging with intervening heating, thermomechanical stress in BEOL layers and TSVs, the copper extrusion in TSVs, issues in the carbon nanotube interconnects, and so on. Process developers are often plagued by the contradictory results during the assembly process or test at end of line. During the assembly of 2.5D package with chip-on-wafer-on-substrate technology, chip-on-chip (CoC) first, or the CoC on substrate (CoC-oS) technology, warpage of package with TSV interposer can be convex or concave, which is difficult to explain. Only after repeated trials without success, the process engineers start to understand and treat the issues seriously by seeking help from the experienced researchers with modeling experience and skills.

Development of highly efficient analysis models and submodels with proved cases is crucial to the successful development of electronic packaging. Undoubtedly, there exist challenges in the multiphysics modeling for electronic packaging which couples electrical, thermal, and mechanical fields in the assembly processes, end-of-line, and reliability tests. The 2.5D/3D IC package may have good electrical performance but is bottlenecked by the poor thermal performance of interstra bonding layers. A comprehensive understanding of the new packages is important to the researchers from both industry and universities.

This book is primarily concerned with the modeling, analysis, design, and test of electronic packaging for heterogeneous integration in the era beyond Moore's law. It mainly focuses on the 2.5D/3D packaging with TSVs and with fine-line RDL interconnects. Fundamental theories and new advances in electrical, thermal, and thermomechanical or thermohygromechanical fields are covered. The electrically and thermally induced issues such as electromagnetic interference and hot spots are related to package signal and thermal integrity, whereas the thermomechanical stress is related to the package structural integrity. Not only the basic principles in electrical, thermal, and mechanical areas will be presented but also the modeling and analysis techniques related to practical technologies such as wafer level packages and 2.5D/3D packages will be analyzed. Some of the new material modeling and testing methodologies are also included to address the challenges related to the new package development.

This book is intended for design engineers, processing engineers, and application engineers as well as for postgraduate students majored in electronics packaging. Exemplified case studies are also demonstrated to satisfy the interests of engineers and researchers specialized in this domain.

All the authors have been engaged in the electronics packaging industry, research institutes, and universities for ten years and above, with some authors for more than 20 years. The authors have been fully in charge of and support of a number of package development projects with in-depth experience and knowledge. More than 80% of the content is based on the authors' personal experience and professions.

There have been several books in English are available to readers in the MtM electronics packaging. However, most of them are limited to the conceptualization of architecture, electrical, and process evaluation, without detailed modeling and verification in the thermal and electrical performances and thermomechanical stress. Their scope is also limited to a certain discipline. This book addresses the 2.5D/3D packages from the three major disciplines with validated models, design, and tests, in hope to stimulate the development of new packaging technologies, predictive models, testing methods, and even a new system.

The content of this book is arranged in seven chapters.

Chapter 1 introduces the background of this book It covers the historical evolution and basic performance metrics of IC packaging, contributed by all the authors and compiled by Dr. Hengyun Zhang. The challenges and solutions are also covered. Chapters 2–5 cover the different aspects of packaging. In Chapter 2, the fundamental of electrical design and modeling is introduced, and the 3D package and integrated passives are being discussed, contributed by Dr. Wensheng Zhao. Chapter 3 addresses the basic heat transfer theories and analysis model from thermal viewpoints, contributed by Dr. Hengyun Zhang. Both in-package and on-package thermal solutions are analyzed, tested, and discussed. Applications of new materials such as carbon nanotube and graphene are also highlighted. Dr. Faxing Che contributes to Chapter 4, in which the stress and reliability analysis of packaging interconnects are discussed. Stress–strain relationship, nonlinear material behavior including viscoplastics, creep, and viscoelastic material models, and characteristic tests are covered. Design for reliability using finite element analysis and simulation has been demonstrated for stacked-die packages with case studies. Chapter 5 presents an analysis on the package failures related to thermal, thermomechanical, and thermohygroswelling aspects, as contributed by Dr. Tingyu Lin. Both Dr. Zhang and Dr. Che contribute to the last two chapters, with Chapter 6 covering the package and material thermal and mechanical test techniques suitable for electronic packaging and Chapter 7 the system-level design and modeling.

The present edition is co-published with Elsevier Inc., which follows the typesetting of Elsevier's edition, including, but not limited to, the fonts, siesz, subscripts, superscripts, normal or italic letters as a courtesy. Metric units have been used throughout this book, though a few common imperial units appear only on very limited occasions. The readers can refer to Appendix 2 for the conversion factors between imperial units and metric units.

Hengyun Zhang, Ph.D.
Eastern Scholar Professor
Shanghai University of Engineering Science,
Shanghai, China

Faxing Che, Ph.D.
Research Scientist
Institute of Microelectronics,
A*STAR, Singapore

Tingyu Lin, Ph.D.
General Manager Assistant
National Center for Advanced Packaging,
Wuxi, China

Wensheng Zhao, Ph.D.
Assoc. Professor
Hangzhou Dianzi University,
Hangzhou, Zhejiang, China

Acknowledgments

It has been a painstaking yet fulfilling journey to document our practitioner experience turning into an officially printed book. Initiation, planning, and preparation of *Modeling, Analysis, Design and Test for Electronics Packaging beyond Moore* were facilitated by the editors at Chemical Industry Press, Elsevier Publisher, coauthors, and colleagues all over us. We would like to thank them all, with special acknowledgments to Ms. Gang Wu and Ms. Yufeng Li from Chemical Industry Press. This book would not have been completed without their professional dedication and inspiration.

The materials in this book have been derived from many publishing organizations, for which we have published some materials previously and reused them in this book. We attempt to acknowledge here for the help and support. They are Institute of Electrical and Electronic Engineers (IEEE), American Society of Mechanical Engineers (ASME), and Elsevier Publisher for the conference proceedings and journal publications including IEEE Transactions on Components and Packaging Technology, IEEE Transactions on Device and Materials Reliability, ASME Transactions on Journal of Electronic Packaging, ASME Transactions on Journal of Heat Transfer, Applied Thermal Engineering, Microelectronics Reliability, Journal of Alloys and Compounds, International Journal of Refrigeration, and Journal of Electronic Materials. Important conferences such as Electronic Components and Technology Conference, Intersociety Conference on Thermal and Thermomechanical Phenomena in Electronic Systems, Electronics Packaging Technology Conference, and International Conference on Electronic Packaging Technology & High Density Packaging are also appreciated for authorizing the reproduction of some of publication materials.

We would also like to acknowledge those who helped review some chapters and sections of this book. They are Professor Yaling He from Xi'an Jiaotong University, Professor Johan Liu from Shanghai University, Professor Xinwei Wang from Iowa State University, Dr. Xiaowu Zhang from Institute of Microelectronics, Singapore, Professor Andrew A.O. Tay from Singapore University of Technology and Design, and Professor John Pang from Nanyang Technological University, Singapore. We would like to thank them for their insightful suggestions and comments benefiting to this book. The help and support from the ex-colleagues including Y.C. Mui, D. Pinjala, M. Rathin, W.H. Zhu, B.S. Xiong, L.H. Xu, D.Y.R. Chong, M.B. Yu, O.K. Navas, G.Y. Tang, B.L. Lau, L. Bu, Daniel M.W. Rhee, and Y. Han are appreciated. We also thank the anonymous reviewers for their constructive review and comments.

The first author would like to thank Shanghai University of Engineering Science for providing the working environment to make this book possible. The first author also would like to express thanks to the postgraduate students for their work, which helped prepare the materials in this book. They are Yang Sui, Le Jiang, Zehua Zhu, Xiaoyu Wu, and Yuchen Deng.

We would like to acknowledge the support of the funding agencies in the past years such as National Natural Science Foundation of China (51876113, 61874038), Institute of Microelectronics from Agency for Science, Technology, and Research, Singapore, and Nanyang Technological University, Singapore.

Hengyun Zhang, Ph.D.
Eastern Scholar Professor
School of Mechanical and Automotive Engineering
Shanghai University of Engineering Science
Shanghai, China

Faxing Che, Ph.D.
Research Scientist
Institute of Microelectronics, A*STAR
Singapore

Tingyu Lin, Ph.D.
General Manager Assistant
National Center for Advanced Packaging
Wuxi, China

Wensheng Zhao, Ph.D.
Assoc. Professor
School of Electronics and Information
Hangzhou Dianzi University
Hangzhou, Zhejiang, China

About the authors

Hengyun Zhang is a professor in Shanghai University of Engineering Science and Eastern Scholar Professor conferred from Shanghai Education Commission. He obtained Bachelor degree and Master degree from University of Science and Technology of China in 1994 and 1997, respectively, and PhD degree from Nanyang Technological University in 2001. After graduation, he has worked in industries and research institutes for 18 years, including Institute of Microelectronics (Singapore), Advanced Micro Devices and Dow Corning. He worked on nearly 20 projects as project and technical leaders. He won the awards of ITHERM Best Poster Paper Award in 2004, AMD Engineering Excellence Award in 2008, and Icept Best Paper Award in 2010. He has more than 80 publications in international journals and conferences and more than 30 patents granted/filed in the United States, Singapore, and China. He also served as committee members for a number of international conferences (IEEE ITHERM, ICEPT, EPTC, ASME MNHMT). He has been reviewers for nearly twenty journal publications. He was one of the founder members in the Thermal Management Association of China Electronics Committee.

Faxing Che is a research scientist with Institute of Microelectronics (IME), Agency for Science, Technology and Research, Singapore. He received BE degree in mechanical engineering and ME degree in solid mechanics from University of Science and Technology, Beijing, China, in 1995 and 1999, respectively, and PhD degree in engineering mechanics from Nanyang Technological University, Singapore, in 2006. He has 13 years' experience in the design-for-reliability for advanced electronics packaging. He worked with the United Test and Assembly Centre, STMicroelectronics, and Infineon Technologies Singapore. He has authored and coauthored more than 120 technical papers in refereed journals and conferences, and the citations given by Google Scholar are close to 2000. Dr. Che serves as a peer reviewer for more than 10 journals. He was the recipients of Best Poster Paper Award from EPTC 2013, Best Poster Paper Award from ITHERM 2006, Best Paper Award from ICEPT 2006, and Outstanding Student Paper Award from EPTC 2003.

Tingyu Lin is General Manager Assistant of National Centre for Advanced Packaging (NCAP), China. He has more than 20 years' experiences in design, process, assembly, reliability, and equipment development in electronics packaging, semiconductor, consumer electronics, and thermal and aerospace industry. He received BE degree in thermal engineering from Tsing Hua University and ME degree in aerospace engineering from the Ministry of Aerospace, China, in 1990 and PhD in microelectronics from National University of Singapore (NUS) in 1997. He was a Motorola (Google)-certified Six Sigma Black Belt in 2012. From 2012–2013, he was a program director of Institute of Microelectronics, Singapore (IME), and responsible for 2.5D, fan-out and WLP development. From 1997–2012, he worked as a senior manager in Philips, Lucent Technologies, the former AT&T Bell Lab, and Motorola Electronics, respectively. Until now, he has been engaged in more than 100 product design and process development in mobile modules, IC components, and consumer electronics, etc. He has published 150 papers and filed more than 10 patents, obtained many best paper awards and gave keynote talks in several conferences, and led two international ASTM standards for TSV development.

Wen sheng Zhao is an associate professor with Hangzhou Dianzi University. He obtained Bachelor degree from Harbin Institute of Technology in 2008 and PhD degree from Zhejiang University in 2013. He was a visiting PhD student at National University of Singapore from 2010 to 2013 and a visiting scholar at Georgia Institute of Technology from 2017 to 2018. His current research interests include interconnects design and simulation, carbon nanoelectronics, and multiphysics simulation. He has published over 70 papers in international journals and conferences (more than 20 IEEE papers). He served as a peer reviewer for more than 10 international journals. He was the receipt of Best Student Paper Award from IEEE EDAPS 2015.

Introduction

1.1 Evolution of integrated circuit packaging

1.1.1 Evolution of integrated circuit devices and applications

Integrated circuit, shortened as IC, by its name, is the combination of more than one electric device such as transistors and capacitors onto a semiconductor chip made of silicon or other materials to fulfill certain functions. Electronic packaging is referred to as the electrical and mechanical connections of IC chip to a system board to form a standalone electronic product or can be interconnected with other packages to form an electronic product. A system may consist of either standalone or interconnected electronic products with the needed functions. The electronic product can be analogized with a human body. With the brain as an IC chip, the skull acts as mechanical protection part of the packaging, and the blood circulation and nervous system act as the power/ground connections and signal transmission lines. Thermal cooling of the human brain is implemented through the natural convection from the head as well as the heat exchange through the dual function of the blood circulation to the body. All the sensory organs work as sensors on the package to communicate with the outer world.

The development of IC devices has been witnessed by the increasing number and decreasing gate size of transistors in the IC chip and packaging. The first transistor was invented in 1947, but the first IC device was developed only after 12 years. In 1959, Jack Kilby from Texas Instruments integrated two transistors and a resistor in silicon with a large spacing of 7/16 inch, which was considered as the birth of modern IC. Ever since, the IC has been evolving in the scaling of transistor gate size as predicted by Moore's law. Today, it has come to a stage that there are more than 5 billion transistors in the latest processor of a mobile phone and even more for a computer processor. In 2016, AlphaGo became the first computer Go program to beat a human professional Go player without handicaps on a full-sized 19×19 board. Behind this sensational human−machine competition were 50 tensor processing units specifically developed for machine learning.

The downscaling to the nanoscale electronics brings fundamental and application issues such as electrical crosstalk, thermal, and mechanical stress management. Fundamental transport processes in the nanoscale devices are playing a role different from that in the macroscopic scale. New designs, processes, and materials are being pushed to the limits to meet the challenges in the packaging from the back end of line of wafer foundry to the packaging and assembly house.

Because of the physical limitations, the device scaling could stop at the 7−5 nm technology node. On the other hand, more and more attentions are being paid to the heterogeneous and hybrid integration of multiple chips and devices in one package. Dissimilar chips such as central processing units, memory, graphic processing units (GPUs), RF chips, microelectromechanical systems sensors, and optical devices are to be integrated into one package to form a higher level of system-in-package assembly with enhanced functionality and improved operation characteristics. This is often called as more than Moore (MtM) or beyond-Moore technologies.

Various applications such as mobile communication, cloud computing, big data, Internet of things, energy storage and conversion, automatic driving, and, more recently, artificial intelligence are becoming the driving forces for the innovation of electronic devices and systems. Many of the applications are highly related to the high-performance electronic packaging and systems with heterogeneous integration instead of the More Moore miniaturization. Such technology crossroad is illustrated in Figure 1.1.1. The vertical axis in Figure 1.1.1 indicates the trend of miniaturization with the Moore's law, whereas the horizontal axis indicates the trend of MtM technology [1]. Typical applications are listed in Figure 1.1.2 [2].

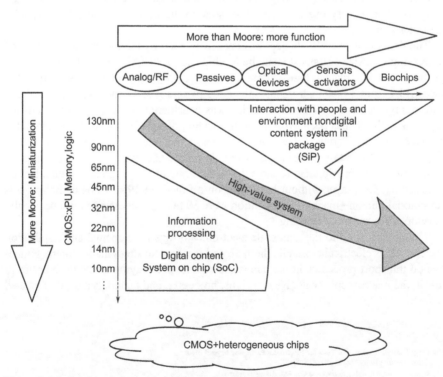

Figure 1.1.1 Semiconductor technology road map on the heterogeneous integration.

Introduction

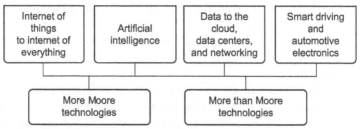

Figure 1.1.2 Applications to drive the electronic products.

1.1.2 Evolution of integrated circuit packaging

The device scaling also affects the interconnects at the packaging and system level. To connect the transistors with downsized scale to the outer world, more intermediate layers are required to facilitate the connections, more I/Os are required to distribute the power and signal lines, and more efficient yet complicated cooling technologies are required to reject the excessive heat from the IC and interconnects as well. The evolution of the electronic packages is illustrated in Figure 1.1.3 [2].

As shown in Figure 1.1.3, semiconductor packages are experiencing five major patterns during last five decades, which are lead-frame, organic substrate with solder balls (BGA), fan-in wafer-level chip-scale package, fan-out wafer-level packaging (FOWLP), and through-silicon via (TSV) technology for 2.5D/3D IC stacking, respectively.

DIP (dual in-line package) lead-frame packages dominated in 1950–1970, followed by QFP, PLCC, and SOIC during 1980. TSSOP and TQFP were much finer pitched lead-frame packages in 1990. At the same time, BGA packages appeared and gradually

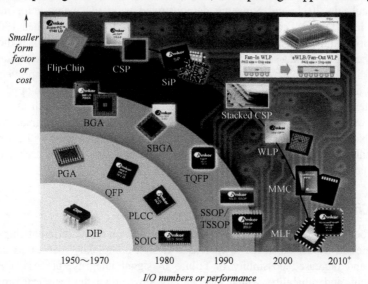

Figure 1.1.3 Semiconductor packaging trend.

dominated in the market. After 2000, there were many packages showing off in the form of flip-chip BGA, flip-chip CSP, etc. Wafer-level packages (WLPs) were major package format due to low cost. Stacked CSP, fan-in WLPs, and fan-out WLPs are gradually burgeoning in the market. Especially fan-out WLP is becoming a mature technology after 2010.

In the trend of shifting from the conventional planar 2D chip packaging to the 3D packaging, advanced electronic packaging formats are to be conceptualized, designed, and manufactured to meet the various application scenarios. Among them, the 2.5D and 3D IC packaging and their hybrid combination are viewed as the most promising technologies. FOWLP provides a low-cost alternative for heterogeneous integration solution ahead of the 3D chip stacking. The initial package-on-package came after 2000 and gradually developed to a 3D package stack in different forms such as wire bonding and mix of flip-chip.

Many efforts have been made on the development of 2.5D and 3D packages based on TSV technologies with accumulated experience and knowledge in the heterogeneous areas [3–5]. The general technology trend toward the 3D IC packaging has been shown in Fig. 1.1.3. A typical 3D IC package relies on the TSV technology as the enabling interconnects, driven by the memory stack, wide I/O, logic + memory, and field-programmable gate array (FPGA) package.

The first commercial product in 2.5D package as shown in Figure 1.1.4a was released by Xilinx in 2010, with FPGA chips on a passive TSV interposer (TSI) after several years of internal development. Many other companies also announced the TSV technology for product execution. For example, TSMC revealed 3D IC design based on silicon interposer in June 2010. Elpida, PTI, and UMC announced the stacks of logic and DRAM with 28 nm technology in 2011, Micron developed the hyper memory cube in 2011, and Samsung demonstrated wide I/O memory for mobile products. In 2015, AMD announced their GPU card, which has more than 190,000 microbumps and the TSV interposer has more than 25,000 C4 bumps. This is the first GPU on the interposer at a size of 1011 mm^2, with the most dies in a single package as illustrated in Figure 1.1.4b, which also took several generations of improvements before its announcement. The 3D-stacked packages are also being developed, with much faster processing speed and reduced total power dissipation, such as for the wide-bandwidth memory. The stacking of logic dies together is also of great interest recently. Nonetheless, most of the work is limited to the research level without being commercialized due to the technical difficulties in areas such as processability and thermal management.

1.1.3 Challenges and solutions

As it has been demonstrated, the heterogeneous integration of the different chips and technologies in a package leads to increased design complexity, decreased design margins, and increased possibility of failure in signal integrity, process, thermal management, and reliability, which may greatly slow down the technical progress and design time. With the increasing demand for packaging architecture and concepts in more applications, it is important to foster the design methodology, modeling and

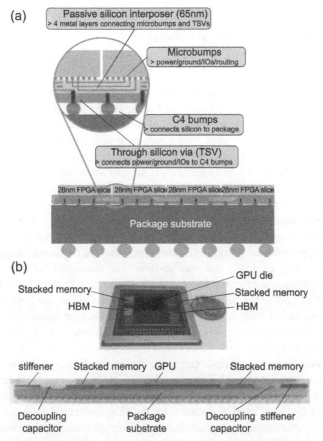

Figure 1.1.4 (a) Xilinx first 2.5D through-silicon via (TSV) digital field-programmable gate array (FPGA) and (b) AMD Radeon graphic card with HBM surrounding the graphic processing unit (GPU) in a single 2.5D package.

analysis tools, manufacturing, and reliability test capacity associated with the 2.5D/3D IC package development.

The corresponding modeling, simulation, and analysis methodologies describing the process, performance, and reliability are urged to develop, though difficult, at the same pace to address the new issues and to predict and guide the design and process. The modeling and analysis methodologies should be physically robust and validated with engineering accuracy and efficiency. A proper use of the modeling and simulation tool can greatly save design time, minimize the defects, and shorten the time to market. Emphasis should be put on the electrical and thermal modeling of interconnects and TSVs, the mechanical stress, the coupled multiphysics interactions during process, and the reliability or field tests. The effects of microscale and nanoscale on the devices, dielectrics, and interfacial materials should also be considered.

1.2 Performance and design methodology for integrated circuit packaging

1.2.1 Electrical performance and design methodology

With the rapid development and downscaling of semiconductor technology, the number of failures caused by the signal integrity problems in the integrated system is on the rise. Signal integrity is referred to as a broad set of IC design issues related to interconnects, such as reflection, crosstalk noise, and electromagnetic interference [6]. The goal of signal integrity analysis is to ensure high-speed data transmission to guarantee reliable system operation. Power integrity problem is another set of problems along the power and ground paths. The objective of the power integrity analysis is to ensure the desired voltage or current from the source to the destination. Obviously, worse signal and power integrity would result in performance degradation and even cause systems to fail.

The impedance, which is defined as the ratio of the voltage to the current, is the key parameter in the electrical design. It is commonly used in the signal and power integrity evaluation. For example, the way to minimize the signal reflection due to the impedance mismatch is essential to keep the impedance at the same level throughout the net. With respect to the power distribution network (PDN), the impedance is used as the criterion. To keep the voltage ripple lower than the specification, it is essential to reduce the PDN impedance below the target impedance at all the frequencies supported by the system.

In a 3D IC package, multiple chips are stacked vertically with the signal transmitted through TSVs. Although TSVs provide shorter interconnect length and thereby reduce power consumption and area, they have to be electrically isolated from the bulk silicon. The capacitive loading induced by the thin insulator layer and lossy effect would cause a considerable degradation in the signal quality. Moreover, the noise coupling between TSVs and between TSVs and active devices becomes more critical with the increasing TSV density, as shown in Figure 1.2.1. The power delivery in 3D integrated system is also a major concern as TSVs produce additional inductance effects,

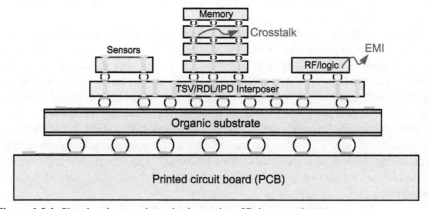

Figure 1.2.1 Signal and power integrity issues in a 3D integrated system.

leading to high peaks of PDN impedance. With the increase in stacked dies, it is predicted that the power density per square area of the 3D IC increases by a factor of $N^{0.5}$, where N is the number of stacked dies, indicating that more power delivery is needed with the increasing number of dies [7].

To evaluate the signal and power integrity, circuit modeling can be carried out to provide an in-depth physical understanding. In this book, the equivalent circuit model of the TSVs is developed with the consideration of the MOS capacitance effect. The influence of the floating silicon substrate on the TSV capacitance is also considered. To guarantee signal integrity, the schemes of coaxial structure and differential signaling are investigated. The PDN impedance is also studied based on the equivalent circuit model, which could provide some useful guidance for the design of 3D IC packaging.

1.2.2 Thermal performance and design methodology

During electronic operation, heat is generated in the chip and along the interconnects due to ohmic heating and power dissipation due to high-frequency switching. In contemporary portable electronic products, the electrochemical heat generation from the bulky batteries may not be ignored and should be considered in the system design. The objective of thermal design is to ensure that the temperatures of the components in a system are maintained within their functional temperature range. Within the temperature range, a component is expected to meet its specified performance without excessive performance degradation and or even thermal runway [8], which is likely to damage the system.

In the design of electronic packaging, the thermal design power (TDP) is used as the target of thermal design. This is the maximum power that the chip and package can dissipate during operation, which defines the upper limit of power dissipation. Given the TDP, the junction temperature should be maintained within a certain temperature level with respect to the prescribed ambient condition. In a processor chip, the projected power dissipation is around 130W, leveling off from the transistor scaling as shown in Figure 1.2.2.

In the absence of appropriate thermal design, logic errors or even thermal runaway would occur, which may cause permanent failure in chip and package. One example of the damaged package due to thermal runaway [8] is indicated in Figure 1.2.3. The target of thermal design is to ensure that the components operate within the prescribed temperature limit based on the ambient condition at the given TDP. In other words, the thermal resistance should be kept within the specified thermal resistance, which is defined by the difference of the junction temperature to reference temperature over power dissipation. The following two thermal resistances are commonly used in package thermal characterization and are therefore introduced to describe the thermal performance. The junction to ambient thermal resistance R_{ja} indicates the heat dissipation capacity from the chip to the surrounding environment, whereas the junction to case thermal resistance R_{jc} reflects the package thermal performance. The lower the thermal resistance is, the more easily the heat can be dissipated and the longer life the package can survive.

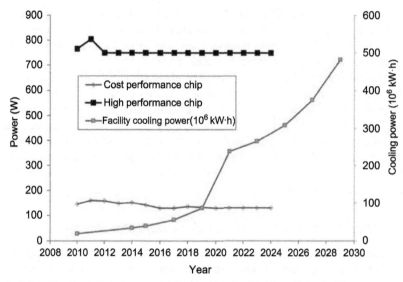

Figure 1.2.2 The thermal design power (W) for the single chip and the facility cooling power for the data center racks.

Figure 1.2.3 Thermal runaway of a package under stressed test.

The thermal design can be implemented through approaches such as thermally enhanced package materials, structural improvement, and heat sink design and optimization. The position of thermal packaging engineer is thus created to cover the thermal design, modeling and analysis, materials characterization, and qualification along with the electrical engineer and materials and process engineer.

With the device scaling and heterogeneous integration, a 2.5D/3D package integrated with multiple chips is being developed as the next generation of package. In a 2.5D package, dissimilar chips can be assembled on a common interposer to fulfill the operations through the microlines and TSV interconnects, whereas a 3D package requires the multiple chips being vertically stacked, with TSVs and microbumps as interconnects. TSVs serve as a common interconnect technology in 2.5D and 3D packages for shortened signal transmission, less delay, and more functionality. The shift to the heterogeneous integration leads to new thermal management issues such as thermal crosstalk and thermal blocking along the thermal path.

Proper evaluation and characterization of the thermal transport properties of 3D chips and the interconnects are the first step to the successful thermal design of the package. On the other hand, the determination of thermal performance is complicated with various design factors including various package elements, geometrical structure, and material types, which may not be solved by a single analysis approach.

In this book, different levels of thermal analysis and design methods are to be addressed. Analytical technique based on porous media theory is wisely sought to resolve the thermal package issues with various repeated or random structures and material compositions. On the other hand, numerical tools are commonly utilized to resolve the thermal package integrated in the system with different levels of package elements and boundary conditions. Nonetheless, package submodels play an important role and should be developed in the numerical model to simplify the computational efforts and time.

Last but not the least, the thermal design and analysis are to be compared and verified with the experimental tests, which constitutes a solid step in utilizing the numerical analysis as optimization tool for implementing the package design.

1.2.3 Stress and reliability issues

Electronic products used both in industries and by consumers are expected to function reliably in service and user environment. Electronic packaging is a technology for fabricating and assembling IC chips into electronic products. Electronic packaging provides not only electrical interconnections but also the mechanical support for protecting the electronic circuits [9]. It is important to ensure the electronic products designed with functions and performances during the specific period of user environment, especially for security and safety demanding applications such as aerospace, automobiles, medical equipment, and military electronics equipment. The reliability of a packaged microelectronic system is related to the probability that will be operational within acceptable limits for a given period of time [10].

Reliability of semiconductor devices and products may depend on assembly process, field application, and environmental conditions. Stress factors affecting device/product reliability include voltage, current density, temperature, humidity, mechanical or thermomechanical stress, vibration, shock and impact, gas, dust, salt, contamination, radiation, pressure, and intensity of magnetic and electrical fields [11]. When the applied stress is higher than strength of material itself, material will degrade at a faster pace, exhibiting ductile or brittle failures. Even when the applied stress is less than the strength of material, the material may exhibit fatigue failure when subjected to cyclic loads such as temperature cycling, power cycling, and cyclic bending. In reality, material strength is not a constant and may degrade with time under high temperature or high humidity condition, which usually results in interfacial delamination or cracking failure.

The methodology of design-for-reliability (DFR) contributes to high quality and reliable electronic assemblies, including the 2.5D and 3D packages. Electronic assemblies are tested at higher stress conditions to accelerate failure for quick understanding of product reliability [12]. The accelerated life test methods shorten the cycle time for reliability design. Reliability tests such as thermal cycling (TC) and thermal shock (TS), cyclic bending, power cycling, high-temperature storage, vibration tests, drop impact, highly accelerated temperature and humidity stress test, unbiased autoclave, and combined testing can provide useful failure assessment methods for reliability of electronic assembly and product. In situ measurement is essential to determine failure time, which is usually achieved through daisy chain or other structure design to measure resistance or capacitance. Sample size should follow the requirement of the test standards. Reliability data can be analyzed by reliability statistical distribution model, such as normal distribution, exponential distribution, and the commonly used Weibull distribution model.

In reliability study, failure analysis (FA) is important to find failure locations and reveal failure modes for understanding failure mechanism. The FA of microelectronic products involves the use of various FA tools and techniques. Scanning acoustic microscope (SAM) is one nondestructive FA tool in detecting cracks, voids, and delaminations within microelectronic packages and assemblies. To localize the failure location, the cross-sectioning technique combined with the focused ion beam etching is used for sample preparation, and then optical microscope or SAM, even transmission electron microscope, is used for taking failure images to determine failure location and failure mode accompanied with a high cost. Dye penetration method is a widely applied and low-cost method to locate fatigue cracks in microelectronic assembly, such as solder joint cracking under repeat impact drop test or TC test.

In addition to failure location identification, stress analysis is essential to better understand failure mechanism. Finite element analysis (FEA) provides a powerful tool in numerical modeling and simulation of stress in electronics packages. FEA modeling and simulation can help reproduce the failure process, reduce the assessment time, and provide a comprehensive failure assessment.

For the advanced packages, the solder joint is often the weakest link, and solder joint reliability becomes even more important with further minimization of electronic assemblies. The solder joint is particularly prone to fatigue failure due to power and TC loading [13]. TC and vibration tests are commonly used for characterization of solder joint failure assessment. TC loads (high-strain, low-cycle fatigue) induce viscoplastic deformation in the solder joints. Vibration loads (low-stress, high-cycle fatigue) primarily induce elastic or elastic–plastic deformation in the solder joints. Drop impact test is used to investigate impact reliability of electronic assemblies for portable electronic products.

FEA simulation provides important understanding on the microdeformation responses in solder joint subjected to reliability test conditions. The combination of FEA simulation and reliability tests provides an effective tool for DFR application in existing and new electronic packaging designs. It should be kept in mind that a multidisciplinary approach is required for electronic assembly failure assessments. Figure 1.2.4 shows the detailed analysis and design methodology in DFR microelectronic assembly.

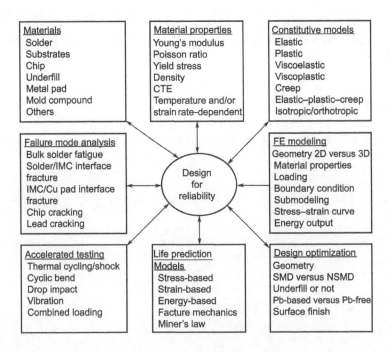

Figure 1.2.4 The multidisciplinary approach in design-for-reliability methodology.

References

[1] Heterogeneous integration, ITRS 2.0, www.itrs.net (accessed Oct. 1st 2017).
[2] The Development Trend and Challenges of Communication Products, Cai Xi, Huawei Technologies Co., Ltd., 2014.
[3] J. Lau, TSV interposer: the most cost-effective integrator for 3D IC integration, in: ASME Interpack, 2011, 52189.
[4] L.L. Yin, C.K.K. Keng, G. Tan, 2.5D/3D device package level defect localization with the use of multiple curve tracings and repeated thermal emission analyses, in: Proc. EPTC, 2015, pp. 287–290.
[5] C.C. Lee, C.P. Hung, et al., An overview of the development of a GPU with integrated HBM on silicon interposer, in: Proc. ECTC, 2016, pp. 1439–1444.
[6] E. Bogatin, Signal and Power Integrity —Simplified, Pearson Education, New York, 2010.
[7] N.H. Khan, S.M. Alam, S. Hassoun, Power delivery design for 3-D ICs using different through-silicon via (TSV) technologies, IEEE Trans. Very Large Scale Integr. Syst. 19 (4) (2011) 647–658.
[8] A. Vassighi, O. Semenov, M. Sachdev, CMOS IC technology scaling and its impact on burn-in, IEEE Trans. Device Mater. Reliab. 4 (2) (2004) 208–220.

[9] M. Pecht, Handbook of Electronic Package Design, Marcel Dekker, Inc., New York, 1991.
[10] R.R. Tummala, Fundamentals of Microsystems Packaging, McGraw-Hill, New York, 2001.
[11] G. Di Giacomo, Reliability of Electronic Packages and Semiconductor Devices, McGraw-Hill, New York, 1996.
[12] H.A. Chan, P.J. Englert, Accelerated Stress Testing Handbook-Guide for Achieving Quality Products, IEEE Press, 2001.
[13] J.H. Lau, Thermal Stress and Strain in Microelectronics Packaging, Van Nostrand Reinhold, New York, 1993.

Further reading

W. Chen, HIR Roadmap Workshop Presentation, 2017. http://cpmt.ieee.org/technology/heterogeneous-integration-roadmap.html.

Electrical modeling and design

2.1 Fundamental theory

2.1.1 Electrical analysis for advanced packaging

Electronic packages provide semiconductor integrated circuit chips, or other electrical devices, with signal and power distribution, capability to remove heat, and protection from mechanical damage and electrostatic discharge [1]. The electrical functions of a package include signal distribution, power allocation, and electromagnetic interference (EMI) protection. In general, the signal passes from the driver circuit to receiver through bonding structures, transmission lines (TLs), and vias, as shown in Fig. 2.1.1. Based on these signal paths, chips can exchange data with each other or with the outside world.

To generate signal, the electrical power should be supplied to the active devices through the package. However, the parasitic inductance of the bonding structures and power distribution network (PDN) has negative impact on the power supply, and therefore, it should be minimized [2,3]. To satisfy the target impedance over a wide frequency range, the decoupling capacitors are usually employed as charge storage components to reduce simultaneous switching noise (SSN). Moreover, as the operating frequency increases, a significant radiated emission would be induced by the high-frequency noise of the clock signal. So special attention should be paid to the EMI-free design of packaging [4].

2.1.2 Signal distribution

Interconnects serve as a bridge for transmitting signals and/or power between the driver and receiver circuits, as shown in Fig. 2.1.2. The circuit generating signal is called driver, whereas the circuit receiving the signal is receiver. The basic element of the circuit is the complementary metal−oxide−semiconductor (CMOS) inverter, which is composed of n-channel and p-channel metal−oxide−semiconductor field-effect transistors (MOSFETs), as shown in Fig. 2.1.3a. The MOSFET is a three-terminal device consisting of a source, drain, and gate. It can be viewed as a switch with an infinite off-resistance and a finite on-resistance R_{on}. When the input voltage is low and equal to 0, the n-channel metal−oxide−semiconductor (MOS) (NMOS) transistor is off, while the p-channel MOS (PMOS) transistor is on. As shown in Fig. 2.1.3b, the output port is connected to the power supply voltage V_{dd}. On the contrary, when the input voltage is high, the NMOS transistor is open while the PMOS transistor is closed, resulting in a steady-state value of zero output voltage. The cross section of a CMOS inverter and its voltage transfer characteristic are illustrated in

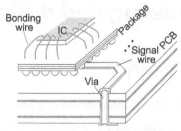

Figure 2.1.1 Electronic packaging.

Figure 2.1.2 Schematic of an interconnect model.

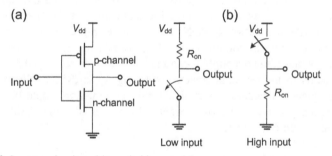

Figure 2.1.3 Inverter circuit and its switching models.

Fig. 2.1.4a and 2.1.4b, respectively. It is evident that the input voltage determines the output one, i.e., whether the interconnect charges to V_{dd} or discharges to the ground node.

Traditionally, the interconnect was modeled as a parasitic capacitance C_{int}, as shown in Fig. 2.1.5. C_L, which is called load capacitance, denotes the input of the receiver circuit. As the switch is closed, the current flows into the interconnect capacitance C and load capacitance C_L through the on-resistance R_{on}. Based on the RC charging circuit shown in Fig. 2.1.5, a time constant is defined as the time required to charge the capacitor by ~63.2% of its initial value and final value, i.e.,

$$\tau = R_{on}(C_{int} + C_L) \tag{2.1.1}$$

The voltage of the load capacitance V_L can be written as

$$V_L(t) = V_{dd}\left(1 - e^{-t/\tau}\right) \tag{2.1.2}$$

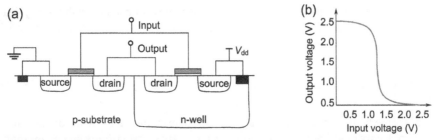

Figure 2.1.4 (a) Complementary metal–oxide–semiconductor (CMOS) inverter and (b) its voltage transfer characteristic.

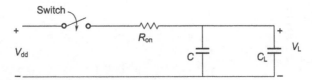

Figure 2.1.5 Charging circuit of an interconnect.

In general, the time period for the capacitor to fully charge is about 5 time constants, i.e., 5τ, as shown in Fig. 2.1.6. In circuit, a signal generating by the driver needs time to reach the receiver. So the time delay is usually defined as the time required for the output to reach 50% of its steady-state value. Based on the charging circuit shown in Fig. 2.1.6, the delay equation can be given as

$$\tau_{50\%} = 0.69 R_{on}(C + C_L) \tag{2.1.3}$$

It is worth noting that the above analysis is based on the lumped model, and the delay equation is derived by using the Kirchhoff's voltage law. However, as higher system bandwidth is desired, it is essential to increase the operating frequency, and the Kirchhoff's laws do not always work at higher frequencies. This is because that

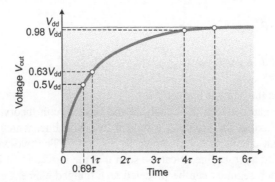

Figure 2.1.6 Transient voltage response.

with the increasing frequency, the wavelength becomes comparable with the physical dimension of the circuit. The wavelength is the ratio of the speed of light in the material to the frequency, i.e.,

$$\lambda = \frac{c}{f\sqrt{\mu_r \varepsilon_r}} \tag{2.1.4}$$

where f is the frequency, $c = 1/\sqrt{\varepsilon_0 \mu_0} = 3 \times 10^8 \mathrm{m/s}$, μ_0 and ε_0 are the permeability and permittivity in vacuum, and μ_r and ε_r are the relative permeability and permittivity of the surrounding material. Under such circumstance, the electrical behavior of the interconnect should be described by the Maxwell's equations, which are given as

$$\nabla \cdot \vec{E} = \frac{\rho}{\varepsilon} \tag{2.1.5}$$

$$\nabla \cdot \vec{B} = 0 \tag{2.1.6}$$

$$\nabla \times \vec{E} + \frac{\partial \vec{B}}{\partial t} = 0 \tag{2.1.7}$$

$$\nabla \times \vec{B} - \mu \left(\vec{J} + \varepsilon \frac{\partial \vec{E}}{\partial t} \right) = 0 \tag{2.1.8}$$

where $\varepsilon = \varepsilon_0 \varepsilon_r$, $\mu = \mu_0 \mu_r$, $\vec{E}$ and $\vec{B}$ are the electric field and magnetic flux density, respectively, ρ is the electric charge density, and $\vec{J}$ is the electric current density. Assuming an $e^{j\omega t}$ time dependent, the Maxwell's equations in the frequency domain can be written as

$$\nabla \cdot \vec{E} = \frac{\rho}{\varepsilon} \tag{2.1.9}$$

$$\nabla \cdot \vec{B} = 0 \tag{2.1.10}$$

$$\nabla \times \vec{E} + j\omega \vec{B} = 0 \tag{2.1.11}$$

$$\nabla \times \vec{B} - \mu (\vec{J} + j\omega \varepsilon \vec{E}) = 0 \tag{2.1.12}$$

where $\omega = 2\pi f$ is the angular frequency.

To bridge the gap between field analysis and basic circuit theory, TL theory was introduced [5]. Taking, for example, a typical two-wire line, which is the simplest TL, the equivalent circuit model is shown in Fig. 2.1.7. The two-wire line consists of a signal line and a return line. It can be seen that the interconnect is meshed along its length, and each segment can be modeled as a lumped-element circuit. Theoretically, the length of the segment is much smaller than one 20th part of the wavelength λ,

Electrical modeling and design

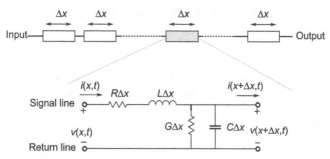

Figure 2.1.7 Transmission line model.

i.e., $\Delta x < \lambda/20$. In the figure, R, L, C, and G are the series resistance, series inductance, shunt capacitance, and shunt conductance per unit length, respectively.

Based on the circuit of Fig. 2.1.7, the following equations can be derived by using the Kirchhoff's laws:

$$v(x,t) - R\Delta x i(x,t) - L\Delta x \frac{\partial i(x,t)}{\partial t} - v(x+\Delta x, t) = 0 \tag{2.1.13}$$

$$i(x,t) - G\Delta x v(x+\Delta x,t) - C\Delta x \frac{\partial v(x+\Delta x,t)}{\partial t} - i(x+\Delta x, t) = 0 \tag{2.1.14}$$

Taking the limit as $\Delta x \to 0$, the above equations can be written as the telegrapher equations:

$$\frac{\partial v(x,t)}{\partial x} = -Ri(x,t) - L\frac{\partial i(x,t)}{\partial t} \tag{2.1.15}$$

$$\frac{\partial i(x,t)}{\partial x} = -Gv(x,t) - C\frac{\partial v(x,t)}{\partial t} \tag{2.1.16}$$

Similarly, assuming an $e^{j\omega t}$ time dependent, the telegrapher equations can be written as

$$\frac{dV(x)}{dx} = -(R+j\omega L)I(x) \tag{2.1.17}$$

$$\frac{dI(x)}{dx} = -(G+j\omega C)V(x) \tag{2.1.18}$$

After some mathematical treatments, the telegrapher equations can be combined as

$$\frac{d^2V}{dx^2} = (R+j\omega L)(G+j\omega C)V^2 \tag{2.1.19}$$

By solving (2.1.19), the signal voltage can be derived as

$$V = V_0^+ e^{-\gamma x} + V_0^- e^{\gamma x} \tag{2.1.20}$$

where V_0^+ and V_0^- are the magnitudes of the harmonic waves, and they are determined by the boundary conditions. γ is the propagation constant, and it is given by

$$\gamma = \sqrt{(R+j\omega L)(G+j\omega C)} \tag{2.1.21}$$

Similarly, the current can be determined as

$$I = I_0^+ e^{-\gamma x} + I_0^- e^{\gamma x} \tag{2.1.22}$$

The propagation constant γ can be written as $\alpha + j\beta$, and α and β are the attenuation constant and phase constant, respectively. The ratio of the amplitudes of the voltage and current is defined as the characteristic impedance Z_0, i.e.,

$$Z_0 = \frac{V_0^+}{I_0^+} = \sqrt{\frac{R+j\omega L}{G+j\omega C}} \tag{2.1.23}$$

For an ideal (or called lossless) TL, the resistance and conductance are set as zero, and the telegrapher equations become

$$\frac{dV(x)}{dx} = -j\omega L I(x) \tag{2.1.24}$$

$$\frac{dI(x)}{dx} = -j\omega C V(x) \tag{2.1.25}$$

The propagation constant and the characteristic impedance of the lossless TL are given by

$$\gamma = j\beta = j\omega\sqrt{LC} \tag{2.1.26}$$

$$Z_0 = \sqrt{\frac{L}{C}} \tag{2.1.27}$$

while the propagation velocity can be written as $v_p = \omega/\beta = 1/\sqrt{LC}$.

Here, some typical package interconnects are listed in the following. As shown in Fig. 2.1.8a, the effective permittivity and the characteristic impedance of the microstrip line with $w < H$ are given by Ref. [6].

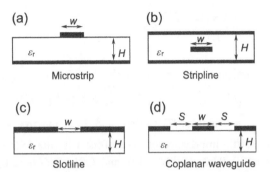

Figure 2.1.8 Typical package interconnects: (a) microstrip; (b) stripline; (c) slotline; and (d) coplanar waveguide.

$$\varepsilon_{\text{eff}} = \varepsilon_0 \left\{ \frac{\varepsilon_r + 1}{2} + \frac{\varepsilon_r - 1}{2} \left[\frac{1}{\sqrt{1 + 12H/w}} + 0.04 \left(1 - \frac{w}{H} \right)^2 \right] \right\} \quad (2.1.28)$$

$$Z_0 = \frac{1}{2\pi} \sqrt{\frac{\mu}{\varepsilon_{\text{eff}}}} \ln \left(\frac{8H}{w} + \frac{w}{4H} \right) \quad (2.1.29)$$

while for microstrip line with $w > H$ are given by

$$\varepsilon_{\text{eff}} = \varepsilon_0 \left(\frac{\varepsilon_r + 1}{2} + \frac{\varepsilon_r - 1}{2} \frac{1}{\sqrt{1 + 12H/w}} \right) \quad (2.1.30)$$

$$Z_0 = \sqrt{\frac{\mu}{\varepsilon_{\text{eff}}}} \left[\frac{w}{H} + 1.393 + \frac{2}{3} \ln \left(1.444 + \frac{w}{H} \right) \right]^{-1} \quad (2.1.31)$$

For stripline in Fig. 2.1.8b, the characteristic impedance is given by Ref. [4].

$$Z_0 = \frac{30\pi}{\sqrt{\varepsilon_r}} \frac{H}{w_{\text{eff}} + 0.441H} \quad (2.1.32)$$

with

$$w_{\text{eff}} = \begin{cases} w, & w > 0.35H \\ w - \left(0.35 - \frac{w}{H} \right)^2 H, & w < 0.35H \end{cases} \quad (2.1.33)$$

For slotlines shown in Fig. 2.1.8c, it should be noted that high permittivity substrate should be employed to minimize the radiation. The effective permittivity of the slotline is given by Ref. [7].

$$\varepsilon_{\text{eff}} = \varepsilon_0 \left(1 + \frac{\varepsilon_r - 1}{2} \frac{K(k_a)}{K(k'_a)} \frac{K(k'_b)}{K(k_b)}\right) \qquad (2.1.34)$$

where $K(k)$ is the complete elliptic integral of the first kind, $k_a = \sqrt{2a/(1+a)}$, $k_b = \sqrt{2b/(1+b)}$, $k'_a = \sqrt{1-k_a^2}$, $k'_b = \sqrt{1-k_b^2}$, $a = \tanh\{(\pi/2)\cdot(w/H)/[1+0.0133/(\varepsilon_r+2)\cdot(\lambda_0/H)^2]\}$, $b = \tanh[(\pi/2)\cdot(w/H)]$, and λ_0 is the free space wavelength. Further, the characteristic impedance of the slotline with the following range of parameters ($9.7 \leqslant \varepsilon_r \leqslant 20$, $0.02 \leqslant w/H \leqslant 1.0$, and $0.01 \leqslant H/\lambda_0 \leqslant 0.25/\sqrt{\varepsilon_r-1}$) is given by Ref. [8].

$$Z_0 = \frac{1}{2}\sqrt{\frac{\mu}{\varepsilon_{\text{eff}}}} \frac{K(k_a)}{K(k'_a)} \qquad (2.1.35)$$

Finally, as shown in Fig. 2.1.8d, the effective permittivity and the characteristic impedance of the coplanar waveguide, which is usually called CPW, are given by Ref. [9].

$$\varepsilon_{\text{eff}} = \varepsilon_0 \left[1 + \frac{\varepsilon_r - 1}{2} \frac{K(k_1)}{K(k'_1)} \frac{K(k'_0)}{K(k_0)}\right] \qquad (2.1.36)$$

$$Z_0 = \frac{1}{4}\sqrt{\frac{\mu}{\varepsilon_{\text{eff}}}} \frac{K(k'_0)}{K(k_0)} \qquad (2.1.37)$$

where $k_0 = w/(w+2S)$, $k_1 = \sinh[\pi w/(4H)]/\sinh[\pi(w+2S)/(4H)]$, $k'_0 = \sqrt{1-k_0^2}$, and $k'_1 = \sqrt{1-k_1^2}$.

It is evident that the characteristic impedance is irrelevant to the length and is determined by the geometry and surrounding material of the TL. In real-world applications, the cross-sectional dimensions of the interconnect vary along the length, thereby resulting in changed characteristic impedance. For example, as shown in Fig. 2.1.9, the interconnect width becomes smaller at the discontinuity point. The characteristic impedances of the left and right TLs are denoted by Z_1 and Z_2, respectively. As $Z_1 \neq Z_2$, the incident signal encounters a change in the instantaneous impedance

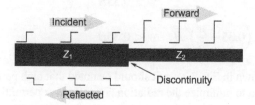

Figure 2.1.9 Interconnect discontinuity.

of the interconnect. Assuming the right line is infinitely long, the reflection coefficient Γ, which is defined as the amplitude of the reflected voltage wave V_0^+ normalized to the amplitude of the incident voltage wave V_0^-, can be calculated by

$$\Gamma = \frac{V_0^-}{V_0^+} = \frac{Z_2 - Z_1}{Z_2 + Z_1} \tag{2.1.38}$$

and the transmission coefficient T is

$$T = 1 + \Gamma = \frac{2Z_2}{Z_2 + Z_1} \tag{2.1.39}$$

It can be seen from Eq. (2.1.38) that the reflection coefficient can be zero as $Z_2 = Z_1$, which means that the reflections are eliminated by impedance matching. As shown in Fig. 2.1.10, the source and load resistances, and the characteristic impedance of the interconnect, are set to be 50Ω. In the figure, l is the interconnect length. As the switch closes, the voltage wave reaches the load at time of l/v_p. However, the voltage across the load is only half of V_{dd}. Therefore, the series termination, which is suitable for CMOS logic due to its low power demand, can be employed [4]. In such scheme, the load end is unterminated, i.e., $R_L \to \infty$, thereby leading to a unit reflection coefficient $\Gamma = 1$. So the voltage at the load can reach V_{dd} at time of $l/v_p = l\sqrt{LC} = l\sqrt{\mu_r \varepsilon_r}/c$, which represents the time delay of the TL. For general case, there exist multiple reflections at the source and load ends, as shown in Fig. 2.1.11. Assuming that Z_0 is the characteristic impedance of the interconnect, the reflection coefficients at the load and source ends are $\Gamma_L = (R_L - Z_0)/(R_L + Z_0)$ and $\Gamma_s = (R_{on} - Z_0)/(R_{on} + Z_0)$, respectively. Based on the bounce diagram, the transient voltage waveform can be captured, as shown in Fig. 2.1.12.

In high-speed, high-density circuits, the crosstalk is inevitable and can limit the performance of the electronic system. Fig. 2.1.13a and 2.1.13b show the coupled interconnects above a ground plane and its equivalent circuit model. Because of the capacitive and inductive couplings, which are denoted as mutual capacitance C_m and mutual inductance L_m, respectively, the signal transmitted on the aggressor line would create an undesirable effect in the victim line [10].

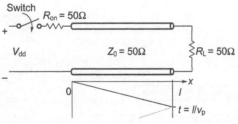

Figure 2.1.10 Impedance matching.

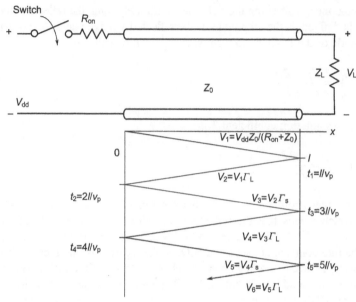

Figure 2.1.11 Bounce diagram.

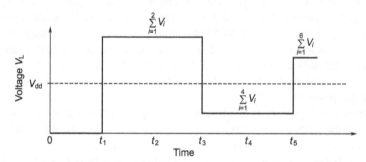

Figure 2.1.12 Transient voltage waveform.

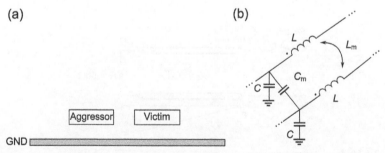

Figure 2.1.13 (a) Coupled interconnects above a ground plane and its (b) equivalent circuit model.

Electrical modeling and design

It is worth noting that various loss mechanisms exist and should be considered in real-world applications. For instance, with the increasing operating frequency, the current would accumulate at the conductor surface, thereby decreasing the effective conductive area. Therefore, the interconnect resistance increases with the frequency, which is called the skin effect. The proximity effect and surface roughness also have significant effect on the resistive loss. Considering the lossy dielectric, the permittivity has real and imaginary parts, i.e., $\varepsilon = \varepsilon' - j\varepsilon''$. By substituting complex permittivity and $\vec{J} = \sigma \vec{E}$ into Eq. (2.1.12), the Maxwell's curl equation can be written as

$$\nabla \times \vec{B} = j\omega\mu\varepsilon' \vec{E} + \mu(\sigma + \omega\varepsilon'')\vec{E} \tag{2.1.40}$$

where σ is the conductivity. It is evident that the imaginary part of permittivity gives rise to energy loss and is indistinguishable from the loss. The dielectric loss is characterized by the electric loss tangent, i.e.,

$$\tan \delta_e = \frac{\varepsilon''}{\varepsilon'} \tag{2.1.41}$$

Similarly, the magnetic loss tangent can be given by

$$\tan \delta_m = \frac{\mu''}{\mu'} \tag{2.1.42}$$

where μ' and μ'' are the real and imaginary parts of the permeability. In particular, for redistribution interconnects, the silicon substrate loss should be treated appropriately. At high frequencies, the eddy current loss has a negative effect on the electrical performance of the interconnects. To reduce the eddy current, some shielding structures, such as partial ground shields and floating shields, can be employed [11].

2.1.3 Power allocation

The interconnects are used not only for transmitting signals but also for power supply. Fig. 2.1.14 shows the power delivery system, and it is evident that the voltage across the load terminal would be decreased by

$$\Delta V = I(R_p + R_g) + (L_p + L_g)\frac{dI}{dt} = IR + L\frac{dI}{dt} \tag{2.1.43}$$

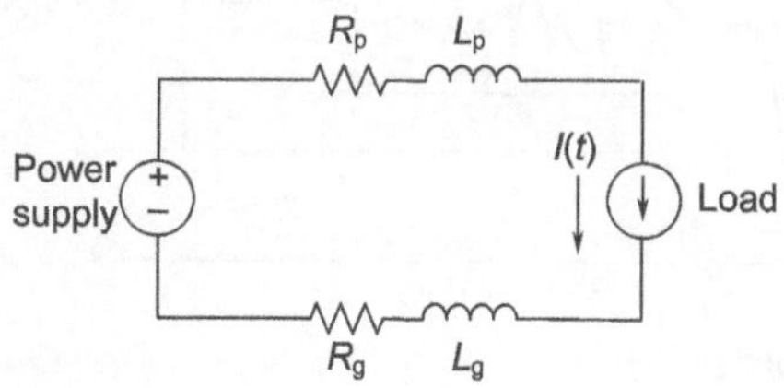

Figure 2.1.14 Power supply system.

where $R_p(R_g)$ and $L_p(L_g)$ represent the parasitic resistance and inductance of the power (ground) lines, respectively, and the variable current source $I(t)$ denotes the power load.

As the direct current passes through the interconnects, the parasitic resistance of the interconnects results in a voltage drop, which is given by the Ohm's law $V = IR$ and is called *IR* voltage drop. To reduce such voltage drop, the simplest way is to increase the interconnect dimensions. More importantly, as a time-varying current passes through the interconnects, the inherent inductive property of the interconnects would lead to a voltage oscillation with time. The direct currents cause large *IR* voltage drop, and the fast transient current produces large inductive voltage drop. These two issues, i.e., *IR* voltage drop and inductive effect, would cause the false transitioning of the circuit and must be taken into account [12].

In general, the change in supply voltage, which is called power supply noise, should be kept within a certain range around the nominal voltage level. This range is named power noise margin, which would be reduced with the advanced technology node, thereby increasing the design difficulty of PDN. The design goal is to keep the impedance of the PDN below a target value Z_{target}, i.e., the maximum PDN impedance. Z_{target} is calculated by

$$Z_{target} = \frac{V_{dd} \times N_{Ripple}}{I_{load}} \qquad (2.1.44)$$

where N_{Ripple} is the maximum tolerable power noise, V_{dd} is the supply voltage of the power lines, and I_{load} is the load current. If the impedance of the designed PDN is smaller than the target impedance, i.e., $Z \leqslant Z_{target}$, as shown in Fig. 2.1.15, the oscillation of the supplied voltage would be smaller than a specific margin. Based on the power supply system, the imaginary component of the impedance $Z = R + j\omega L$ increases linearly with the increasing frequency. As shown in Fig. 2.1.16, as the frequency exceeds the maximum frequency $f_{max} = Z_{target}/(2\pi L)$, the impedance becomes larger than the target value.

In determining the impedance of the PDN, it is essential to understand the inductive properties of the power and ground interconnects. The inductance of an interconnect denotes its capability to store energy in the form of magnetic field. In this viewpoint,

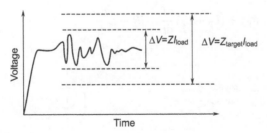

Figure 2.1.15 Supply voltage.

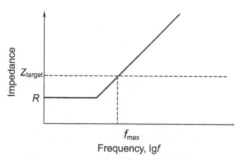

Figure 2.1.16 Impedance of a power distribution network (PDN) without decoupling capacitors.

the self-inductance per unit length of a conductor with the current of I_0 can be defined as

$$L_s = \frac{1}{\mu |I_0|^2} \int_S \vec{B} \cdot \vec{B}^* \, ds \qquad (2.1.45)$$

A more intuitive definition of the inductance, as shown in Fig. 2.1.16, is defined as N, the number of magnetic field line rings around the conductor, over the current, i.e.,

$$L_s = \frac{N}{I_0} \qquad (2.1.46)$$

Note that the inductance is irrelevant to the current and is only determined by the interconnect geometry and surrounding material. Considering two adjacent lines, and supposing a current passes through one line, some number of magnetic field line rings would encircle another line, as shown in Fig. 2.1.17. So the mutual inductance between two adjacent lines can be defined as

$$L_{21} = \frac{\Delta N}{I_1} \qquad (2.1.47)$$

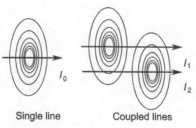

Figure 2.1.17 Definition of the inductance.

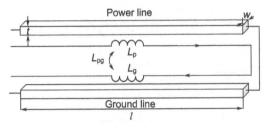

Figure 2.1.18 A complete current loop formed by parallel interconnects.

It is evident that $L_{21} = L_{12}$, and the mutual inductance must be smaller than the self-inductance of any line. Fig. 2.1.18 shows a complete current loop formed by parallel interconnects. Here, the interconnect width, thickness, and length are denoted as w, t, and l, and the pitch between two lines is P. The total inductance of the current loop, which is called loop inductance, can be obtained by

$$L_{\text{loop}} = L_p + L_g - 2L_{pg} \tag{2.1.48}$$

where L_p and L_g represent the self-inductances of the power and ground lines, respectively, and they can be calculated by

$$L_p = L_g = \frac{\mu l}{2\pi}\left(\ln\frac{2l}{w+t} + \frac{1}{2} + 0.2235\frac{w+t}{l}\right) \tag{2.1.49}$$

The mutual inductance between the power lines is

$$L_{pg} = \frac{\mu l}{2\pi}\left(\ln\frac{2l}{P} - 1 + \frac{P}{l}\right) \tag{2.1.50}$$

The high-frequency impedance can be reduced by placing capacitors across the power and ground interconnects. On one hand, the capacitors can provide energy to the load (see Fig. 2.1.19), and on the other hand, they decouple the high-impedance paths of the PDN from the load, i.e., provide low impedance between the power and ground interconnects. Therefore, the capacitors are called decoupling capacitors. Note that the capacitor has not only capacitance but also equivalent series resistance (ESR) and equivalent series inductance (ESL), as shown in Fig. 2.1.20. The impedance of the capacitor can be given by

$$Z = \text{ESR} + j\,\omega\text{ESL} + \frac{1}{j\omega C} \tag{2.1.51}$$

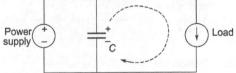

Figure 2.1.19 Power delivery system with decoupling capacitor.

Electrical modeling and design

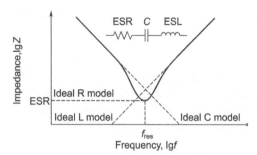

Figure 2.1.20 Equivalent circuit model of the capacitor and its impedance.

At the resonant frequency $f_{res} = 1/(2\pi\sqrt{ESL \cdot C})$, the impedance is equal to ESR. At low frequencies, the capacitor behaves like an ideal capacitance model, i.e., the impedance decreases linearly with the increasing frequency. However, as the frequency exceeds f_{res}, the impedance increases linearly with the frequency. Through carefully designing the number and positions of the decoupling capacitors, the PDN impedance can be suppressed effectively.

2.1.4 Electromagnetic compatibility and interference

With the increase of clock frequency and integration density, EMI would be likely to occur in the electronic systems. Therefore, more attention should be paid to the electromagnetic compatibility (EMC) design, which becomes more and more important. From a designer's viewpoint, the EMC indicates the ability of the electronic systems to operate without degradation, nor be a source of interference, in the electromagnetic environment.

Traditionally, as shown in Fig. 2.1.21, the EMI sources were mainly related to the cables, radio frequency devices, and high-speed digital systems [4]. For example, if a cable shield is connected to the ground, it would act as a dipole antenna and cause significant radiated emission. Moreover, as the operating frequency increases, the

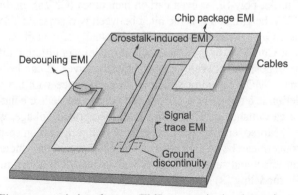

Figure 2.1.21 Electromagnetic interference (EMI) sources in the electronic system.

devices and packages also become the EMI source, thereby significantly increasing the difficulty in the EMI control. To avoid EMC/EMI problems and achieve first-pass design success, it is highly desirable to perform accurate modeling and simulation before real fabrication [13]. Various approaches including analytical and numerical methods have been proposed to model the passive structures for electronic packaging. To suppress the electromagnetic radiated emission, some shielding technologies were developed. For instance, by spraying a conductive material on the package, good shielding effectiveness can be achieved without additional penalty to the package size [14].

2.2 Modeling, characterization, and design of through-silicon via packages

2.2.1 Through-silicon via modeling

Through-silicon vias (TSVs) are high-aspect-ratio vertical interconnects penetrating the silicon substrate and provide electrical connectivity between active layers. With the fabrication technologies of TSVs becoming mature [15,16], it is mandatory to accurately and efficiently evaluate the electrical characteristics of TSVs for performance analysis and design optimization. Basic studies of TSV characterization have been reported by several academic and industrial researchers [17–19]. The numerical techniques such as finite-element method are accurate but slow and memory intensive. Moreover, they have no clear links to the processing and real measurements, which limits the use of simulation significantly. In contrast to the numerical techniques, the equivalent circuit model is intuitive; meanwhile, it can facilitate the measurement and application.

2.2.1.1 Through-silicon via structure

Fig. 2.2.1 shows the schematic of a pair of signal and ground TSVs, which are composed of Cu, Al, poly-Si, or even carbon nanotubes (CNTs). In the figure, H_{TSV} represents the TSV height, and P is the pitch between two adjacent TSVs. In general, TSV is surrounded by a dielectric layer, which is usually silicon dioxide and used for DC isolation. For Cu TSVs, to prevent the Cu diffusion, an ultrathin diffusion barrier layer should be deposited between Cu and dielectric layer. It can be seen that TSV essentially forms a MOS capacitor structure. In accordance with the semiconductor physics, a depletion layer appears as the applied voltage exceeds the flat-band voltage. So the parasitic capacitance of TSV varies with the applied voltage, which is called MOS effect [20]. Moreover, with the increasing signal propagation speed, the highest frequency of concern can be as high as several tens of gigahertz. Under such circumstances, the skin effect and substrate loss, as well as the MOS effect, must be considered in the TSV modeling [21].

Electrical modeling and design

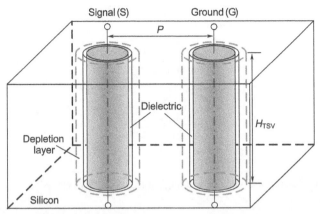

Figure 2.2.1 Schematic of a pair of through-silicon vias (TSVs).

2.2.1.2 Circuit modeling

Fig. 2.2.2 shows the equivalent circuit model of a pair of TSVs. In the figure, V_s is the source voltage, V_{in} and V_{out} represent the input and output voltages, and the impedance of 50Ω is applied at both the input and output ports. R_{TSV} and L_{TSV} are the TSV resistance and inductance, C_{Si} and G_{Si} are the silicon capacitance and conductance, and C_{ox} and C_{dep} are the oxide capacitance and depletion capacitance, respectively.

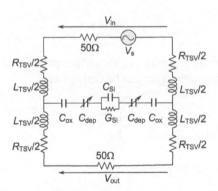

Figure 2.2.2 Equivalent circuit model of a pair of through-silicon vias (TSVs).

2.2.1.2.1 Metal–oxide–semiconductor effect

To clearly analyze the MOS effect, the cross-sectional view of a TSV is plotted in Fig. 2.2.3. In the figure, r_{via} is the radius of the TSV filling conductor, t_{ox} is the thickness of the isolation dielectric layer, and w_{dep} is the thickness of the depletion layer. Analogously with the conventional planar MOS capacitor structure, the TSV MOS capacitor can be calculated by solving 1D Poisson equation in the radial direction [17].

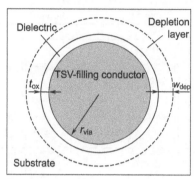

Figure 2.2.3 Cross-sectional view of a through-silicon via (TSV) embedded into a p-type silicon substrate.

$$\frac{1}{r}\frac{\partial}{\partial r}\left(r\frac{\partial \psi}{\partial r}\right) = \frac{qN_a}{\varepsilon_{Si}}, \quad r_{ox} \leqslant r \leqslant r_{dep} \tag{2.2.1}$$

where q is the elementary charge, N_a is the p-type bulk doping concentration, ε_{Si} is the permittivity of silicon, $r_{ox} = r_{via} + t_{ox}$ is the oxide radius, and $r_{dep} = r_{via} + t_{ox} + w_{dep}$ is the depletion radius. The boundary conditions can be expressed as

$$\psi|_{r=r_{dep}} = 0 \tag{2.2.2}$$

$$\left.\frac{\partial \psi}{\partial r}\right|_{r=r_{dep}} = 0 \tag{2.2.3}$$

By applying the boundary conditions, the surface potential at the interface between silicon substrate and dielectric layer can be derived as

$$\psi_s = \frac{qN_a}{2\varepsilon_{Si}}\left(r_{dep}^2 \ln\frac{r_{dep}}{r_{ox}} - \frac{r_{dep}^2 - r_{ox}^2}{2}\right) \tag{2.2.4}$$

The applied voltage can be given by

$$V = V_{FB} + \psi_s + \frac{qN_a(r_{dep}^2 - r_{ox}^2)}{2\varepsilon_{ox}}\ln\frac{r_{ox}}{r_{via}} \tag{2.2.5}$$

where V_{FB} is the flat-band voltage,

$$V_{FB} = \phi_{ms} - \frac{2\pi r_{ox} q q_{ot}}{2\pi\varepsilon_{ox}}\ln\frac{r_{ox}}{r_{via}} \tag{2.2.6}$$

ϕ_{ms} is the work function difference between the filling conductor and the semiconductor, ε_{ox} is the permittivity of the dielectric layer, and q_{ot} is the total oxide charges.

At threshold, the thickness of depletion layer reaches its maximum value, i.e., $r_{dep} = r_{max}$, and the surface potential is fixed at $2\phi_F = 2k_B T \ln (N_a/n_i)/q$,

$$\frac{qN_a}{2\varepsilon_{Si}}\left(r_{max}^2 \ln \frac{r_{max}}{r_{ox}} - \frac{r_{max}^2 - r_{ox}^2}{2}\right) = \frac{2k_B T}{q} \ln \frac{N_a}{n_i} \qquad (2.2.7)$$

where k_B is the Boltzmann's constant, T is the temperature, and $n_i = 9.38 \times 10^{19} (T/300)^2 \exp(-6884/T)$ is the intrinsic carrier concentration of silicon [22]. By solving Eq. (2.2.7), the maximum depletion radius r_{max} can be calculated, and the threshold voltage can be subsequently determined by

$$V_{Th} = V_{FB} + \frac{2k_B T}{q} \ln \frac{N_a}{n_i} + \frac{qN_a(r_{max}^2 - r_{ox}^2)}{2\varepsilon_{ox}} \ln \frac{r_{ox}}{r_{via}} \qquad (2.2.8)$$

Fig. 2.2.4 shows the TSV MOS capacitance versus the applied voltage. There are three distinct regions including accumulation region ($V < V_{FB}$), depletion region ($V_{FB} \leqslant V \leqslant V_{Th}$), and inversion region ($V > V_{Th}$). In the accumulation region, only the oxide capacitance exists, i.e.,

$$C_{TSV} = C_{ox} = \frac{2\pi \varepsilon_{ox} H_{TSV}}{\ln(r_{ox}/r_{via})} \qquad (2.2.9)$$

When the voltage exceeds the flat-band voltage, the depletion layer appears, and r_{dep} can be determined by Eq.(2.2.5). The depletion capacitance is obtained by

$$C_{dep} = \frac{2\pi \varepsilon_{Si} H_{TSV}}{\ln(r_{dep}/r_{ox})} \qquad (2.2.10)$$

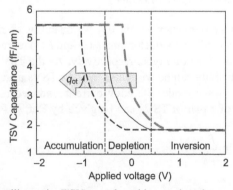

Figure 2.2.4 Through-silicon via (TSV) metal—oxide—semiconductor (MOS) capacitance versus the applied voltage.

and the total TSV capacitance is a series combination of the oxide and depletion capacitances (see Fig. 2.2.2)

$$C_{TSV} = \left(\frac{1}{C_{ox}} + \frac{1}{C_{dep}}\right)^{-1} = 2\pi H_{TSV}\left(\frac{1}{\varepsilon_{ox}}\ln\frac{r_{ox}}{r_{via}} + \frac{1}{\varepsilon_{Si}}\ln\frac{r_{dep}}{r_{ox}}\right) \quad (2.2.11)$$

With the increase of applied voltage, r_{dep} increases, thereby leading to decreased TSV capacitance. As the applied voltage becomes larger than the threshold one, the depletion layer thickness reaches its maximum value, which can be determined by Eq. (2.2.7). Therefore, the TSV capacitance is kept as its minimum value in the inversion region.

It is worth noting that the oxide charge will be induced by plasma damage during via-hole etching and dielectric deposition and is inevitable in the real-world applications [23]. As shown in Fig. 2.2.4, the increase in the oxide charge leads to a leftward shift in the TSV capacitance−voltage curve. This is because that the flat-band and threshold voltages decrease with the increasing oxide charge. To reduce the TSV parasitic capacitance, Katti et al. [24] proposed that proper process can introduce a high amount of oxide charge, thus ensuring the minimum TSV capacitance in the desired operating voltage range. However, the depletion layer thickness and the corresponding minimum TSV capacitance are susceptible to the temperature variation. To accurately extract the temperature-dependent TSV capacitance, an iterative method should be employed [25], as summarized by the flowchart in Fig. 2.2.5. The temperature rise will increase the intrinsic carrier concentration, thereby reducing the maximum depletion radius and increasing the TSV capacitance.

2.2.1.2.2 RLCG parameters
For each TSV, the internal impedance can be computed by Ref. [26].

$$Z_{inter} = \frac{H_{TSV}\sqrt{j\omega\mu_0/\sigma}}{2\pi r_{via}}\frac{I_0(r_{via}\sqrt{j\omega\mu_0\sigma})}{I_1(r_{via}\sqrt{j\omega\mu_0\sigma})} \quad (2.2.12)$$

where ω is the angular frequency, μ_0 is the permeability of free space, σ is the conductivity of the TSV filling conductor, and $I_0(\cdot)$ and $I_1(\cdot)$ represent the 0th and first order Bessel functions of the first type, respectively. As the TSV filling conductor is usually Cu, the conductivity can be written as σ_{Cu}. As TSVs usually operate at quasi-TEM mode and slow-wave mode, the eddy current−induced resistance is ignored. The loop inductance of a pair of TSVs can be given by Ref. [27].

$$L_{loop} = \frac{\mu_0 H_{TSV}}{\pi}\ln\frac{P}{r_{via}} \quad (2.2.13)$$

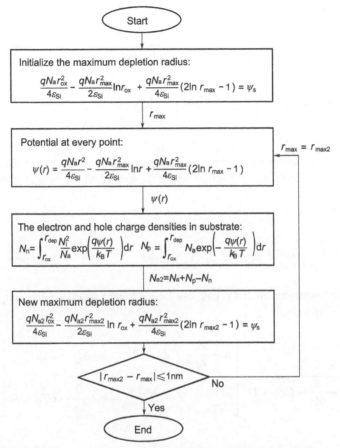

Figure 2.2.5 Flowchart for extracting depletion radius at a given temperature.

Therefore, the resistance and inductance of a pair of TSVs can be calculated by

$$R_{TSV} = \text{Re}(Z_{inter}) \tag{2.2.14}$$

$$L_{TSV} = \frac{\text{Im}(Z_{inter} + j\omega L_{loop})}{j\omega} \tag{2.2.15}$$

The silicon capacitance and conductance can be computed by Ref. [28].

$$C_{Si} = \frac{\pi \varepsilon_{Si} H_{TSV}}{\text{arccosh}\left(\dfrac{P}{2r_{dep}}\right)} \tag{2.2.16}$$

$$G_{Si} = \frac{\sigma_{Si}}{\varepsilon_{Si}} C_{Si} \qquad (2.2.17)$$

where σ_{Si} is the silicon conductivity and can be expressed as [29].

$$\sigma_{Si} = qN_a \left[\mu_{min} + \frac{\mu_{max} - \mu_{min}}{1 + (N_a/N_{ref})^{0.76}} \right] \cdot \left(\frac{T}{300} \right)^{-1.5} \qquad (2.2.18)$$

$\mu_{max} = 495 \text{ cm}^2 \cdot \text{V}^{-1} \cdot \text{s}^{-1}$, $\mu_{min} = 47.7 \text{ cm}^2 \cdot \text{V}^{-1} \cdot \text{s}^{-1}$, and $N_{ref} = 6.3 \times 10^{16} \text{cm}^{-3}$.
In general, the circuit model of the signal/ground TSV pair can be simplified as a TL model [18]. The impedance and admittance of the simplified TL model can be given as

$$Z_{TSV} = (R_{TSV} + j\omega L_{TSV}) \qquad (2.2.19)$$

$$Y_{TSV} = \left(\frac{2}{j\omega C_{TSV}} + \frac{1}{G_{Si} + j\omega C_{Si}} \right)^{-1} \qquad (2.2.20)$$

The S-parameter of the TSV pair can be computed based on the simplified TL model as follows [30].

$$S = \frac{1}{A + \frac{B}{Z_0} + CZ_0 + D} \cdot \begin{bmatrix} A + \frac{B}{Z_0} - CZ_0 - D & AD - BC \\ 2 & -A + \frac{B}{Z_0} - CZ_0 + D \end{bmatrix} \qquad (2.2.21)$$

with $Z_0 = 50\Omega$, and the *ABCD* transmission matrix is

$$T = \begin{bmatrix} A & B \\ C & D \end{bmatrix} = \begin{bmatrix} \cosh(\gamma H_{TSV}) & Z \sinh(\gamma H_{TSV}) \\ \sinh(\gamma H_{TSV})/Z & \cosh(\gamma H_{TSV}) \end{bmatrix} \qquad (2.2.22)$$

The characteristic impedance and the propagation constant are given by

$$Z = \sqrt{Z_{TSV}/Y_{TSV}} \qquad (2.2.23)$$

$$\gamma = \sqrt{Z_{TSV} Y_{TSV}}/H_{TSV} \qquad (2.2.24)$$

2.2.1.3 Floating substrate Effect

As aforementioned, the TSV capacitance is kept as its minimum value due to the MOS effect. These analyses are based on the premise that the silicon substrate is well grounded with bulk contacts, but such prerequisite cannot be fulfilled at any time due to imperfect grounding. So the electrical analyses of TSVs in the passive silicon substrate were carried out by neglecting the MOS effect [31,32]. In a floating silicon substrate, however, the electric fields of the single TSVs terminate not at the substrate but at the ground TSVs [33, 73].

Fig. 2.2.6 shows the capacitance–voltage characteristic of the signal/ground TSVs in the floating silicon substrate. It can be seen that at low frequency (<10 MHz), the capacitance of the signal TSV decreases as the voltage exceeds the flat-band one and then increases and approaches to the oxide capacitance with the increasing voltage. In contrast, at high frequency (>10 MHz), the capacitance of the signal TSV remains at its minimum value when the voltage becomes larger than the threshold one. This is because that the generation of inversion carriers cannot follow the voltage change rate at high frequency [34]. The capacitances of the signal and ground TSVs are symmetrical to the zero voltage as the same but opposite charges are accumulated at the signal and ground TSVs, respectively. The increase in temperature mainly affects the TSV capacitance at high frequency as the temperature rise increases the intrinsic carrier concentration and the minority electron concentration.

Because of the floating substrate effect, the depletion layer thicknesses of TSVs will be different. The silicon capacitance between two TSVs with different depletion layer thicknesses can be calculated using the conformal mapping method [35].

$$C_{Si} = \frac{2\pi\varepsilon_{Si}H_{TSV}}{\ln\left[\frac{P^2 - r_{dep,s}^2 - r_{dep,g}^2}{2r_{dep,s}r_{dep,g}} + \sqrt{\left(\frac{P^2 - r_{dep,s}^2 - r_{dep,g}^2}{2r_{dep,s}r_{dep,g}}\right)^2 - 1}\right]} \qquad (2.2.25)$$

where $r_{dep,s}$ and $r_{dep,g}$ represent the depletion radius of the signal and ground TSVs in the floating silicon substrate, respectively. The results of silicon capacitance are shown in Fig. 2.2.7 for the signal/ground TSVs in the floating silicon substrate at different temperatures. The impact of different depletion layer thicknesses of signal and ground TSVs has been considered appropriately. As shown in the middle of Fig. 2.2.7, the capacitance at low and high frequencies overlap with each other, while the difference occurs in the inversion region. As the temperature rise mainly influences

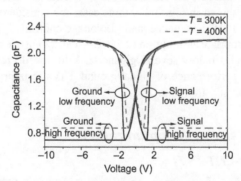

Figure 2.2.6 Capacitance of single through-silicon via (TSV) in signal/ground TSVs in the floating silicon substrate.

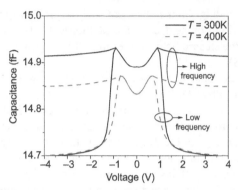

Figure 2.2.7 Silicon capacitance between signal/ground through-silicon vias (TSVs) in the floating silicon substrate.

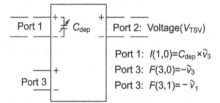

Figure 2.2.8 Symbolically defined device (SDD) block for modeling the voltage-controlled depletion capacitance.

the depletion layer thickness, the silicon capacitance is unchanged when the TSVs operate at the accumulation region. In the depletion and inversion regions, the silicon capacitance decreases with the increasing temperature.

To accurately capture the electrical performance of TSVs in the floating silicon substrate, as shown in Fig. 2.2.8, the symbolically defined device (SDD) block in ADS software is employed for modeling the voltage-controlled depletion capacitance [36]. In the SDD block, $\tilde{v}_i$ is the voltage of the ith port, while $\tilde{v}_2 = V_{TSV}$. Based on the equivalent circuit model, the electrical characteristics of the signal/ground TSVs can be captured. As shown in Fig. 2.2.9, three cases are considered: (1) oxide capacitance; (2) minimum TSV capacitance; and (3) voltage-controlled TSV capacitance. It is evident that inappropriate assumption of the TSV capacitance leads to significant inaccuracy at frequencies below several gigahertz. With the decreasing TSV capacitance, the electrical performance of signal/ground TSVs is improved, which is due to the reduced silicon substrate loss.

2.2.2 Through-silicon via optimization

2.2.2.1 CNT/Cu-CNT TSVs

CNT, since its first discovery in 1991 [37], has attracted a lot of research interests due to its extraordinary physical properties, such as high ampacity and long mean free path

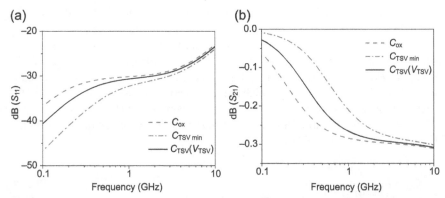

Figure 2.2.9 S-parameters of the signal/ground through-silicon vias (TSVs). (a) dB(S_{11}) and (b) dB(S_{21}).

(MFP) [38]. Depending on the diameter dimensions, CNTs can be divided into two categories: single-walled CNT (SWCNT) and multiwalled CNT (MWCNT). SWCNTs can be either semiconducting or metallic, while MWCNTs are always metallic. In general, CNTs are grown vertically, and it is intuitive to consider CNTs as TSV-filling conductor. It is expected that the implementation of CNT TSVs can avoid some problems such as voiding, extrusion, and electromigration [39]. Moreover, as the CNTs have ultrahigh thermal conductivity, the thermal issues of 3D ICs can be alleviated by using CNT TSVs [40]. Fig. 2.2.10a shows the schematic of the silicon substrate embedded with TSV array. A heating power q is input on the top of the substrate, while the bottom of the substrate is fixed at 300 K [41]. By using the COMSOL Multiphysics software, the temperature rise ΔT can be captured, and the equivalent thermal conductivity of the substrate can be calculated by $k_{eq} = qH_{TSV}/\Delta T$. The thermal conductivity of Cu is $k_{Cu} = 400 \text{ W} \cdot \text{m}^{-1} \cdot \text{K}^{-1}$, while for CNTs, the thermal conductivity is in the range of [1750, 5800] $\text{W} \cdot \text{m}^{-1} \cdot \text{K}^{-1}$. As shown in Fig. 2.2.10b, the

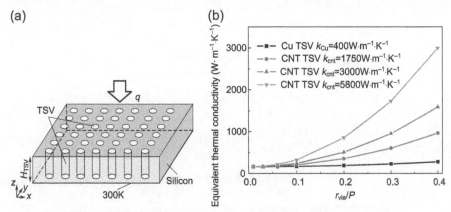

Figure 2.2.10 (a) Schematic of silicon substrate embedded with TSV array and (b) its equivalent thermal conductivity.

implementation of CNT TSVs, even with a pessimistic CNT thermal conductivity k_{cnt} of 1750 W·m^{-1}·K^{-1}, can significantly improve the equivalent thermal conductivity.

To consider the impact of CNT kinetic inductance, Xu et al. proposed the concept of "effective complex conductivity" and performed the electrical characterization of CNT TSVs [18]. However, most of the literatures on modeling of CNT TSVs are based on the ideally packed CNTs [42,43], which are still no possible even with the state-of-the-art fabrication process. As shown in Fig. 2.2.11, assuming that N_{cnt} identical CNTs are uniformly distributed in the TSV, a CNT filling ratio is defined as the proportion of CNT area to total area of TSV, i.e.,

$$f_{cnt} = N_{cnt} \cdot \frac{D_{cnt}^2}{4 r_{via}^2} \tag{2.2.26}$$

where D_{cnt} is the diameter of an isolated CNT. As the mixed CNT bundle behaves like a bundle of identical CNTs with the mean diameter [44], it is neglected in the following. So the effective complex conductivity of the CNT TSV is given by

$$\sigma_{cnt} = f_{cnt} \sigma_{cnt,ideal} \tag{2.2.27}$$

where $\sigma_{cnt,ideal}$ is the conductivity of the ideally packed CNTs, i.e.,

$$\sigma_{cnt,ideal} = \frac{4 H_{TSV}}{\pi D_{cnt}^2} \cdot Fm \cdot \frac{f_{cnt}}{Z_{cnt}} \tag{2.2.28}$$

where Fm is the fraction of metallic CNTs in the bundle and equals 1 for the MWCNT TSVs. Z_{cnt} is the intrinsic self-impedance of an isolated CNT. For SWCNT, one has [18].

$$Z_{cnt} = R_{mc} + \frac{h}{2q^2 N_{ch}} \left(1 + \frac{H_{TSV}}{\lambda_{eff}} + j\omega \frac{H_{TSV}}{2 v_F} \right) \tag{2.2.29}$$

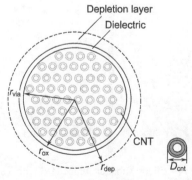

Figure 2.2.11 Cross-sectional view of a carbon nanotube (CNT) through-silicon via (TSV).

where R_{mc} represents the imperfect contact resistance and is neglected here as it highly depends on the fabrication process. h is the Planck's constant, λ_{eff} is the effective MFP, N_{ch} is the number of conducting channels, and v_F is the Fermi velocity [45]. As an MWCNT can be viewed as a coaxial assembly of cylinders of SWCNTs, the intrinsic self-impedance of an MWCNT is a shunt combination of each shell impedance.

Although CNTs have the potential to achieve superior performance over the Cu counterpart, they are still far from the real-world applications. For example, even for an ultrahigh-density SWCNT bundle reported in Ref. [46], the maximum value of the conductivity is only 1.98×10^7 S/m, which is still smaller than that of Cu. Recently, a Cu-CNT composite was fabricated in Ref. [47], and it was demonstrated that such Cu-CNT composite can provide a more practical way to achieve balance between the performance and reliability. Further, the sample of Cu-CNT TSVs was successfully realized in Ref. [48]. The effective conductivity of Cu-CNT TSVs can be given by

$$\sigma_{eff} = (1 - f_{cnt})\sigma_{Cu} + f_{cnt}\sigma_{cnt} \tag{2.2.30}$$

Note that there is a separation of 0.31 nm between the carbon and Cu atoms. So the CNT diameter D_{cnt} should be replaced with (D_{cnt}+0.31 nm) in the modeling of Cu-CNT composite TSVs.

By substituting $Z_{cnt} = R_{cnt} + j\omega L_{cnt}$ into Eq. (2.2.30), the effective conductivity of CuCNT can be expressed as [49].

$$\sigma_{eff} = (1 - f_{cnt})\sigma_{Cu} + \frac{4H_{TSV}}{\pi D_{cnt}^2} \cdot Fm \cdot \frac{f_{cnt}^2}{R_{cnt}^2 + \omega^2 L_{cnt}^2} \cdot (R_{cnt} - j\omega L_{cnt}) \tag{2.2.31}$$

The real and imaginary parts of the effective conductivity are given as

$$Re(\sigma_{eff}) = (1 - f_{cnt})\sigma_{Cu} + \frac{4H_{TSV}}{\pi D_{cnt}^2} \cdot Fm \cdot \frac{f_{cnt}^2}{R_{cnt}^2 + \omega^2 L_{cnt}^2} \cdot R_{cnt} \tag{2.2.32}$$

$$Im(\sigma_{eff}) = -\frac{4H_{TSV}}{\pi D_{cnt}^2} \cdot Fm \cdot \frac{f_{cnt}^2}{R_{cnt}^2 + \omega^2 L_{cnt}^2} \cdot \omega L_{cnt} \tag{2.2.33}$$

Fig. 2.2.12 and 2.2.13 show the real and imaginary parts of the effective conductivity of CNT and Cu-CNT composite TSVs with different values of f_{cnt}. $f_{cnt} = 0.2$ is adopted from the experiment in Ref. [46], while $f_{cnt} = 1$ represents the ideally packed CNTs. For SWCNTs and Cu-SWCNT composite, Fm is set as 1/3. It can be seen that $Re(\sigma_{eff})$ decreases with the increasing frequency. As the kinetic inductance has significant influence on the electrical characteristics of MWCNT, R_{cnt}/L_{cnt} of an MWCNT is much larger than that of an SWCNT. Therefore, $Re(\sigma_{eff})$ is kept almost unchanged for SWCNTs, while it decreases for MWCNTs. After codepositing Cu with CNTs, $Re(\sigma_{eff})$ can be increased to be close to that of Cu, i.e., $\sigma_{Cu} = 5.8 \times 10^7$ S/m.

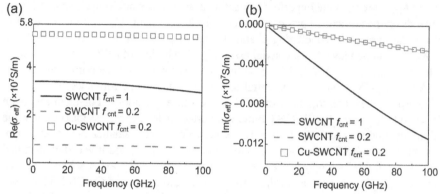

Figure 2.2.12 Effective conductivity of the single-walled carbon nanotube (SWCNT) and Cu-SWCNT through-silicon vias (TSVs).

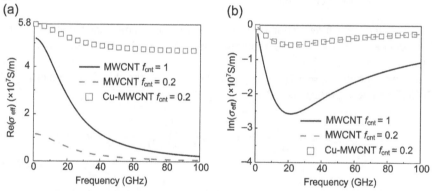

Figure 2.2.13 Effective conductivity of the multiwalled carbon nanotube (MWCNT) and Cu-MWCNT through-silicon vias (TSVs).

According to Eq. (2.2.33), with the increasing frequency, Im(σ_{eff}) drops at first and then rises as the frequency exceeds $f_0 = R_{\text{cnt}}/(2\pi L_{\text{cnt}})$, which is equal to 269.8 GHz for SWCNTs and to 21.1 GHz for MWCNTs, respectively. As the highest frequency plotted in the figure is only 100 GHz, it seems that Im(σ_{eff}) decreases almost linearly in Fig. 2.2.12b.

Moreover, it was experimentally observed that the CNT kinetic inductance may be 15 times larger than its theoretical value [50]. The impacts of CNT kinetic inductance variation on the electrical performance of MWCNT and Cu-MWCNT TSVs are investigated based on the equivalent circuit model (see Fig. 2.2.2). Here, the silicon substrate is assumed to be well grounded, and therefore, the TSV capacitance is kept as its minimum value. As shown in Fig. 2.2.14, when the CNT kinetic inductance is $L_K = 8$ nH/μm, the electrical performances of both MWCNT TSVs and Cu-MWCNT TSVs are comparable with that of Cu TSVs. As the CNT kinetic inductance increases to 60 nH/μm, the performance of MWCNT TSVs is degraded significantly,

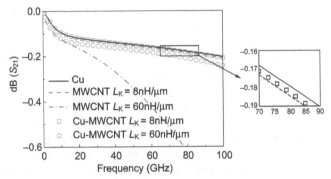

Figure 2.2.14 S_{21}-parameters of Cu, multiwalled carbon nanotube (MWCNT), and Cu-MWCNT composite through-silicon vias (TSVs).

which is mainly attributed to the impedance mismatch. The Cu-MWCNT TSVs, however, are less susceptible to the CNT kinetic inductance variation, implying that Cu-MWCNT TSVs have better stability than pure MWCNT TSVs.

2.2.2.2 Coaxial through-silicon vias

As TSVs penetrate the silicon substrate in 3D ICs, mitigation of substrate noise coupling is crucial. To solve such issue, a new TSV structure with the self-shielding function, i.e., coaxial TSV, was proposed and explored [51]. Fig. 2.2.15a shows the geometry of coaxial TSV, which is composed of a central via and an outer shielding

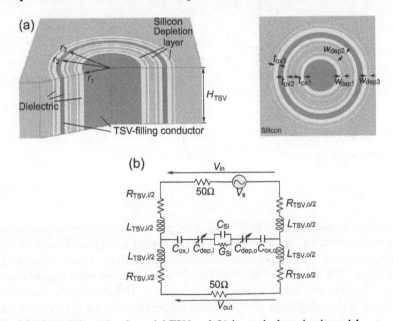

Figure 2.2.15 (a) Schematic of coaxial TSV and (b) its equivalent circuit model.

shell. In the figure, H_{TSV} is the TSV height, r_1 is the radius of the central via, l_{io} is the distance between the central via and the inner surface of the shielding shell, and t_{sh} is the thickness of the shielding shell. There are three dielectric layers in the coaxial TSV, and their thicknesses are t_{ox1}, t_{ox2}, and t_{ox3}. The corresponding depletion layer widths are w_{dep1}, w_{dep2}, and w_{dep3}. The following geometrical parameters are defined: $r_2 = r_1 + l_{io}$ and $r_3 = r_2 + t_{sh}$.

Because of the self-shielding function, the outer silicon substrate has negligible influence on the electrical characteristics of the coaxial TSV. Therefore, the parasitic capacitance of the outer surface of the shielding shell can be neglected in the circuit model. That is, the coaxial TSV can be modeled with the equivalent circuit model in Fig. 2.2.15b. Here, the subscripts i and o represent the respective corresponding quantities of the inner via and outer shielding shell. The resistance R_{TSV} and inductance L_{TSV} of the coaxial TSV can be extracted by either numerical methods such as partial element equivalent circuit (PEEC) method [52] or closed-form expressions [53]. The oxide and the deletion capacitances are given by

$$C_{ox,i} = 2\pi\varepsilon_{ox}H_{TSV}/\ln\left(\frac{r_1 + t_{ox1}}{r_1}\right) \qquad (2.2.34)$$

$$C_{dep,i} = 2\pi\varepsilon_{Si}H_{TSV}/\ln\left(\frac{r_1 + t_{ox1} + w_{dep1}}{r_1 + t_{ox1}}\right) \qquad (2.2.35)$$

$$C_{ox,o} = 2\pi\varepsilon_{ox}H_{TSV}/\ln\left(\frac{r_2}{r_2 - t_{ox2}}\right) \qquad (2.2.36)$$

$$C_{dep,o} = 2\pi\varepsilon_{Si}H_{TSV}/\ln\left(\frac{r_2 - t_{ox2}}{r_2 - t_{ox2} - w_{dep2}}\right) \qquad (2.2.37)$$

The silicon capacitance and conductance are

$$C_{Si} = 2\pi\varepsilon_{Si}H_{TSV}/\ln\left(\frac{r_2 - t_{ox2} - w_{dep2}}{r_1 + t_{ox1} + w_{dep1}}\right) \qquad (2.2.38)$$

$$G_{Si} = \frac{\sigma_{Si}}{\varepsilon_{Si}}C_{Si} \qquad (2.2.39)$$

Moreover, for coaxial TSVs embedded in the passive interposer or with no substrate contact on the inner silicon, the floating substrate effect should be considered. Under such circumstance, the charges will accumulate alternatively at the surfaces of the central via and the shielding shell [54]. Fig. 2.2.16 shows the low-frequency capacitance (<10 MHz) and high-frequency capacitance (>10 MHz) of the coaxial TSV with electrically floating inner silicon. Different from the cylindrical TSVs, the coaxial TSV with electrically floating inner silicon possesses asymmetrical MOS capacitances.

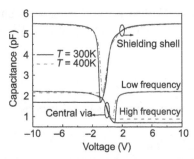

Figure 2.2.16 Capacitance of the central via and the shielding shell of the coaxial through-silicon via (TSV) with electrically floating inner silicon.

Based on the equivalent circuit model, the time-domain analysis is carried out for the coaxial TSV. The source voltage is a clock-like signal with a fundamental frequency of 2 GHz, rising/falling time of 50ps, and amplitude from −2 to 2V. The transient waveform of the output voltage is shown in Fig. 2.2.17a. Three cases are considered: 1) no depletion (i.e., maximum capacitance C_{ox}); 2) full depletion (i.e., minimum capacitance $C_{TSV,min}$); and nonlinear depletion (i.e., voltage-controlled capacitance $C_{TSV}(V_{TSV})$). Fig. 2.2.17b shows the total capacitance of the coaxial TSV for three cases. It is evident that ignoring the floating substrate effect results in inaccuracy.

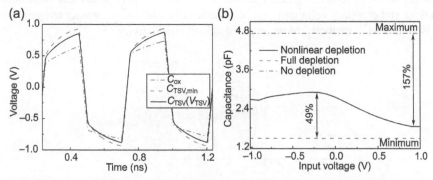

Figure 2.2.17 (a) Transient waveform of the output voltage and (b) total capacitance of the coaxial through-silicon via (TSV) with electrically floating inner silicon for different cases.

2.2.3 Through-silicon via signal/power integrity

2.2.3.1 Crosstalk Effect

With the increasing demand for high bandwidth, the operating frequency and the TSV density are continually increased. Under such circumstance, the noise coupling issue is becoming a major concern in the 3D ICs. Considering a uniformly distributed TSV array, the parasitic capacitance and the resistance of each TSV can be computed by Eqs. (2.2.11) and (2.2.12), respectively. The other circuit parameters can be obtained based on the multiconductor TL theory [27]. Taking one ground TSV as the reference TSV, the loop inductance of the ith TSV and ground TSVs can be obtained by

$$L_{ii} = \frac{\mu H_{TSV}}{\pi} \ln \frac{P_i}{r_{via}} \tag{2.2.40}$$

and the mutual inductance between two loops is given by

$$L_{ij} = \frac{\mu H_{TSV}}{\pi} \ln \frac{P_i P_j}{P_{ij} r_{via}}, \quad i \neq j \tag{2.2.41}$$

where P_i (P_j) is the pitch between the ith (jth) TSV and ground TSV and P_{ij} is the pitch between the ith and jth TSVs. Then, the inductance matrix $[L]$ can be established, and $[L_{Si}]$ can be formed by replacing r_{via} with r_{dep}. The silicon capacitance matrix can be obtained by $[C_{Si}] = \mu \varepsilon_{Si} H_{TSV}^2 [L_{Si}]^{-1}$, while the silicon conductance matrix $[G_{Si}] = \sigma_{Si}[C_{Si}]/\varepsilon_{Si}$ [26].

Take for example the typical three-TSV array, namely signal-ground-signal TSVs (see Fig. 2.2.18). The impedance of 50Ω is added to each port, and the source voltage is applied to the input port between S1 and ground TSVs. Fig. 2.2.19 shows magnitudes of S_{31} and S_{41} parameters of the three-TSV array with considering the voltage-controlled capacitance. Further, the time-domain crosstalk analysis is carried out for different temperatures, as shown in Fig. 2.2.20. It is found that both near- and far-end crosstalk voltages decrease significantly with the increasing temperature, which is mainly attributed to the reduced silicon conductivity.

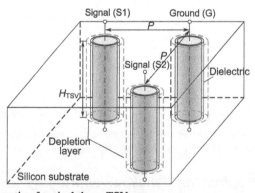

Figure 2.2.18 Schematic of typical three-TSV array.

2.2.3.2 Differential signaling

The continuous demand for higher bandwidth performance necessitates higher operating frequency, which makes the signal integrity problems worse. To guarantee signal integrity, differential signaling is essential for real-world applications of 3D ICs [55]. Fig. 2.2.21 shows the configuration of the ground-signal-signal-ground TSVs for

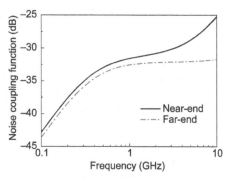

Figure 2.2.19 Noise coupling function in three-TSV array.

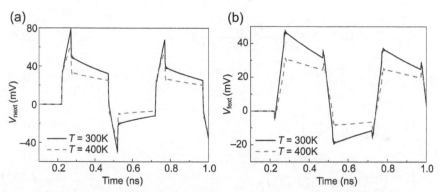

Figure 2.2.20 (a) Near-end crosstalk voltage V_{next} and (b) far-end crosstalk voltage V_{fext} in three-TSV array.

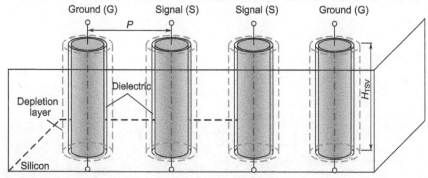

Figure 2.2.21 Schematic of ground-signal-signal-ground through-silicon vias (TSVs) for differential signaling.

differential signaling. In the figure, two signal TSVs are placed in the middle of two ground TSVs. In general, the pitch between adjacent TSVs is six times larger than the TSV radius, and therefore, the differential TSVs are weakly coupled. Under such circumstance, the silicon admittance between adjacent pair of TSVs $Y_{Si}(=G_{Si}+j\omega C_{Si})$ can be computed by Eqs. (2.2.16) and (2.2.17). Using the Δ-Y-Δ transformation method, the circuit model can be simplified into the equivalent circuit model shown in Fig. 2.2.22. The admittance between signal and ground TSVs can be given by

$$Y = G + j\omega C = \frac{1}{2Y_{Si}}\left(\frac{1}{2Y_2 Y_{Si}} + \frac{1}{2Y_1 Y_{Si} + Y_1^2}\right)^{-1} \tag{2.2.42}$$

while the admittance between two signal TSVs is given by

$$Y_m = G_m + j\omega C_m = \frac{Y_2^2 Y_{Si}}{2Y_1 Y_{Si} + 2Y_2 Y_{Si} + Y_1^2} \tag{2.2.43}$$

where

$$Y_1 = \left(\frac{1}{Y_{Si}} + \frac{1}{j\omega C_{TSV}}\right)^{-1} \tag{2.2.44}$$

$$Y_2 = \left(\frac{1}{2Y_{Si} + Y_1} + \frac{1}{j\omega C_{TSV}}\right)^{-1} \tag{2.2.45}$$

Based on the current directions in the differential mode and common mode, the frequency-dependent impedance $Z_{TSV}(=R_{TSV}+j\omega L_{TSV})$ can be extracted using the PEEC method [56]. Based on the circuit model, the odd mode and even mode propagation constants and characteristic impedances can be obtained by

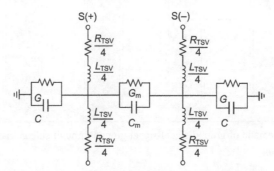

Figure 2.2.22 Equivalent circuit model of the differential through-silicon vias (TSVs).

$$\gamma_{\text{odd}} = \frac{1}{H_{\text{TSV}}} \cdot \sqrt{\frac{Z_{\text{TSV}}(Y + 2Y_m)}{2}} \qquad (2.2.46)$$

$$Z_{\text{odd}} = \sqrt{\frac{Z_{\text{TSV}}}{2(Y + 2Y_m)}} \qquad (2.2.47)$$

$$\gamma_{\text{even}} = \frac{1}{H_{\text{TSV}}} \cdot \sqrt{\frac{Z_{\text{TSV}} Y}{2}} \qquad (2.2.48)$$

$$Z_{\text{even}} = \sqrt{\frac{Z_{\text{TSV}}}{2Y}} \qquad (2.2.49)$$

Then, the *ABCD* transmission matrixes and *S*-parameter matrixes of the differential TSVs can be obtained by substituting Eqs. (2.2.42)– (2.2.45) into Eq. (2.2.1) and Eq. (2.2.22) [57].

Fig. 2.2.23 shows the scattering parameters of the differential TSVs. It can be seen that the forward transmission coefficient magnitude in the common mode is larger than that in the differential mode as the frequency is below 35 GHz, as a virtual ground exists between two signal TSVs in the differential mode. However, because of the negative mutual inductance between two signal TSVs in the differential mode, the differential mode inductance is smaller than the common mode counterpart. Therefore, in the high-frequency region, the electrical performance in the differential mode is better than that in the common mode.

It is well known that the annular TSVs have easier fabrication process and can alleviate the reliability issues due to the reduced thermal expansion mismatch between the TSV filling conductor and the silicon substrate. As shown in Fig. 2.2.24, by replacing the conventional cylindrical TSVs with annular TSVs, the insertion loss in the

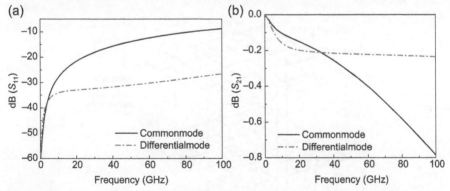

Figure 2.2.23 *S*-parameters of the differential through-silicon vias (TSVs). (a) dB(S_{11}) and (b) dB(S_{21}).

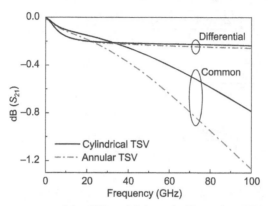

Figure 2.2.24 S_{21}-parameters of the differential through-silicon vias (TSVs).

common mode is aggravated significantly, while it is almost unchanged in the differential mode, implying that the annular TSVs are more suitable for the applications of transmitting differential signals than the conventional cylindrical TSVs.

2.2.3.3 Power distribution network impedance

Besides the signal integrity issues, the TSV-based 3D ICs have another major challenge: how to deliver clean power to the switched devices, in particular on the topmost chip? Therefore, special attention should be paid to the design and optimization of PDN in the 3D ICs, as shown in Fig. 2.2.25. Generally speaking, the low-frequency PDN impedance is dominated by the off-chip PDN, while the high-order resonances beyond several tens of gigahertz are induced by the chip PDN alone [58]. However, the inductance produced by TSVs will resonate with the chip PDNs and thereby lead to antiresonance peak in the PDN impedance at the midfrequency [59].

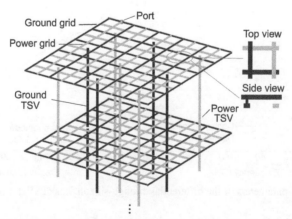

Figure 2.2.25 Schematic of a multistacked chip power distribution network (PDN) connected by through-silicon vias (TSVs).

It is essential to mitigate the antiresonance peaks to keep PDN impedance under the target impedance over the whole operating frequency of devices. Therefore, some necessary measures, such as decoupling capacitor placement [60], should be taken in the PDN design.

To facilitate the PDN design, the primary objective is to develop a circuit model to capture the PDN impedance. An equivalent circuit model of the 3D IC PDN was proposed in Ref. [61], and the circuit elements were calculated analytically. Based on the proposed circuit model, the electrical performance of 3D IC PDN can be investigated. As aforementioned, CNTs can be utilized as the filling conductor for TSV applications, and it is essential to investigate the impacts of CNT TSVs on the characteristics of 3D IC PDN [62]. Fig. 2.2.26 shows the self-impedance of 2-stacked chip PDNs connected by a pair of power and ground TSVs. It is shown that the PDN impedance decreases linearly as the frequency is below several gigahertz, implying that chip PDN behaves as a capacitive element in the low-frequency region. The first two resonances are formed by the TSV inductance and chip PDN capacitances. As the frequency exceeds 20 GHz, the resonances are only induced by the chip PDNs.

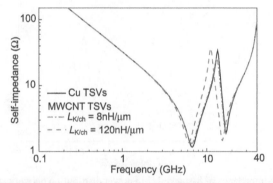

Figure 2.2.26 Self-impedance of 2-stacked chip power distribution networks (PDNs) connected by P/G through-silicon via (TSV) pair.

Further, the Cu TSVs are replaced with MWCNT TSVs with $f_{cnt} = 0.2$ and $L_K = 8$ nH/μm. It is shown that with the implementation of MWCNT TSVs, the impedance oscillation is suppressed and the resonances are shifted down slightly. If the CNT kinetic inductance is increased to 60 nH/μm, which is 15 times larger than the theoretical value, the inductance of MWCNT TSVs will be increased significantly, thereby decreasing the antiresonance frequency. Fig. 2.2.27 shows the PDN impedance of the 8-stacked chips connected by the power and ground TSVs. It can be seen multiple resonances are induced by the TSV inductances and chip PDN capacitances. Analogously, the presence of MWCNT TSVs suppresses the impedance oscillation, and large CNT kinetic inductance will reduce the frequency range of the 3D IC PDNs. To avoid the influence of CNT TSVs on the PDN performance, the fabrication process should be further optimized to densify CNTs.

Finally, the impact of temperature on the self-impedance of the 3D IC PDN with MWCNT TSVs is characterized, as shown in Fig. 2.2.28. The temperature dependence

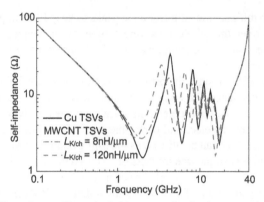

Figure 2.2.27 Self-impedance of 8-stacked chip power distribution networks (PDNs) connected by P/G through-silicon via (TSV) pairs.

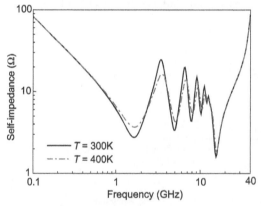

Figure 2.2.28 Self-impedance of 8-stacked chip power distribution networks (PDNs) connected by P/G multiwalled carbon nanotube (MWCNT) through-silicon via (TSV) pairs at different temperatures.

of TSV capacitance has been considered [25]. As the temperature rises, the resistance and the inductance of the MWCNT TSVs are increased and decreased, respectively. Therefore, it can be seen that the resonances are shifted slightly, and the impedance oscillation is suppressed.

2.2.4 TSV/IPD interposer

To guarantee high yield rate, a minimum TSV density rule should be met in the real-world fabrication. For example, Tezzaron requires that there must be at least one TSV in every 250×250 μm^2 window [63]. A lot of dummy TSVs need to be inserted to satisfy the rule over the entire chip area, thereby increasing the area overhead. It is intuitive to make use of these dummy TSVs to realize passive components as integrated passive devices (IPDs).

Basic studies of TSV-based components have been reported by several academic researchers Ref. [64,65]. In Ref. [64], TSVs were proposed to build substrate-integrated waveguide, which behaves as a high-pass filter. They were also employed to form a hairpin band pass filter to reduce the footprint. It is evident that the electrical performances of these TSV-based components are limited by the dielectric loss due to the lossy silicon substrate. To alleviate these problems, high resistivity silicon substrate or polymer-enhanced technology can be utilized [66]. Another interest is in the application of dummy TSVs toward on-chip inductors, which are critical for various microelectronic applications, such as voltage control oscillators and radiofrequency circuits. In this aspect, Tida, et al. systematically studied the TSV-based solenoid inductors [67]. They found that the substrate loss can be suppressed dramatically by utilizing the microchannel technique. By combining two TSV-based solenoid inductors, a high-performance 3D transformer was developed [68]. In addition, an LC resonator was fabricated by integrating TSV-based solenoid inductors with an interdigitated capacitor [65].

Although there are multiple loss mechanisms for TSV-based solenoid inductors, it can still be modeled with the conventional π model. Fig. 2.2.29a shows the structure of TSV-based solenoid inductor, which is composed of TSVs and redistribution lines. The simple π circuit model is illustrated in Fig. 2.2.29b [69]. A RL ladder is used to model frequency-dependent parameters. The series and shunt admittances can be given as

$$Y_{\text{series}} = \left(\frac{1}{R_{s0}} + \frac{1}{R_{s1} + j\omega L_{s1}}\right)^{-1} + j\omega L_{s0} \qquad (2.2.50)$$

$$Y_{\text{shunt}} = \left(\frac{1}{j\omega C_{\text{ox}}} + \frac{1}{G_{\text{Si}} + j\omega C_{\text{Si}}}\right)^{-1} \qquad (2.2.51)$$

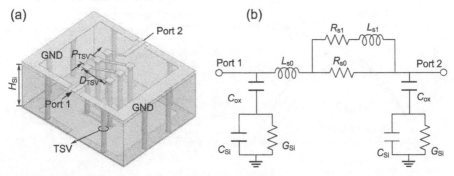

Figure 2.2.29 (a) Schematic of through-silicon vias (TSV)−based solenoid inductor and its (b) circuit model.

where C_{ox} is the oxide capacitance and G_{Si} and C_{Si} are the silicon conductance and capacitance, respectively. Based on the measured scattering parameters, the admittance matrix of the TSV-based solenoid inductor can be obtained.

$$Y = \begin{bmatrix} Y_{11} & Y_{12} \\ Y_{21} & Y_{22} \end{bmatrix} \tag{2.2.52}$$

At low frequencies, $Y_{series} = -Y_{12}$ and $Y_{shunt} = Y_{11} + Y_{12}$. Therefore, the circuit parameters in Fig. 2.2.29b can be extracted as follows:

$$R_{s1} = a_0 - \frac{a_0^2 a_2}{a_1^2} \tag{2.2.53}$$

$$R_{s0} = \frac{a_0 R_{s1}}{R_{s1} - a_0} \tag{2.2.54}$$

$$L_{s1} = \frac{1}{R_{s1}} \sqrt{a_1 (R_{s0} + R_{s1})^3} \tag{2.2.55}$$

$$L_0 = L_{dc} - \frac{R_{s0}^2 L_{s1}}{(R_{s0} + R_{s1})^2} \tag{2.2.56}$$

$$C_{ox} = -\frac{1}{\omega \text{Im}\left[(Y_{11} + Y_{12})^{-1}\right]} \tag{2.2.57}$$

$$G_{Si} = b_0 \tag{2.2.58}$$

$$C_{Si} = \sqrt{b_0 b_1} \tag{2.2.59}$$

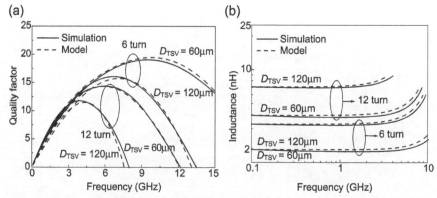

Figure 2.2.30 (a) Schematic of through-silicon via (TSV)–based solenoid inductor and its (b) circuit model.

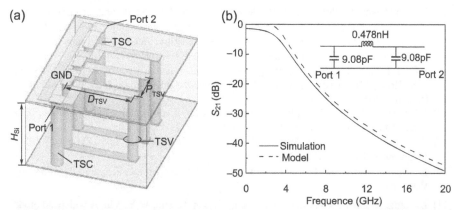

Figure 2.2.31 (a) Schematic of through-silicon vias (TSV)—based Butterworth filter and its (b) filtering characteristics.

The coefficients a_0, a_1, a_2, b_0, and b_1 can be obtained by fitting $-\text{Re}(1/Y_{\text{series}}) = a_0 + a_1\omega^2 + a_2\omega^4$ and $1/\text{Re}(1/Y_{\text{shunt}}) = b_0 + b_1\omega^2$ [70]. Here, the simulated results are employed for the model validation. Fig. 2.2.30 plots the quality factor and effective inductance, which can be obtained by

$$\text{Quality factor} = -\frac{\text{Im}(Y_{11})}{\text{Re}(Y_{11})} \tag{2.2.60}$$

$$L_{\text{eff}} = \frac{1}{\omega}\text{Im}\left(\frac{1}{Y_{11}}\right) \tag{2.2.61}$$

It can be seen that the modeled results agree well with the measurements over a wide frequency range.

Further, a compact third-order Butterworth filter is designed based on the TSV technology. As shown in Fig. 2.2.31a, the filter is composed of one TSV-based solenoid inductor and two through-silicon capacitors [71]. The equivalent circuit model of the ideal third-order Butterworth filter is given in the inset of Fig. 2.2.31b. It can be seen that the simulated and modeled results match well with each other. The discrepancy is mainly due to the parasitic capacitance and series resistance [72].

References

[1] J.H. Lau, C.P. Wong, W. Nakayama, J.L. Prince, Electronic Packaging: Design, Materials, Process, and Reliability, McGraw-Hill, New York, 1998.
[2] M. Swaminathn, J. Kim, I. Novak, J.P. Libous, Power distribution networks for system-on-packages: Status and challenges, IEEE Trans. Adv. Packag. 27 (2) (2004) 286–300.

[3] R. Jakushokas, M. Popovich, A.V. Mezhiba, S. Kose, E.G. Friedman, Power Distribution Networks with On-Chip Decoupling Capacitors, second ed., Springer, Boston, 2010.
[4] R.R. Tummala, Fundamentals of Microsystems Packaging, McGraw-Hill, New York, 2001.
[5] D.M. Pozar, Microwave Engineering, John Wiley & Sons, Hoboken, 2009.
[6] I. Bahl, D.K. Trivedi, A designer's guide to microstrip line, Microwaves (1977) 174–182.
[7] J. S. Hong, Microstrip Filters for RF/Microwave Applications, John Wiley & Sons, Hoboken, 2011.
[8] J. Svacina, Analysis of multilayer microstrip lines by a conformal mapping method, IEEE Trans. Microw. Theory Tech. 40 (4) (1992) 769–772.
[9] R.N. Simons, Coplanar Waveguide Circuits, Components, and Systems, Wiley-Interscience, New York, 2001.
[10] E. Bogatin, Signal and Power Integrity—Simplified, second ed., Pearson Education New York, 2010.
[11] K. Kang, X. Cao, Y. Wu, Z. Gao, Z. Tang, Y. Ban, L. Sun, W.Y. Yin, A wideband model for on-chip interconnects with different shielding structures, IEEE Trans. Compon. Packag. Manuf. Technol. 7 (10) (2017) 1702–1712.
[12] M. Swaminathan, K.J. Han, Design and Modeling for 3-D ICs and Interposers, World Scientific, 2014.
[13] E.P. Li, X.C. Wei, A.C. Cangellaris, E.X. Liu, Y.J. Zhang, M. D'Amore, J. Kim, T. Sudo, Progress review of electromagnetic compatibility analysis technologies for packages, printed circuit boards, and novel interconnects, IEEE Trans. Electromagn Compat. 52 (2) (2010) 248–265.
[14] N. Karim, J. Mao, J. Fan, Improving electromagnetic compatibility of packages and SiP modules using a conformal shielding solution, in: Asia-Pacific Symposium on Electromagnetic Compatibility, Beijing, China, 2010.
[15] M. Motoyoshi, Through-silicon via (TSV), Proc. IEEE 97 (1) (2009) 43–48.
[16] J.P. Gambino, S.A. Adderly, J.U. Knickerbocker, An overview of through-silicon-via technology and manufacturing challenges, Microelectron. Eng. 135 (2015) 75–106.
[17] G. Katti, M. Stucchi, K. De Meyer, W. Dehaene, Electrical modeling and characterization of through silicon via for three-dimensional ICs, IEEE Trans. Electron Devices 57 (1) (2010) 256–262.
[18] C. Xu, H. Li, R. Suaya, K. Banerjee, Compact AC modeling and performance analysis of through-silicon vias in 3-D ICs, IEEE Trans. Electron Devices 57 (12) (2010) 3405–3417.
[19] B. Xie, M. Swaminathan, FDFD modeling of signal paths with TSVs in silicon interposer, IEEE Trans. Compon. Packag. Manuf. Technol. 4 (4) (2014) 708–717.
[20] T. Bandyopadhyay, K.J. Han, D. Chung, R. Chatterjee, M. Swaminathan, R. Tummala, Rigorous electrical modeling of through silicon vias (TSVs) with MOS capacitance effects, IEEE Trans. Compon. Packag. Manuf. Technol. 1 (6) (2011) 893–903.
[21] E.X. Liu, E.P. Li, W.B. Ewe, H.M. Lee, T.G. Lim, S. Gao, Compact wideband equivalent-circuit model for electrical modeling of through-silicon via, IEEE Trans. Microw. Theory Tech. 59 (6) (2011) 1454–1460.
[22] A.B. Sproul, M.A. Green, Improved value for the silicon intrinsic carrier concentration from 275 to 300 K, J. Appl. Phys. 70 (2) (1991) 846–854.
[23] D.W. Hess, Plasma-assisted oxidation, anodization, and nitridation of silicon, IBM J. Res. Dev. 43 (1/2) (1999) 127–145.
[24] G. Katti, M. Stucchi, J. Van Olmen, K. De Meyer, W. Dehaene, Through-silicon-via capacitance reduction technique to benefit 3-D IC performance, IEEE Electron. Device Lett. 31 (6) (2010) 549–551.
[25] G. Katti, M. Stucchi, D. Velenis, B. Soree, K. De Meyer, W. Dehaene, Temperature-

dependent modeling and characterization of through-silicon via capacitance, IEEE Electron. Device Lett. 32 (4) (2011) 563−565.
[26] A.E. Engin, S.R. Narasimhan, Modeling of crosstalk in through silicon vias, IEEE Trans. Electromagn Compat. 55 (1) (2013) 149−158.
[27] C.R. Paul, Analysis of Multiconductor Transmission Lines, Wiley, NY, 2008.
[28] D.M. Pozar, Microwave Engineering, second ed., Wiley, NY, 1998.
[29] W.S. Zhao, W.Y. Yin, X.P. Wang, X.L. Xu, Frequency- and temperature-dependent modeling of coaxial through-silicon vias for 3-D ICs, IEEE Trans. Electron Devices 58 (10) (2011) 3358−3368.
[30] W.R. Eisenstadt, Y. Eo, S-parameter-based IC interconnect transmission line characterization, IEEE Trans. Comp. Hybr. Manufac. Technol. 15 (4) (1992) 483−490.
[31] J. Kim, J.S. Pak, J. Cho, E. Song, J. Cho, H. Kim, T. Song, J. Lee, H. Lee, K. Park, S. Yang, M.S. Suh, K.Y. Byun, J. Kim, High-frequency scalable electrical model and analysis of a through silicon via (TSV), IEEE Trans. Compon. Packag. Manuf. Technol. 1 (2) (2011) 181−195.
[32] I. Ndip, B. Curran, K. Lobbicke, S. Guttowski, H. Reichl, K.D. Lang, H. Henke, High-frequency modeling of TSVs for 3-D chip integration and silicon interposers considering skin-effect, dielectric quasi-TEM and slow-wave modes, IEEE Trans. Compon.Packag. Manuf. Technol. 1 (10) (2011) 1627−1641.
[33] R. Fang, X. Sun, M. Miao, T. Jin, Characteristics of coupling capacitance between signal-ground TSVs considering MOS effect in silicon interposers, IEEE Trans. Electron Devices 62 (12) (2015) 4161−4168.
[34] D.A. Neamen, Semiconductor Physics and Devices: Basic Principle, fourth ed., McGraw-Hill, New York, 2011.
[35] L. Yu, Research into TSV Interconnect Characterization for 3D Integrated Circuit Design, PhD dissertation, UCAS, Beijing, China, 2012.
[36] S. Piersanti, F. de Paulis, A. Orlandi, J. Fan, Impact of frequency-dependent and nonlinear parameters on transient analysis of through silicon vias equivalent circuit, IEEE Trans. Electromagn Compat. 57 (3) (2015) 538−545.
[37] S. Ijima, Helical microtubules of graphitic carbon, Nature 354 (1991) 56−58.
[38] B.Q. Wei, R. Vajtai, P.M. Ajayan, Reliability and current carrying capacity of carbon nanotubes, Appl. Phys. Lett. 79 (2001) 1172−1174.
[39] D. Jiang, W. Mu, S. Chen, Y. Fu, K. Jeppson, J. Liu, Vertically stacked carbon nanotube-based interconnects for through silicon via application, IEEE Electron. Device Lett. 36 (5) (2015) 499−502.
[40] W.S. Zhao, L. Sun, W.Y. Yin, Y.X. Guo, Electrothermal modeling and characterisation of submicron through-silicon carbon nanotube bundle vias for three-dimensional ICs, Micro & Nano Lett. 9 (2) (2014) 123−126.
[41] H.Y. Zhang, Y.S. Wang, W.H. Zhu, T. Lin, Effective thermal conductivity model for TSVs with insulation layer as contact resistance, in: IEEE 16th International Conference on Electronic Packaging Technology, Changsha, China, 2015, pp. 125−131.
[42] A.G. Chiariello, A. Maffucci, G. Miano, Electrical modeling of carbon nanotube vias, IEEE Trans. Electromagn Compat. 54 (1) (2012) 158−166.
[43] Y.F. Liu, W.S. Zhao, Z. Yong, Y. Fang, W.Y. Yin, Electrical modeling of three-dimensional carbon-based heterogeneous interconnects, IEEE Trans. Nanotechnol. 13 (3) (2014) 488−496.
[44] W.S. Zhao, J. Zheng, L. Dong, F. Liang, Y. Hu, L. Wang, G. Wang, Q. Zhou, High-frequency modeling of on-chip coupled carbon nanotube interconnects for millimeter-

wave applications, IEEE Trans. Compon. Packag. Manuf. Technol. 6 (8) (2016) 1226–1232.

[45] A. Naeemi, J.D. Meindl, Compact physical models for multiwall carbon-nanotube interconnects, IEEE Electron. Device Lett. 27 (5) (2006) 338–340.

[46] G. Zhang, J.H. Warner, M. Fouquet, A.W. Robertson, B. Chen, J. Robertson, Growth of ultrahigh density single-walled carbon nanotube forests by improved catalyst design, ACS Nano 6 (4) (2012) 2893–2903.

[47] C. Subrammniam, T. Yamada, K. Kobashi, A. Sekiguchi, D.N. Futaba, M. Yumura, K. Hata, One hundred fold increase in current carrying capacity in a carbon nanotube-copper composite, Nat. Commun. 4 (2013) 2202.

[48] S. Sun, W. Mu, M. Edwards, D. Mencarelli, L. Pierantoni, Y. Fu, K. Jeppson, J. Liu, Vertically aligned CNT-Cu nano-composite material for stacked through-silicon-via interconnects, Nanotechnology 27 (2016) 335705.

[49] W.S. Zhao, J. Zheng, Y. Hu, S. Sun, G. Wang, L. Dong, L. Yu, L. Sun, W.Y. Yin, High-frequency analysis of Cu-carbon nanotube composite through-silicon vias, IEEE Trans. Nanotechnol. 15 (3) (2016) 506–511.

[50] J.J. Plombon, K.P. O'Brien, F. Gstrein, V.M. Dubin, Y. Jiao, High-frequency electrical properties of individual and bundled carbon nanotubes, Appl. Phys. Lett. 90 (6) (2007) 063106.

[51] S.W. Ho, S.W. Yoon, Q. Zhou, K. Pasad, V. Kripesh, J.H. Lau, High RF performance TSV silicon carrier for high frequency application, in: Proc. Electronic Components and Technology Conference, Lake Buena Vista, FL, USA, 2008, pp. 1946–1952.

[52] W.S. Zhao, X.P. Wang, W.Y. Yin, Electrothermal effects in high density through silicon via (TSV) arrays, Prog. Electromagn. Res. 115 (2011) 223–242.

[53] F. Liang, G. Wang, D. Zhao, B.Z. Wang, Wideband impedance model for coaxial through-silicon vias in 3-D integration, IEEE Trans. Electron Devices 60 (8) (2013) 2498–2504.

[54] W.S. Zhao, J. Zheng, J. Wang, F. Liang, F. Wen, L. Dong, D. Wang, G. Wang, Modeling and characterization of coaxial through-silicon via with electrically floating inner silicon, IEEE Trans. Compon. Packag. Manuf. Technol. 7 (6) (2016) 936–943.

[55] J. Kim, J. Cho, J. Kim, J.M. Yook, J.C. Kim, J. Lee, K. Park, J.S. Pak, High-frequency scalable modeling and analysis of a differential signal through-silicon via, IEEE Trans. Compon. Packag. Manuf. Technol. 4 (4) (2011) 697–707.

[56] W.S. Zhao, J. Zheng, F. Liang, K. Xu, X. Chen, G. Wang, Wideband modeling and characterization of differential through-silicon vias for 3-D ICs, IEEE Trans. Electron Devices 63 (3) (2016) 1168–1175.

[57] Q. Lu, Z. Zhu, Y. Yang, R. Ding, Electrical modeling and characterization of shield differential through-silicon vias, IEEE Trans. Electron Devices 62 (5) (2015) 1544–1552.

[58] H. He, J.J.Q. Lu, Modeling and analysis of PDN impedance and switching noise in TSV-based 3-D integration, IEEE Trans. Electron Devices 62 (4) (2015) 1241–1247.

[59] J.S. Pak, J. Kim, J. Cho, K. Kim, T. Song, S. Ahn, J. Lee, H. Lee, K. Park, J. Kim, PDN impedance modeling and analysis of 3D TSV IC by using proposed P/G TSV array model based on separated P/G TSV and chip-PDN models, IEEE Trans. Compon. Packag. Manuf. Technol. 1 (2) (2011) 208–219.

[60] E. Song, K. Koo, J.S. Pak, J. Kim, Through-silicon-via-based decoupling capacitor stacked chip in 3-D ICs, IEEE Trans. Compon. Packag. Manuf. Technol. 3 (9) (2013) 1467–1480.

[61] C.H. Cheng, T.Y. Cheng, C.H. Du, Y.C. Lu, Y.P. Chiou, S. Liu, T.L. Wu, An equation-based circuit model and its generation tool for 3-D IC power delivery networks with an

emphasis on coupling effect, IEEE Trans. Compon. Packag. Manuf. Technol. 4 (6) (2014) 1062—1070.

[62] J. Jin, W.S. Zhao, D.W. Wang, H.S. Chen, E.P. Li, W.Y. Yin, Investigation of carbon nanotube-based through-silicon vias for PDN applications, IEEE Trans. Electromagn Compat. 60 (3) (2018).

[63] D.H. Kim, K. Athikulwongse, M. Healy, M. Hossain, M. Jung, I. Khorosh, G. Kumar, Y.J. Lee, D. Lewis, T.W. Lin, C. Liu, S. Panth, M. Pathak, M. Ren, G. Shen, T. Song, D.H. Woo, X. Zhao, J. Kim, H. Choi, G. Loh, H.H. Lee, S.K. Lim, 3D-MAPS: 3D massively parallel processor with stacked memory, in: 2012 IEEE International Solid-State Circuits Conference Design of Technical Papers, ISSCC, San Francisco, CA, USA, 2012, pp. 188—190.

[64] S. Hu, L. Wang, Y.Z. Xiong, T.G. Lim, B. Zhang, J. Shi, X. Yuan, TSV technology for millimeter-wave and terahertz design and applications, IEEE Trans. Compon. Packag. Manuf. Technol. 1 (2) (2011) 260—267.

[65] W.A. Vitale, M. Fernandez-Bolanos, A. Klumpp, J. Weber, P. Ramm, A.M. Ionescu, in: Ultra Fine-Pitch TSV Technology for Ultra-dense High-Q RF Inductors, 2015 Symposium on VLSI Technology, Kyoto, Japan, 2015, pp. 52—53.

[66] P.A. Thadesar, M.S. Bakir, Fabrication and characterization of polymer-enhanced TSVs, inductors, and antennas for mixed-signal silicon interposer platforms, IEEE Trans. Compon. Packag. Manuf. Technol. 6 (3) (2016) 455—463.

[67] U.R. Tida, R. Yang, C. Zhuo, Y. Shi, On the efficacy of through-silicon-via inductors, IEEE Trans. Very Large Scale Integr. Syst. 23 (7) (2015) 1322—1334.

[68] Z. Feng, M.R. Lueck, D.S. Temple, M.B. Steer, High-performance solenoidal RF transformers on high-resistivity silicon substrates for 3D integrated circuits, IEEE Trans. Microw. Theory Tech. 60 (7) (2012) 2066—2072.

[69] J. Zheng, D.W. Wang, W.S. Zhao, G. Wang, W.Y. Yin, Modeling of TSV-based solenoid inductors for 3-D integration, in: 2015 IEEE MTT-S International Microwave Workshop Series on Advanced Materials and Processes for RF and THz Applications (IMWS-AMP), Suzhou, China, 2015.

[70] H.H. Chen, H.W. Zhang, S.J. Chung, J.T. Kuo, T.C. Wu, Accurate systematic model-parameter extraction for on-chip spiral inductors, IEEE Trans. Electron Devices 55 (11) (2008) 3267—3273.

[71] K. Dieng, P. Artillan, C. Bermond, O. Giller, T. Lacrevaz, S. Joblot, G. Houzet, A. Farcy, A.L. Perrier, Y. Lamy, B. Flechet, Modeling and frequency performance analysis of through silicon capacitors in silicon interposer, IEEE Transactions on Components, Packaging and Manufacturing Technology 7 (4) (2017) 477—484.

[72] F. Wang, N. Yu, An ultracompact Butterworth low-pass filter based on coaxial through-silicon vias, IEEE Trans. Very Large Scale Integr. Syst. 25 (3) (2017) 1164—1167.

[73] W.S. Zhao, J. Zheng, S. Chen, X. Wang, G. Wang, Transient analysis of through-silicon vias in floating silicon substrate, IEEE Trans. Electromagn Compat. 59 (1) (2017) 207—216.

Thermal modeling, analysis, and design

3.1 Principles of thermal analysis and design

Waste heat is generated during the operation of electronic devices and systems, which is to be transferred to the ambient in a timely manner to avoid overheating and thermal failures. The thermal analysis and design reply on the heat transfer theories as documented in Refs. [1–3], which are summarized in this section. In addition, the thermal interface resistance related to different types of thermal interface materials (TIMs) encountered in electronic packaging is elaborated in this section.

3.1.1 Principles of thermal analysis

There exist three modes of heat transfer: heat conduction, heat convection, and heat radiation. Heat conduction is referred to the heat transfer that occurs across a medium due to the activities of atoms, molecules, or electrons, whereas heat convection is referred to heat transfer that occurs between a surface and a moving fluid at different temperatures. The third mode of heat transfer is called thermal radiation in the form of electromagnetic waves. In the absence of an intervening medium, there is net heat transfer by radiation between two surfaces at different temperatures. Heat transfer processes can be quantified in terms of the heat transfer laws, which are used to determine the amount of energy being transferred at the macroscopic level.

3.1.1.1 Heat conduction

As shown in Figure 3.1.1, the heat conduction rate across a plane wall is governed by Fourier's law. Namely

$$q_x = -kA\frac{dT}{dx} \qquad (3.1.1)$$

Here A is the area perpendicular to the surface, q_x is the heat flow (W) defining the heat transfer rate in the x-direction per unit area perpendicular to the direction of transfer, which is proportional to the temperature gradient, dT/dx, in this direction. k is the thermal conductivity (unit: $W \cdot m^{-1} \cdot K^{-1}$) and is a characteristic property of the conductive material. The minus sign in Eq. (3.1.1) is due to the fact that heat is transferred in the direction of decreasing temperature. Under the steady-state heat transfer

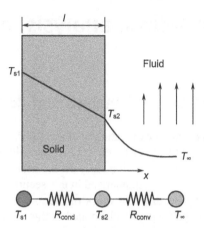

Figure 3.1.1 The steady-state heat transfer process and the thermal resistance network, with heat conduction from the hot wall with temperature T_{s1} to the cold wall with temperature T_{s2} and heat convection from cold wall to the ambient air with temperature T_∞.

as illustrated in Figure 3.1.1, the temperature gradient is linear across a homogeneous material, and the heat flow can be expressed as

$$q_x = kA\frac{T_{s1} - T_{s2}}{l} = kA\frac{\Delta T}{l} \tag{3.1.2}$$

where T_{s1} and T_{s2} are the temperatures of the hot and cold surfaces of the wall, respectively, l is the thickness of the wall, and A is the area normal to the direction of heat transfer.

The thermal conductivity is a key material property, varying with material types, temperature, and even size at the micro- to nanoscale. The range of thermal conductivity for different states of matter is shown in Figure 3.1.2 at normal temperatures and pressure. Among them, graphene has a thermal conductivity up to $5000 \text{W} \cdot \text{m}^{-1} \cdot \text{K}^{-1}$, whereas the lowest thermal conductivity matters include both the gaseous CO_2 ($0.0166 \text{W} \cdot \text{m}^{-1} \cdot \text{K}^{-1}$) and the solid aerogel ($0.013 \text{W} \cdot \text{m}^{-1} \cdot \text{K}^{-1}$). From the gases to the highly thermally conductive graphene, there is a difference of nearly 6 orders of magnitude. The thermal conductivity is also a function of temperature. The temperature dependence of thermal conductivity for selected materials is shown in Figure 3.1.3 for selected materials used in electronic packaging, including the major semiconductor materials such as silicon, silicon carbide, and gallium nitride as well as the fluid coolants applicable in electronic cooling.

3.1.1.2 Micro- and nanoscale effect on heat conduction

In the preceding section, the thermal conductivity for bulk materials is discussed. In applications such as electronics packaging, the material's characteristic dimensions can be on the order of micrometers or nanometers, and care should be taken to account

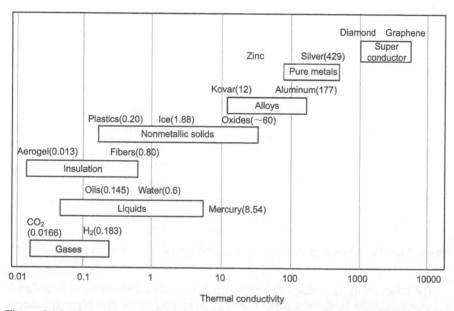

Figure 3.1.2 Thermal conductivity of different states of matter at normal temperature and pressure.

for the possible reduction of thermal conductivity that can occur with size reduction. For example, the energy conversion at the drain of a nanoscale MOSFET device would lead to a temperature much higher than what is predicted based on the classical Fourier's law, resulting in a much faster thermal failure [2]. As the film dimensions are much larger than the mean free path, or l_{mfp}, namely, $l_x/l_{mfp} \gg 1$, the scattering effect of the boundaries on reducing the average energy carrier path length is minor, and conduction heat transfer behaves as the bulk material. However, as the film becomes thinner, the physical boundaries of the material can decrease the average net distance traveled by the energy carriers, typically electrons and phonons. The development of aerogel as insulation makes use of air with the size less than its l_{mfp} to achieve a thermal conductivity even lower than that for the bulk air.

For $l_x/l_{mfp} > 1$, the predicted thermal conductivity components perpendicular to the film plane and in parallel to the film plane, k_z and k_x, may be estimated to be within 20% from the following expression.

$$\frac{k_z}{k} = 1 - \frac{l_{mfp}}{3l_z} \qquad (3.1.3a)$$

$$\frac{k_x}{k} = 1 - \frac{2l_{mfp}}{3\pi l_x} \qquad (3.1.3b)$$

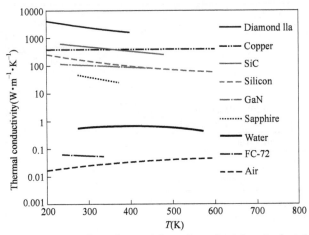

Figure 3.1.3 The temperature dependence of thermal conductivity of selected materials.

The critical film thicknesses, below which microscale effects must be considered, $l_{crit,z}$ perpendicular to the film plane and $l_{crit,x}$ in parallel to the film plane are shown in Table 3.1.1, together with the mean free path for several materials at temperature $T \sim 300$ K [1,3].

Table 3.1.1 Mean free path and critical film thickness for various materials at $T \sim 300$K [1,4].

Material	l_{mfp} (nm)	$l_{crit,z}$ (nm)	$l_{crit,x}$ (nm)
Aluminum oxide	5.08	36	22
Diamond (IIa)	315	2200	1400
Gallium arsenide	23	160	100
Gold	31	220	140
Copper	40	280	180
Silicon	43	290	180
Silicon dioxide	0.6	4	3

3.1.1.3 Heat convection

The convection heat transfer mode comprises two mechanisms, the energy transfer due to random molecular motion or diffusion and the energy transferred by the bulk motion of the fluid. The convection heat transfer mode is sustained both by random molecular motion and by the bulk motion of the fluid within the boundary layer. Convection heat transfer can be classified into forced convection and natural convection (also called free convection). The forced convection is called so that the flow is caused by external means, such as by a fan or a pump. In contrast, the natural convection is induced by buoyancy forces, due to factors such as density differences associated with temperature variations in the fluid. An example is the natural convection heat transfer that occurs from hot components on a vertical array of circuit boards in air. Air that makes contact with the components experiences an increase in temperature and hence a reduction in density. As the heated air is lighter than the surrounding air, buoyancy forces induce a vertical motion for which warm air ascending from the boards is replaced by an inflow of cooler ambient air. In special cases, the phase change takes place involving the heat exchange and convection such as melting, boiling, and condensation.

The convection heat transfer is usually described by the Newton's law of cooling, namely,

$$q = hA(T_{s2} - T_\infty) \tag{3.1.4}$$

where h in $W \cdot m^{-2} \cdot K^{-1}$ is the convective heat transfer coefficient, A is the surface area, T_{s2} is the surface temperature, and T_∞ is the bulk temperature of the fluid. It is noted that h values vary significantly with the fluid types, flow speed, and geometrical configurations. For natural convection, h is in the range of $5-15 W \cdot m^{-2} \cdot K^{-1}$ in gases and $50-100 W \cdot m^{-2} \cdot K^{-1}$ in liquids. For forced convection, h values would reach $15-250 W \cdot m^{-2} \cdot K^{-1}$ in gases and of $100-2000 W \cdot m^{-2} \cdot K^{-1}$ in liquids. The heat transfer correlations for different regimes and specific configurations are tabulated in Table 3.1.2 [1,2].

3.1.1.4 Thermal radiation

Thermal radiation is energy emitted by matter that is at a nonzero temperature, transported in the form of electromagnetic waves (or alternatively, photons). The thermal radiation power from surface 1 to surface 2, as shown in Figure 3.1.4, is expressed by

$$q = \varepsilon \sigma A F_{12} (T_{s1}^4 - T_{s2}^4) \tag{3.1.5}$$

where T_{s1} and T_{s2} are the absolute temperature (K) of the surfaces 1 and 2, respectively, and σ is the Stefan–Boltzmann constant, $\sigma = 5.67 \times 10^8 \, W \cdot m^{-2} \cdot K^{-4}$, ε is the surface emissivity of the surface 1, and F_{12} is the view factor between surfaces 1 and 2. A surface with $\varepsilon = 1$ is called the blackbody, an ideal radiator. With ε values in the range $0-1$, this property provides a measure of efficiency for a surface emitting energy relative to the blackbody.

The thermal radiation may be neglected in the presence of forced convection in electronic cooling. Nonetheless, it may be significant when the natural convection is the major heat transfer mode.

3.1.1.5 Thermal resistance

An analogy exists between the diffusion of heat and electrical charge. Just as an electrical resistance is associated with the conduction of electricity as depicted in

Table 3.1.2 Empirical correlations for the heat transfer correlations under different conditions.

Regimes of convection	Correlations for the heat transfer coefficient
Natural convection from isothermal vertical surface with length l	$h = C\left(\frac{k}{l}\right)\left(\frac{\rho^2 \alpha g c_p \Delta T^3}{\mu k}\right)^n$ $C = 0.59, n = 1/4$ for $1 < \mathrm{Ra} < 10^9$ $C = 0.10, n = 1/3$ for $10^9 < \mathrm{Ra} < 10^{14}$
Natural convection from isoflux vertical surface with length l	$h = 0.631\left(\frac{k}{l}\right)\left[\frac{c_p \rho^2 g \alpha q''^4}{\mu k^2}\right]^{1/5}$
Natural convection on isothermal horizontal surface with length l	$h = C\left(\frac{k}{l}\right)\left(\frac{\rho^2 \alpha g c_p \Delta T^3}{\mu k}\right)^n$ $C = 0.54, n = 1/4$ for $10^4 \leqslant \mathrm{Ra}_l \leqslant 10^7$ $C = 0.15, n = 1/3$ for $10^7 \leqslant \mathrm{Ra}_l \leqslant 10^{11}$
Natural convection on vertical cylinder with length l [5]	$h = \left(\frac{k}{l}\right)\mathrm{Ra}^{0.25}\left(0.59 + 0.52\mathrm{Ra}^{-0.25}\frac{l}{D}\right)$
Natural convection in rectangular cavities with cold and hot vertical walls with width l	$h = 0.22\left(\frac{k}{l}\right)\left(\frac{\rho \alpha g c_p \Delta T_g}{k}\right)^n$ $2 < H/l < 10, \mathrm{Pr} < 10^5, 10^3 < \mathrm{Ra}_l < 10^{10}$
Forced convection on isothermal flat plate with length l	$h = C\left(\frac{k}{l}\right)\left(\frac{\rho u l}{\mu}\right)^n\left(\frac{\mu c_p}{k}\right)^{1/3}$ Laminar: $C = 0.0664, n = 1/2$ Turbulent: $C = 0.0296, n = 4/5$

Table 3.1.2 (continued)

Regimes of convection	Correlations for the heat transfer coefficient
Forced convection on isoflux flat plate with length l	$h = C\left(\frac{k}{l}\right)\left(\frac{\rho u l}{\mu}\right)^n \left(\frac{\mu c_p}{k}\right)^{1/3}$ Laminar: $C = 0.453, n = 1/2$ Turbulent: $C = 0.0308, n = 4/5$
Laminar forced convection in circular tube with diameter D and length l	$h = \frac{k}{D}\left(3.66 + \frac{0.0668(D/l)\text{Re}_D\text{Pr}}{1+0.04[(D/l)\text{Re}_D\text{Pr}]^{2/3}}\right) \text{Re}_D = \left(\frac{\rho u D}{\mu}\right), \text{Pr} = \left(\frac{\mu c_p}{k}\right) \text{ for } \text{Re}_D < 2000$
Turbulent forced convection in circular tube with diameter D	$h = 0.023\left(\frac{k}{D}\right)\text{Re}_D^{0.8}\text{Pr}^n (\text{Re}_D > 2000)$ $n = 0.4$ for heating, $n = 0.3$ for cooling

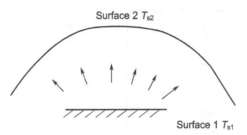

Figure 3.1.4 Heat transfer from a surface at temperature T_{s1} to a hemisphere ambient at temperature T_{s2} by thermal radiation.

Ohm's law, the thermal resistance can be associated with the transfer of heat. Defined as the ratio of a driving potential to the corresponding transfer rate, the thermal resistance R_{cond} for conduction in a planar wall, following Eq.(3.1.2), is expressed as

$$R_{cond} = \frac{T_{s1} - T_{s2}}{q_x} = \frac{l}{kA} \tag{3.1.6a}$$

Or the specific thermal resistance per unit area R''_{cond} is

$$R''_{cond} = \frac{T_{s1} - T_{s2}}{q_x} A = \frac{l}{k} \tag{3.1.6b}$$

Note that R_{cond} and R''_{cond} have the units of K/W and K·cm²/W, respectively. The thermal resistances take different forms for the different modes of heat transfer. The thermal resistance for convection is then

$$R_{conv} = \frac{T_{s2} - T_\infty}{q_x} = \frac{1}{hA} \tag{3.1.7}$$

The specific thermal resistance for convection is simply the inverse of heat transfer coefficient. For thermal radiation, the thermal resistance is expressed by

$$R_{conv} = \frac{1}{\varepsilon \sigma A F_{12}(T_{s1} + T_{s2})(T_{s1}^2 + T_{s2}^2)} \tag{3.1.8}$$

Several thermal resistances based on the junctions of semiconductor chips have been widely adopted in designing and characterizing various thermal problems in electronic packaging and systems, which are to be elaborated in the later sections.

Thermal resistance network representations provide a useful tool in both conceptualizing and describing heat transfer problems quantitatively. The equivalent thermal resistance network for the plane wall with convection surface conditions has been shown in Figure 3.1.1. The overall heat transfer rate may be determined from the overall temperature difference over the addition of the thermal resistance in series

Thermal modeling, analysis, and design

connection. For the thermal network shown in Figure 3.1.1, the heat flow can be expressed in term of thermal resistance.

$$q_x = \frac{T_{s1} - T_{s2}}{R_{cond}} = \frac{T_{s2} - T_\infty}{R_{conv}} = \frac{T_{s1} - T_\infty}{R_{cond} + R_{conv}} \tag{3.1.9}$$

3.1.1.6 Heat conduction through radial systems

The heat conduction for a concentric cylindrical system with the inner radius r_1, outer radius r_2, and length of l can be solved by the following equation.

$$\frac{1}{r}\frac{d}{dr}\left(kr\frac{dT}{dr}\right) = 0 \tag{3.1.10}$$

Thus, the heat flow along the radial direction is

$$q_r = \frac{2\pi l k (T_{s1} - T_{s2})}{\ln(r_2/r_1)} \tag{3.1.11}$$

The thermal resistance between the inner surface and outer surface is obtained as follows:

$$R_{cond} = \frac{\ln(r_2/r_1)}{2\pi l k} \tag{3.1.12}$$

3.1.2 Thermal interfacial resistance

Various interfaces exist in electronic packages and the temperature drop across the interface between materials can be appreciable. This temperature drop could be large in view of the contact resistance because of the surface asperity, trapped air voids, and microscale heat transfer effect in between the matching surfaces. A description of the interfacial resistance with air voids can be found in the reference such as Ref. [6]. To minimize the contact resistance, TIMs have been used to fill in the gaps between the two surfaces to result in a reduced thermal resistance. This is known as the thermal interfacial resistance. The contact resistances can be minimized to some degree but are never zero-valued due to the boundaries and possible imperfect wetting between the TIM and the two solid surfaces. The concept of the thermal interfacial resistance is schematically shown in Figure 3.1.5. The thermal resistance and specific thermal resistance are defined respectively as follows:

$$R_{tim} = \frac{T_1 - T_2}{q_x} \tag{3.1.13a}$$

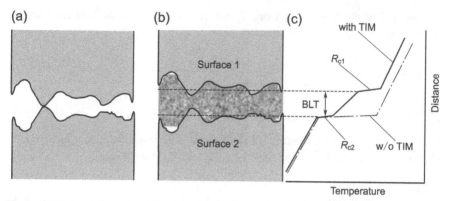

Figure 3.1.5 The schematic of the thermal interfacial resistance. (a) Surface contact without thermal interface material (TIM); (b) surface contact with interface material; and (c) the temperature gradient along the interface distance with and without TIM.

$$R''_{\text{tim}} = \frac{T_1 - T_2}{q''_x} \qquad (3.1.13b)$$

The thickness increase due to the adding of TIM is called bond line thickness (BLT), which can be used to determine the bulk thermal resistance related to the TIM. The thermal resistance for the thermal interface filled in with conductive materials can be obtained based on the experimental method as described in the ASTM 5470D standard [7]. Nonetheless, the in situ evaluation tests of TIM in a package is of more practical interest, which include power cycling test, thermal cycling test, transient test, cold plate test, and so on. The thermal interfacial resistances vary for different TIM materials, depending on the material properties, reliability, and application scenarios. Common TIMs include thermal grease, thermal gel, thermal adhesive, elastomer, and thin foils made of metal or graphite sheets.

The thermal grease usually has excellent wetting and thermal performance with minimum contact resistance when being first applied package surface. Made of silicone oil with thermally conductive fillers, the thermal grease has excellent ability to conform to mating surfaces and performs well at low operational temperatures, such as temperatures below 90 °C. However, thermal grease degrades significantly during temperature cycling or power cycling tests due to the pumping out mechanism. The CTE mismatch of the lid and package tends to squeeze out the silicone oil during the test, which is difficult to apply at the interfaces from the deformable package, such as the Type 1 thermal interface material (TIM1) between the silicon chip and the metallic lid.

Thermal gel material is a gel-like TIM with a low modulus but a high elongation rate, which has been successfully developed for on-chip application (TIM1) between lid and flip-chip. Filled with thermally conductive fillers, thermal gel possesses polydimethylsiloxane as polymer matrix. The material is normally supplied as liquid paste format, and the storage shear modulus G' is less than the loss shear modulus G''

(corresponding to the viscosity for liquids). At elevated curing temperature, vinyl groups in the base polymer react with the crosslinking agent to cure into a skin−soft polymer network with the aid of a platinum catalyst. This curing makes $G'>G''$ and thus makes it endure the pump out to maintain its gelation form during the various reliability tests without significant degradation (say 0.02−0.03 K·cm^2/W increase in thermal resistance for typical processor applications). Intel has successfully characterized the material properties and applied as TIM1 material between the lid and the chip [8−10]. The typical specific interfacial resistance varies from 0.05K·cm^2/W to 0.2K·cm^2/W.

Thermal adhesive is referred to the fully cross-linked resin type TIM filled with thermally conductive fillers. Because of its high modulus and hardness, the material BLT is designed larger in purpose than the thermal gel of the same polymer system to resist the excessive thermomechanical stress during operation. The contact resistance between TIM is also larger due to the use of coarser fillers. In general, the thermal adhesive cannot be reworked or repaired once assembled at the end of line.

The elastomeric TIM (or elastomer) is a solid-like material, precured in factory and then supplied to assembly house. The elastomer can be used as the TIM for a secondary die in a multiple chip package in conjunction with other TIMs. One example is to apply at top of a memory stack in a 2.5D package, in which both the processor and memory are assembled on a common interposer. Although with relatively large contact resistance, an elastomer has the advantage of easy assembly with good stress-absorbing compressibility and reworkability.

For high performance packages such as land-grid array (LGA) and pin-grid array (PGA) packages, the indium solder TIM is assembled at the die top by major chip makers such as Intel and AMD to minimize the interfacial thermal resistance. Nonetheless, the die top must be metallized carefully to form the metallurgical indium solder. For ball-grid array (BGA) applications, which require an even higher temperature reflow exceeding the indium melting point, a polymer gel is also used as TIM1 with a plain die without metallization.

Graphite sheet can also work as TIM in the test of the package. The thermal interface resistance for graphite foil TIM materials is shown in Figure 3.1.6 used in the system-level testing (SLT) of a logic die assembled with a memory chip in a 2.5D package. The thermal gel is applied on top of the logic die, whereas the elastomer is applied on the memory chip to fill the gap to the lid. The test was conducted by attaching a temperature-controlled thermal head the package top.

Because of the good lubrication, mechanical integrity, and thermal performance at low cost, graphite sheets can endure multiple insertions before mechanical failures, which allows multiple uses in the burn-in and system-level test of packages. Figure 3.1.6 shows the test of a lidded package used in a burn-in test configuration. Figure 3.1.7 show the thermal resistance of graphite foils between the package and the thermal control unit at different pressures [10]. The contact resistance of the graphite sheet as TIM is comparable or higher than the bulk thermal resistance even if the surface is flat. On the contrary, the liquid-based TIM such as thermal grease, due to its excellent wetting characteristics, has a very small contact resistance in comparison with its bulk thermal resistance.

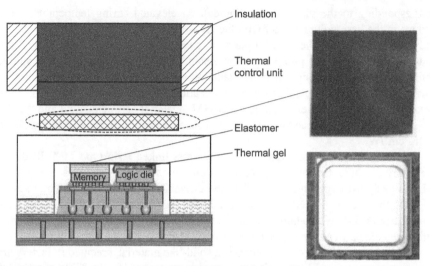

Figure 3.1.6 A burn-in test configuration with the solid thermal interface material (TIM) on a lidded package.

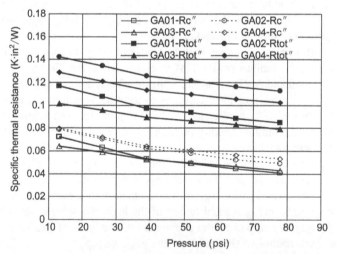

Figure 3.1.7 Contact thermal resistances in comparison with the total interfacial resistances for thermal interface materials (TIMs) made from graphite foils ($1\text{K} \cdot \text{in}^2/\text{W} = 6.45\text{K} \cdot \text{cm}^2/\text{W}$, $1\text{psi} = 6895\text{Pa}$).

The research studies on the development of new TIMs with advanced materials such as carbon nanotube (CNT) and graphene are gaining increasing interest. CNT in the TIMs can be found for example in Ref. [11]. Although the bulk thermal resistance can be reduced slightly, the contact resistance between the CNT and the base polymer remains at a relatively high level. With the birth of the graphene, more efforts have been devoted to develop the TIM with graphene [12,13]. Shahil et al. show a significant enhancement with graphene-based thermal grease [12]. The thermal grease mixed with the multilayer graphene (MLG) flakes goes up to 5.4W·m^{-1}·K^{-1}, which is 28 times of the thermal conductivity of base polymer matrix. They attributed the significant thermal enhancement to the factors such as low Kapitza resistance at the graphene/matrix interface, the flake-like geometrical shape of multiple layer graphene flakes and optimum mix of graphene, and MLG with different thickness and lateral size. It is expected that the research work in this area will surge up in the near future to accelerate the adoption of graphene-based TIMs.

3.1.3 Heat transfer from extended fin surfaces

Finned surfaces are commonly seen in the electronic cooling field. Considering the simplest case of straight rectangular fin of uniform cross section, each fin is attached to a base surface of temperature $T(0) = T_b$ and extends into a fluid of temperature T_∞ with the thermal conductivity of k, as indicated in Figure 3.1.8a. For the prescribed fins with a constant cross-sectional area A_c, the thermal energy equation can be obtained as

$$\frac{d^2 T}{dx^2} - \frac{hP}{kA_c}[T(x) - T_\infty] = 0 \qquad (3.1.14)$$

where P_w is the perimeter of the fin and $P_w = 2(w + t)$, with w the fin width and t the fin thickness. Introducing $\theta(x) = T(x) - T_\infty$, and solving Eq. (3.1.14), one obtains (3.1.15)

$$\frac{\theta}{\theta_b} = \frac{\cosh m(l-x) + (h/mk)\sinh m(l-x)}{\cosh ml + (h/mk)\sinh ml} \qquad (3.1.15)$$

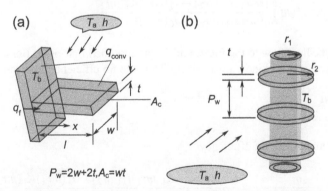

Figure 3.1.8 Straight fins of uniform cross section under forced convective heat transfer coefficient h: (a) rectangular fin and (b) annular fins.

Here

$$m^2 = \frac{hP_w}{kA_c} \tag{3.1.16}$$

Note that the magnitude of the temperature gradient decreases with increasing x. This trend is a consequence of the reduction in the conduction heat transfer $q_x(x)$ with increasing x due to continuous convection loss from the fin surface.

Analysis of fin thermal behavior becomes more complex if the fin is of circular cross section. Consider the annular fin shown in the inset of Figure 3.1.8 b. Although the fin thickness t is uniform (t is independent of r), the cross-sectional area, $A_c = 2\pi rt$, varies with r, the general form of the fin equation reduces to

$$\frac{d^2T}{dr^2} + \frac{1}{r}\frac{dT}{dr} - \frac{2h}{kt}(T - T_\infty) = 0 \tag{3.1.17}$$

with $m^2 = 2\,h/kt$ and $\theta = T - T_\infty$,

$$\frac{d^2\theta}{dr^2} + \frac{1}{r}\frac{d\theta}{dr} - m^2\theta = 0 \tag{3.1.18}$$

If the temperature at the base of the fin is prescribed, $\theta(r_1) = \theta_b$, an adiabatic tip is presumed, and a temperature distribution in the following form is obtained.

$$\frac{\theta}{\theta_b} = \frac{I_0(mr)K_1(mr_2) + K_0(mr)I_1(mr_2)}{I_0(mr_1)K_1(mr_2) + K_0(mr_1)I_1(mr_2)} \tag{3.1.19}$$

where $I_0(\cdot)$ and $K_0(\cdot)$ are modified zero-order Bessel functions of the first and the second types, respectively. It follows that

$$q_f = 2\pi k r_1 t \theta_b m \frac{K_1(mr_1)I_1(mr_2) - I_1(mr_1)K_1(mr_2)}{K_0(mr_1)I_1(mr_2) - I_0(mr_1)K_1(mr_2)} \tag{3.1.20}$$

where q_f is the heat flow to the fin, I_1 and K_1 are modified first-order Bessel functions of the first and second kinds, respectively. Thus the fin efficiency becomes

$$\eta_f = \frac{q_f}{q_{max}} = \frac{q_f}{hA_f\theta_b} \tag{3.1.21a}$$

For the circular fin, fin efficiency is

$$\eta_f = \frac{q_f}{h2\pi(r_2^2 - r_1^2)\theta_b} = \frac{2r_1}{m(r_2^2 - r_1^2)} \frac{K_1(mr_1)I_1(mr_2) - I_1(mr_1)K_1(mr_2)}{K_0(mr_1)I_1(mr_2) - I_0(mr_1)K_1(mr_2)}$$

$$\tag{3.1.21b}$$

In contrast to the fin efficiency η_f, which characterizes the performance of a single fin, the overall surface efficiency η_0 characterizes an array of N fins and the base surface to which they are attached. Typical fin arrays are shown in Figure 3.1.8. In each case, the overall efficiency is defined as

$$\eta_0 = \frac{q_t}{q_{max}} = \frac{q_t}{hA_t\theta_b} \qquad (3.1.22)$$

where q_t is the total heat rate from the surface area A_t associated with both the N fins and the exposed portion of the base, $A_t = NA_f + A_b$.

Thus, the total heat transfer rate from the fins and unfinned area is expressed as

$$q_t = h[N\eta_f A_f + (A_t - NA_f)]\theta_b = hA_t\left[1 - \frac{NA_f}{A_t}(1-\eta_f)\right]\theta_b \qquad (3.1.23)$$

The overall thermal resistance can be determined by the following expression.

$$R_t = \frac{T_b - T_\infty}{q_t} = \frac{1}{h[A_t - NA_f(1-\eta_f)]} \qquad (3.1.24)$$

3.1.4 Heat generation in electronics systems

A common heat generation process involves the conversion from electrical to thermal energy in a current-carrying medium (also called Ohmic, or Joule heating) as well as high-frequency dissipation. The rate at which heating energy is generated by passing a current I through a medium of electrical resistance R_e is

$$q = I^2 R_e \qquad (3.1.25)$$

Heat generation may also occur as a result of the deceleration and absorption of neutrons in the fuel element of a nuclear reactor or exothermic chemical reactions occurring within a medium. For example, lithium-ion batteries are used widely in consumer electronics and mobile devices such as a mobile phone. Endothermic reaction takes place in a lithium-ion battery, which would generate significant portion of waste heat and have adverse effect on the electronic package and system. The simplified heat generation for a lithium-ion can be expressed as follows [14].

$$q = I(V_{oc} - V) + IT\left(-\frac{\partial V_{oc}}{\partial T}\right) \qquad (3.1.26)$$

Here I is the electric current, V_{oc} the open circuit voltage, V the closed circuit voltage, and T the absolute temperature in Kelvin scale (K). The first term on the right hand corresponds to the Ohmic heating due to internal electrical resistance and the second term corresponds to the reversible heat generation during electrochemical reaction depending on the charge and discharge processes. The heat dissipation for a typical

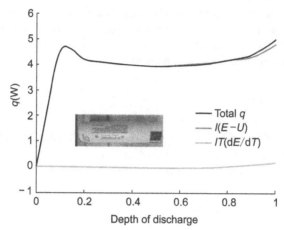

Figure 3.1.9 The curves of reversible, irreversible, and total heat of the 18,650 battery at the 3C discharge rate.

LiCoO$_2$-based 18650 battery with a standard size of 18 mm in diameter and 65 mm in height is displayed in Figure 3.1.9. The discharge rate is 3 times of the nominal current (3C), which shows an average power of around 4W.

3.2 Package-level thermal analysis and design

A conventional planar package such as flip-chip BGA involves a variety of parts including a single chip, solder bumps, a substrate with traces and vias, solder balls, etc. With the development in heterogeneous integration, 2.5D and 3D packages are developed as next-generation packaging technologies integrated with more parts. In a 2.5D package, multiple dissimilar chips are stacked on a common interposer, with through-silicon vias (TSVs) and microbumps to connect to each other or to the substrate, whereas a 3D package requires multiple chips vertically stacked with TSVs and microbumps as interconnects. Dissimilar chips, logic chip, memory, optical devices, etc., can be assembled on a common interposer to fulfill the operations through the microlines with finer on-silicon RDL and TSV processes. Proper modeling and analysis of the thermal properties of TSVs together with other parts of package of increasing complexity is crucial to the successful thermal design of the package. In the following, the submodels of various package parts are to be elaborated based on the principles of thermal analysis as described in Section 3.1.

3.2.1 Analytical solutions for analysis of through-silicon vias

The determination of thermal property of TSVs is not straightforward as it involves a number of design and process factors, including via size and geometry, the surrounding insulation materials, and barrier layer. The TSVs, mostly plated with copper, have higher thermal conductivity than the bulk silicon, which is expected to promote the heat transfer. On the other hand, the electrical insulation material is required to deposit between the TSV and the surrounding semiconductor silicon to minimize the signal loss. Different types of materials such as silicon oxide or polymers made of parylene, benzocyclobutene (BCB), or phenolic aldehyde have been used between silicon and TSV as electrical insulation [15–18]. The insulation material, though much thinner than the copper via, induces additional thermal resistance, adversely affecting the thermal transport capability.

In Liu's group [19], the high aspect ratio CNT TSV has been successfully developed through the CVD process. The multiwalled carbon nanotube (MWCNT) is then transferred to the wafer holes, separated from the silicon with SiO_2 and metallization layers. In addition, the MWCNT TSV can be bonded together without solder or adhesive, which eliminate the thermal resistance of the bonding layer by a conventional solder bump method. The fabricated package on interposer with copper TSV is exemplified in Ref. [19].

In line with the TSV development, thermal analysis of the TSV thermal performance is being called for to address thermal issue. Most of the thermal analysis and characterizations have been conducted by ignoring the insulation effect on the in-plane thermal conductivity [20–26].

Lau et al. [23] conducted numerical simulation of copper TSV effect on the thermal performance. Ma et al. (2014) [24] considered the cylindrical model of the metallic TSVs and build an analytical model for the effective thermal conductivity without taking into account the insulation layer. Oprins et al. [25] conducted an experimental study by cooling the 2.5D/3D package with a cold plate. It was found that the measured chip temperature was obviously higher than the numerically simulated values. They attributed this underestimation of temperature to the crack of TSVs without evidence, which seems unlikely as the package was well fabricated without undergoing reliability test. It is more believed that such an underestimation may be due to the neglect of the insulation layer in the thermal simulation. A few research teams realized that the insulation layer may have an adverse effect on the transverse thermal conductivity. Chen et al. [26] studied the adverse effect of thermal insulation by analyzing the thermal network and optimized the TSV array in the chip. However, the effective thermal conductivity model was numerically obtained without providing detailed TSV data.

Recently, it has been gradually recognized that the insulation material, though small in thickness, would have an unfavorable impact on the thermal performance of high-power chips [26] based on a numerical analysis tool. Effort has been conducted to correlate the effect of the silicon oxide layer on the thermal performance of the interposer with TSV. Nonetheless, the empirically fitted correlation, due to the lack of physical foundation, cannot cover all the scenarios especially when the insulation is much smaller than the copper TSV. The unfavorable effect of TSVs on the thermal performance of the package is also not well addressed for the package under different thermal boundary conditions. Zhang et al. [27] identified that the insulation layer could affect the transverse thermal conductivity significantly, which may also have an impact on the package thermal performance. The analysis starts as follows with a brief review of the models in thermal transport in porous media.

3.2.1.1 Porous media model

Before addressing the TSV thermal model, it is instructive to preview the heat transfer model in porous media. In one way or another, the porous media model can be introduced to simplify the analysis of various geometrical parts in the electronic packaging field. A porous medium consists of several phases in either solid or fluid with either periodic or random structure and layout. Thermal analysis of porous media is available in literature such as the well-known Maxwell–Eucken's equation [28]. Similar to porous media, many parts in electronic packaging exhibit periodic structure. Typical examples include solid bumps filled with underfilled epoxy, substrate vias of different forms, and more recently, the array of TSVs, and redistribution layer (RDL) on the silicon interposer. The TSV array is mostly periodic, at least partially, and can be treated as porous media in whole or partially, so that the key thermophysical properties can be obtained analytically without resorting to time-consuming numerical approach.

The thermal model based on the porous media is to be briefed as follows. In a series composite wall consisting of two phases in heat conduction mode, the effective thermal conductivity can be obtained based on the analysis of series resistance. Assuming that the phase 1 has a length of βl and phase 2 of length $(1-\beta)l$, the effective thermal conductivity through the plane in series is given as

$$k_{\text{eff,series}} = \frac{1}{(1-\beta)/k_1 + \beta/k_2} \quad (3.2.1)$$

Alternatively, the effective thermal conductivity for the parallel composite wall consisting of the phase 1 with a width βw and phase 2 with a width $(1-\beta)w$, as the copper and silicon illustrated in Figure 3.2.2d, can be described by the following equation.

$$k_{\text{eff,parallel}} = \beta k_2 + (1-\beta)k_1 \quad (3.2.2)$$

It is noted that the above correlation can also be developed based on the volume averaging theory in porous media.

Another expression known as the Maxwell–Eucken's equation [1,28] is available for dispersed phase in a continuous composite system for the effective thermal conductivity of phase 1 interspersed with uniformly distributed, noncontacting spherical inclusions phase 2.

$$k_{\text{eff}} = \left[\frac{k_2 + 2k_1 - 2\beta(k_1 - k_2)}{k_2 + 2k_1 + \beta(k_1 - k_2)}\right]k_1 \quad (3.2.3)$$

Here β is referred to the volume fraction of the non-contacting phase 2. In the realistic electronic package with either periodic or random patterns, the above porous media model can be utilized to derive the effective thermal conductivities in one way or another.

3.2.1.2 Submodels of through-silicon vias

In this part, the different analysis models are examined for the modeling of TSVs. In literature, some work has been conducted to obtain the effective thermal conductivity of the TSVs numerically, which is however time-consuming and inconvenient for thermal design. Instead, analytical solutions can be utilized to avoid significant computing effort within acceptable engineering accuracy. The TSV array, either with and without the insulation layer, is shown in Figure 3.2.1.

In the analytical model provided by Ma et al. [24], the insulation layer is neglected and thus the effective thermal conductivity was derived for in-line matrix of TSVs embedded in silicon as shown in Figure 3.2.1. The representative block consists of a metal via with the diameter D embedded in a silicon block with edge length of P. Let $e = D_v/P$, then the effective thermal conductivity is written as

$$k_{\text{eff}//} = \frac{k_{\text{Si}}^2 + k_{\text{Si}}(k_{\text{Cu}} - k_{\text{Si}})e}{k_{\text{Si}} + (k_{\text{Cu}} - k_{\text{Si}})e(1 - e)} \tag{3.2.4}$$

Define the volume ratio of copper and silicon is β, which is expressed by

$$\beta = \frac{v_{\text{Cu}}}{v_{\text{Si}}} = \frac{\pi e^2}{4 - \pi e^2} \tag{3.2.5}$$

Thus, the effective thermal conductivity can be expressed in term of β also, namely,

$$k_{\text{eff}//} = \frac{\pi(1+\beta)k_{\text{Si}}^2 + k_{\text{Si}}(k_{\text{Cu}} - k_{\text{Si}})\sqrt{4\pi\beta(1+\beta)}}{\pi(1+\beta)k_{\text{Si}} + (k_{\text{Cu}} - k_{\text{Si}})\left(\sqrt{4\pi\beta(1+\beta)} - 4\beta\right)} \tag{3.2.6}$$

The above equation can be recast in terms of the TSV diameter to spacing ratio D_v/P, namely,

$$\frac{k_{\text{eff}//}}{k_{\text{Si}}} = \frac{1 + (k_{\text{Cu}}/k_{\text{Si}} - 1)(D_v/P)}{1 + (k_{\text{Cu}}/k_{\text{Si}} - 1)(D_v/P)(1 - D_v/P)} \tag{3.2.7}$$

It is noted that this model does not consider the effect of the insulation layer, which is thin in size but cannot be neglected in the thermal evaluation.

In Zhang et al.'s work [27], a closed-form thermal conductivity model for the TSV chips is developed to capture this insulation effect. An analytical model is proposed by treating the insulation layer as a contact resistance surrounding the TSV. This analytical model is validated with the numerical analysis with reasonable agreement. The in-plane thermal conductivity of the interposer decreases dramatically with the increase in the ratio of the TSV diameter to the pitch, D_v/P. Besides the D_v/P ratio,

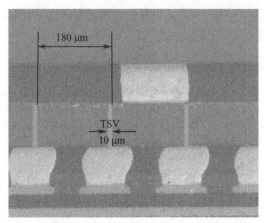

Figure 3.2.1 Fabricated copper through-silicon via (TSV) of 10 μm in diameter in interposer for 2.5D package.

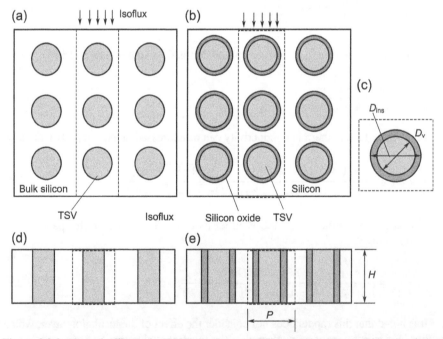

Figure 3.2.2 Through-silicon via (TSV) structure: (a) TSV array in silicon without insulation layer; (b)TSV array with surrounding insulation layer; (c) representative unit cell with insulation; (d) the lateral view of the TSV without insulation layer; and (e) the lateral view of the TSV with insulation layer.

the reduction in effective thermal conductivity depends also on the insulation parameters such as insulation thickness and thermal conductivity.

This work starts with the in-line TSV array without interacting with the heat-generating zones such as logic area and layered metal layers. The plan view of the in-line TSV array with insulation layer is shown in Figure 3.2.2 b, together with the case without insulation layer. Mostly, the insulation layer is made of silicon oxide with a thermal conductivity of $1.3 \text{W} \cdot \text{m}^{-1} \cdot \text{K}^{-1}$, which is two order of magnitude lower than those for the bulk silicon at room temperature ($150 \text{W} \cdot \text{m}^{-1} \cdot \text{K}^{-1}$) and copper TSV ($385 \text{W} \cdot \text{m}^{-1} \cdot \text{K}^{-1}$). Other insulation made of polymer materials such as parylene or BCB has an even lower thermal conductivity around $0.1-0.2 \text{W} \cdot \text{m}^{-1} \cdot \text{K}^{-1}$. In general, the insulation thickness is in the range of $0.1-1$ μm, much smaller in comparison with the TSV diameter from 5 to 50 μm. However, the insulation layer has a significant blockage effect on the in-plane heat conduction, depending on several geometrical factors including insulation thickness, via diameter and via pitch, and thermophysical properties.

Because the insulation layer is much smaller than the TSV diameter, it is reasonable to treat the insulation layer as a contact resistance for the TSV embedded in bulk silicon. Thus, the following model can be used, namely,

$$\frac{k_{\text{eff}//}}{k_{\text{Si}}} = \frac{\left[\frac{k_v}{k_{\text{Si}}}(1+2\gamma)+2\right] + 2v_{\text{ins}}\left[\frac{k_v}{k_{\text{Si}}}(1-\gamma)-1\right]}{\left[\frac{k_v}{k_{\text{Si}}}(1+2\gamma)+2\right] - v_{\text{ins}}\left[\frac{k_v}{k_{\text{Si}}}(1-\gamma)-1\right]} \quad (3.2.8)$$

Here k_v is the thermal conductivity of metal via, v_{ins} the volume fraction of the insulation and metal via material with respect to the volume of the unit cell bulk, and γ the normalized contact resistance which is defined by

$$\gamma = \frac{2R_c k_{\text{Si}}}{D_v} \quad (3.2.9a)$$

The specific contact resistance R_c'' is defined with the thermal resistance across the concentric insulation layer, namely,

$$R_c'' = \frac{D_v \cdot \ln(D_{\text{ins}}/D_v)}{2k_{\text{ins}}} \quad (3.2.9b)$$

Thus, γ is simplified into the following form:

$$\gamma = \frac{k_{\text{Si}}}{k_{\text{ins}}} \cdot \ln\left(\frac{D_{\text{ins}}}{D_v}\right) \quad (3.2.9c)$$

It is noted that Eq. (3.2.8) has been used previously in estimating the contact resistance of nanocomposites, where the contact resistance was utilized to quantify the heat transport at the interface of infinitesimally small scale between nanoparticles and the matrix material [29,30]. The novelty of Zhang et al.'s work is to apply Eq. (3.2.8) in obtaining the overall thermal property with the finite thickness of the insulation as there is little work in analyzing the TSV insulation layer as the contact resistance. Such a treatment may apply to the cases in which the insulation layer thickness is much smaller than the via diameter, which is also in line with the trend for the TSV implementation. To increase the analysis accuracy, the effect of the radial heat conduction for the concentric insulation layer has been taken into account in Eq. (3.2.9b) to obtain the contact resistance R_c. In addition, the volume of the insulation combining with the via is included in the calculation of v_{ins} in Eq. (3.2.8), which is not considered in the previous studies [29,30].

If the insulation layer thickness reduces to zero, γ approaches zero and Eq. (3.2.8) reduce to the case without the insulation layer, which is equivalent to the known Maxwell–Eucken equation, namely,

$$\frac{k_{eff//}}{k_{Si}} = \frac{\frac{k_v}{k_{Si}} + 2\beta_v\left(\frac{k_v}{k_{Si}} - 1\right)}{\frac{k_v}{k_{Si}} + 2 - \beta_v\left(\frac{k_v}{k_{Si}} - 1\right)} \tag{3.2.10}$$

Here, the volume fraction β_v is the ratio of the via volume to the unit cell volume. Eq. (3.2.10) can be used to predict the effective thermal conductivity for a two-phase system such as metal vias in the bulk silicon material.

The effective density and heat capacity of the interposer are given as follow:

$$\rho_{eff} = \frac{\rho_v V_v + \rho_{ins} V_{ins} + \rho_{Si} V_{Si}}{V_v + V_{ins} + V_{Si}} \tag{3.2.11a}$$

$$c_{p,eff} = \frac{\rho_v C_v V_v + \rho_{ins} C_{ins} V_{ins} + \rho_{Si} C_{Si} V_{Si}}{\rho_v V_v + \rho_{ins} V_{ins} + \rho_{Si} V_{Si}} \tag{3.2.11b}$$

where ρ_{eff} and $c_{p,eff}$ are the effective density and specific heat for the bulk silicon with TSVs, respectively.

The analysis model given in Eq. (3.2.8) is compared with existing studies such as Ref. [24]. In the meantime, the numerical simulation is also conducted to examine the effective thermal conductivity. The TSV diameter has been set to be 10 μm and insulation 1 μm with different D_{ins}/P ratios. The results are shown in Figure 3.2.3.

Without the insulation SOX, good agreement is achieved between the present analysis and numerical results. Both are lower than Ma et al.'s results. With the insulation SOX, the present analysis results are close to the numerical results within 10% for $D_{ins}/P \leqslant 0.5$, which shall be sufficient for thermal modeling of most TSV array for $D_{ins}/P \leqslant 0.5$.

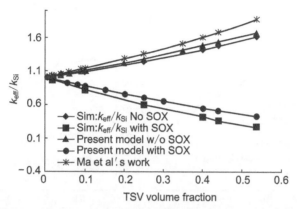

Figure 3.2.3 In-plane thermal conductivity ratios for the cases with and without through-silicon via (TSV) insulation layers at the ratio of silicon oxide diameter to via diameter of 1.1 for the different cases: simulation results, Ma et al.'s work, and present analytical model.

In Fig. 3.2.4, the analysis work from Zhang et al. [27] is also compared with results from available correlation on the TSV insulation layer from Chien et al.'s work [31], where an empirical correlation for TSVs with silicon oxide layer was fitted based on numerically computed results. The fitted correlation takes the following form.

$$k_{\text{eff}//} = (90 t_{\text{ins}}^{-0.3} - k_{\text{Si}}) \left(\frac{D_{\text{ins}}}{P}\right) H^{-0.1} + 160 t_{\text{ins}}^{0.07} \qquad (3.2.12)$$

where t_{ins} is the thickness of silicon oxide, H the via height in microns, both given in microns. It is noted that this correlation is not homogenous, which may provide

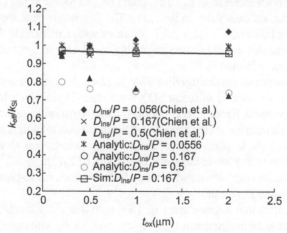

Figure 3.2.4 In-plane thermal conductivity ratios for $D_v = 10$ μm versus the silicon oxide thickness at different D_{ins}/P values.

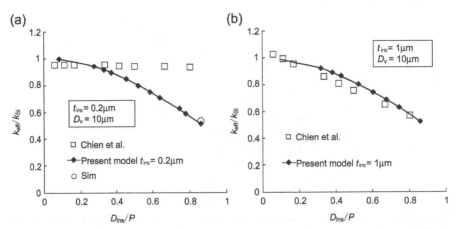

Figure 3.2.5 Calculated results for the present model in comparison with Chien et al.'s model at: (a) $t_{ins} = 0.2$ μm; (b) $t_{ins} = 1$ μm.

inappropriate results at small D_V/P values due to the lack of physical foundation, as indicated in the following discussion. This correlation is limited to the specific bulk silicon, silicon oxide, and copper via structure, which may not be extended to general cases with other types of materials such as polymer insulation with varying thermal conductivity range.

Figure 3.2.5 shows the comparison of the present analytical results with the fitted correlation from Chien et al. [31]. The silicon oxide layer thickness, t_{ins}, is used as the abscissa for convenience of comparison. The TSV diameter has been imposed to be 10 μm. It is observed that the effective thermal conductivity ratio from Ref. [31] increases unexpectedly above the line of $k_{eff}/k_{Si} = 1$ at a small value D_{ins}/P of 0.056. This deviates from the physical rule as the effective thermal conductivity should decrease with increase in t_{ins}. This could be due to the fitting error in the correlation due to limited case study in Ref. [31]. The deviations from the present model are further exemplified in Figure 3.2.5a for cases with small oxide thickness ($t_{ins} < 1$ μm). Such deviations could become negligibly small only at large insulation thickness $t_{ins} \geq 1$ μm, as indicated in Figures 3.2.5b and 3.2.4.

Figure 3.2.6 summarizes the trend of TSV in-plane thermal conductivity with the via diameter to pitch ratio at different SOX thickness. Obviously, a large D_V/P ratio leads to higher volume fraction, accounting for the remarkable reduction in the in-plane thermal conductivity. At the same D_V/P, the insulation thickness also contributes to the reduction of the in-plane thermal conductivity, nonetheless to a lesser degree. Polymer insulation with a lower thermal conductivity (0.2 W·m^{-1}·K^{-1}) is also shown in the same figure for comparison, which may further reduce the in-plane thermal conductivity at the same D_V/P ratio.

Figure 3.2.6 hints that a dense array of TSV with thermally insulated layer might lead to an increase in the junction temperature and TSVs with smaller D_V/P ratios could be used to avoid the decrease in the in-plane thermal conductivity. This can

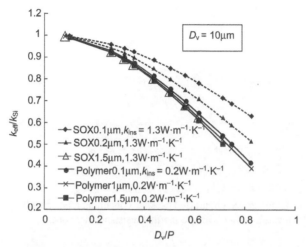

Figure 3.2.6 Comparison of through-silicon via (TSV) thermal conductivities with different insulation layer parameters.

be implemented by either reducing the via diameter or increasing the TSV pitch. On the other hand, a system-level thermal evaluation is to be conducted to examine the effect on the overall thermal metrics such as junction-to-ambient thermal resistance due to the presence of insulated TSVs.

A 2.5D package mounted with a heat generating chip is also examined at the board-level, the structure of which is schematically shown in Figure 3.2.7. It is found that, in the presence of insulated TSVs, the junction-to-ambient thermal resistance is almost unchanged for the low D_v/P ratio of 0.1 but increases by 7% at a large D_v/P ratio of 0.5. Therefore, it is suggested not to neglect the TSV in the design stage if a dense TSV array is used.

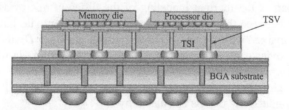

Figure 3.2.7 Schematic of a 2.5D package with through-silicon via (TSV) interconnects.

3.2.2 Submodels for substrate and redistribution layer

The laminated substrate with the typical via microstructure layers is shown in Figure 3.2.8. A detailed discussion on the laminated substrate with multiple trace layers will be shown in Section 3.3. The laminated substrate consists of both the core layer with plated through hole (PTH) vias and the buildup layer with microvias

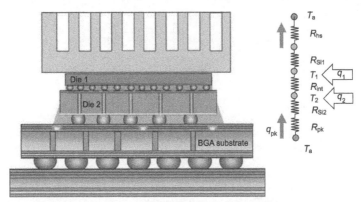

Figure 3.2.8 The 3D integrated circuit (IC) package containing representative package parts for submodel development, together with the thermal resistance network.

of tapering outer dimension. Similarly, the RDL on the interposer also possesses multiple layers of microvias through the dielectric materials such as polyimide. To facilitate the thermal analysis, the PTH vias and tapering vias are represented in thermal submodels based on the heat transfer theory.

The equivalent thermal conductivities in each layer should be computed and incorporated in the thermal modeling at the system level. Otherwise the temperature fields lack of accuracy to guide the design and development work. The substrate model is discussed as follows.

To simplify the submodel, the substrate can be divided into several layers; each layer is represented by effective thermal conductivities in different directions. For the top copper trace layer, the copper traces are interconnected; see also Layer 1 (L1) in Figure 3.3.3. The effective thermal conductivity can be obtained with Eq. (3.2.1) based on the portion of the copper content, which can be obtained from the software during the routing of electrical wiring.

A difficult part is the thermal modeling of the different microvias plated in the buildup layer in between the copper trace layers. The different vias have been shown in Figure 3.2.9, marked with via top and bottom diameters D_1 or D_2 and plating thickness t. The microvias can be divided into three types: the solid via filled fully with copper, the conformal via filled with polymer, and the PTH vias in the core layer filled with resin (through hole via). In each metal layer or via layer, the equivalent through-plane thermal conductivity can be calculated by Eq. (3.2.1).

Nonetheless, the in-plane thermal conductivity has to be simplified to reduce the complexity of the modeling. For example, the trapezoidal microvia can be simplified into a cylindrical shape with same diameter, which is then calculated based on the porous media model. The simplification process is shown in Figure 3.2.9. For the filled via with a shape of circular truncated cone, the via is approximated as a solid cylindrical via of equivalent diameter of $(D_1+D_2)/2$ of the same material, as shown in Figure 3.2.9a. Following Eqs. (3.2.2)–(3.2.3), one obtains the through-plane and in-plant thermal conductivities of the via layer.

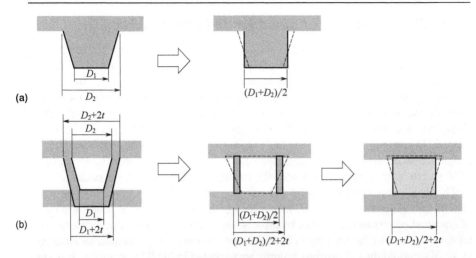

Figure 3.2.9 Different substrate via model: (a) the filled via and its simplified thermal submodel and (b) the conformal via and its simplified thermal submodel.

For the conformal via with a plating thickness t, the via is first approximated as a concentric cylindrical via of the inner diameter of $(D_1+D_2)/2$ and the outer diameter of $(D_1+D_2)/2+2t$, as illustrated in Figure 3.2.9b. Then the concentric cylindrical via is further approximated as a solid via with a reduced thermal conductivity of the same outer diameter based on the Maxwell equation in Eq. (3.2.4). Again one obtains the through-plane and in-plane thermal conductivities of the via layer following Eqs. (3.2.2)–(3.2.3). A concrete example is given as follows for the conformal via indicated in Fig. 3.2.9b. Assuming $t = 15$ μm, $D_1 = 50$ μm, and $D_2 = 80$ μm, the outer and inner diameters for the concentric cylinder are 95μm and 65μm, respectively. The volume fraction of the inner air phase is $\beta = 43.1\%$. The equivalent thermal conductivity for the solid via through the plane is calculated to be $k_\perp = V_y k_{Cu} + V_{air} k_{air} = (95^2-65^2)/95^2 \times 390 + 65^2/95^2 \times 0.026 = 207.4 \text{ W}\cdot\text{m}^{-1}\cdot\text{K}^{-1}$.

The through-plane thermal conductivity is calculated to be

$$k_{//} = \left[\frac{k_{air} + 2k_{Cu} - 2\beta(k_{Cu} - k_{air})}{k_{air} + 2k_{Cu} + \beta(k_{Cu} - k_{air})}\right] k_{Cu} = 168.1 \text{ W}\cdot\text{m}^{-1}\cdot\text{K}^{-1}$$

Although the two thermal conductivities are slightly different in values, both are much larger than the thermal conductivities of internal air voids and the surrounding polymer materials.

The solid ball at the second level can be viewed as cylindrical blocks of the same volume and thermal conductivity. Considering the same two-phase medium made of the cylindrical block and the surrounding air, both the through-plane and in-plane thermal conductivities can be attained in a similar manner.

With thermal submodels of the different parts, the package model can be established through the numerical solver such as Flotherm. In-depth discussion of the numerical methodology and solutions is to be given in Section 3.3.

3.2.3 Thermal model for 3D integrated circuit package

Vertically integrated 3D circuits have attracted significant attention in the recent years due to potential benefits such as the increased device density, reduced signal delay, and enabled new architecture designs and heterogeneous integration. Vertical integration of heterogeneous technologies using 3D integration may offer many advantages over alternatives such as package-on-package (POP) and system-on-single chip. Compared with the conventional 3D POP package, a 3D IC package is viewed as the real 3D integration, in which the logic chips or logic and memory chips are stacked with much faster signal transmission speed due to shortened interconnects. Metal-filled TSV enables communication between the two dies as well as with the package.

The integration technologies for 3D ICs include both face-to-face integration and back-to-face integration. While 3D technology has clearly established benefits in terms of electrical performance, it also exacerbates the severe challenge of microelectronics cooling. Because of the stacking of ICs of already extremely high temperature inherent in the 3D stacked dies, disastrous failures are expected in 3D IC integration. It is suggested to either minimize the power consumption on the die or along the interconnects to mitigate the thermal challenge.

A low power design involving stacking memory on logic was analyzed, and the increase in peak temperature in this case was expected to be limited to a few degrees Celsius [20]. Alternatively, volumetric heat removal techniques are also being pursued to reduce the chip temperature. Convective heat transfer from a multi-die stack using liquid cooling has been also reported in references such as Refs. [32–34].

In this section, a preliminary thermal network analysis of the 3D stacked IC is presented to address the fundamental heat transfer questions critical to the successful implementation of 3D technology. The schematic of a 2-die stack on package and the thermal resistance network with multiple heat generating junctions are shown in Figure 3.2.8. Uniform heat generation in the device planes is assumed, and heat flows only in the direction normal to the device planes. In general, a 3D microelectronics system comprises two or more strata mechanically connected to each other through an appropriate interstrata bonding technology. Each strata and interface is characterized by the thermal resistance of the interstrata bonding layer made of bumps and underfills and the back end of line (BEOL) metal dielectric stack, which is lumped as R_{int} in the thermal network in Figure 3.2.7. As indicated, the two dies are stacked face to face with the heat sink thermal resistance R_{hs} and package thermal resistance R_{pk}.

Assuming the self-heating for the two dies are q_1, and q_2, and the heat flow through the package is q_{pk}, the one-dimensional heat transfer model for a general multilayer 3D integrated circuit is given as follows:

$$T_1 - T_a = (q_1 + q_2 + q_{pk})(R_{hs} + R_{Si1}) \qquad (3.2.13a)$$

$$T_2 - T_a = (q_2 + q_{pk})(R_{int}) \qquad (3.2.13b)$$

$$T_2 - T_a = -q_{pk}(R_{pk} + R_{Si2}) \qquad (3.2.13c)$$

R_{int} can be expressed by

$$R_{int} = 2R_{beol} + R_{bump} \tag{3.2.13d}$$

Thus, the solution for temperature is obtained as

$$T_1 - T_a = \frac{[q_1(R_b + 2R_{beol} + R_{pk} + R_{Si2}) + q_2(R_{Si2} + R_{pk})](R_{hs} + R_{Si1})}{R_{hs} + R_{Si1} + R_{bump} + 2R_{beol} + R_{Si2} + R_{pk}} \tag{3.2.14a}$$

$$T_2 - T_a = \frac{[q_1(R_{hs} + R_{Si1}) + q_2(R_{hs} + R_{Si1} + R_b + 2R_{beol})](R_{Si2} + R_{pk})}{R_{hs} + R_{Si1} + R_{bump} + 2R_{beol} + R_{Si2} + R_{pk}} \tag{3.2.14b}$$

Consider the case that there is no heat flow through the package. This may be relevant for high-power applications where a heat sink removes most of the heat, and the package end is assumed to be insulated. In this case, the temperature solution is given by

$$T_1 - T_a = (q_1 + q_2)(R_{Si1} + R_{hs}) \tag{3.2.15a}$$

$$T_2 - T_a = (q_1 + q_2)(R_{Si1} + R_{hs}) + q_2(R_{bump} + 2R_{beol}) \tag{3.2.15b}$$

As shown in Eqs. (3.2.14a)–(3.2.14b), the relative magnitudes of the various thermal resis-tances in the network play a key role in determining the temperature profile. Because of different bonding and BEOL technology, a certain 3D stack with the same power pattern may not be feasible for another package. In general, the silicon thermal resis tances are much smaller than other thermal resistances, which lead to further simplification of the solutions.

The thermal metrics of the die stack can be optimized by power allocation, depending on the respective thermal resistances along the thermal network. Consider the following 2-die stack as an example. Each die is 300 μm thick with a 10 mm × 10 mm footprint. In Case 1, heat sink and package thermal resistances are assumed to be 2 and 20 K/W, respectively. The uniform heat is input to the two dies, which are varied from 2W to 18W, but the total power is fixed to 20W as a constraint, namely $q_1+q_2 = 20W$. In addition, the thermal design goal is to obtain the lowest temperature for both dies. The interstrata bond layer is assumed to be 20 μm thick, with an effective thermal conductivity of 0.2 $W \cdot m^{-1} \cdot K^{-1}$. As the interstrata layer has a large thermal resistance, the BEOL thermal resistance is neglected in this analysis. Varying the ratio of the die powers, the calculated die temperature rises are shown in Figure 3.2.10. Because of the high performance of the heat sink, it is seen that the minimum die temperature rise of 37K is obtained at the maximum die 1 to die 2 power ratio of 9, with 18W on die 1 and 2W on die 2. In other words, the die next to heat sink should be assigned the highest power to achieve the design goal. When each die heat dissipation is 10W with the

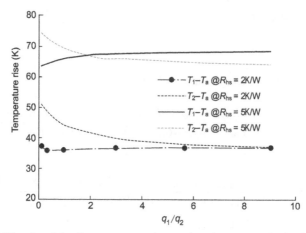

Figure 3.2.10 The plot of the die temperature rises against the power ratio for the 2-die stack at different heat sink and package thermal resistances. The total power is fixed to be 20W.

power ratio of 1, the temperature rises in the two dies are determined 36.15K and 43.95K. Obviously, this is not an optimal design.

In contrast, if the heat sink thermal resistance is increased to 5K/W, whereas the package resistance is improved up to 10K/W, the optimal die power assignment is different. Again fixing the total power dissipation to be 20W, the q_1/q_2 ratios are varied to examine the minimum temperatures for both dies. The die temperature rises are plotted against the power ratio in Figure 3.2.10. It is shown that the minimum temperature rises are obtained at the power ratio around 2, with 13.3W on die 1 and 6.7W on die 2.

With the development of CNT TSVs integrated with the package, it is interesting to estimate the thermal performance of CNT TSVs. Based on the present analysis model, we can estimate how capable CNT TSV can be used in the thermal enhancement of 3D stack with two dies. The thermal conductivity along the MWCNT varies from $300 \text{W} \cdot \text{m}^{-1} \cdot \text{K}^{-1}$ to $3000 \text{W} \cdot \text{m}^{-1} \cdot \text{K}^{-1}$, depending on the fabrication and measurement techniques [35]. It is reasonable to set the thermal conductivity of CNT to be $1000 \text{W} \cdot \text{m}^{-1} \cdot \text{K}^{-1}$ in the present analysis. Compared with a conventional 2-die stack with bonding layer, the use of MWCNT TSVs improves the out-of-plane thermal conductivity of the bulk silicon chip. Another significant thermal benefit is that the CNT TSV can be directly bonded end to end, without introducing the interstrata interconnect thermal resistance.

Let us examine the thermal performance of the 2-die stack as shown in Figure 3.2.7. A densified array of MWCNT TSVs with a diameter of 30 μm and pitch of 60 μm is used here, similar to the MWCNT in Ref. [19]. Thus, for the same die dimensions of 10 mm × 10 mm, the die temperature rises are calculated based on the above equations. The results are shown in Figure 3.2.11. It is seen that the temperature rise for the CNT TSV remains an almost constant temperature rise of 36.55 °C in spite of the power ratio of die 1 to die 2. This will allow more freedom in configuring the logic

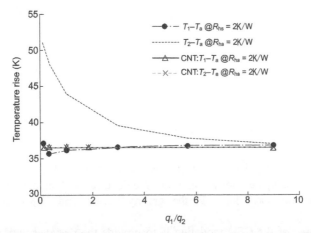

Figure 3.2.11 The plot of the die temperature rises for the carbon nanotube (CNT) through-silicon via (TSV) as against the conventional process ($R_{hs} = 2K/W$, $R_{pk} = 20K/W$). The total power is fixed to be 20W.

die and memory die in a 3D stack. Because the logic die has more I/Os and power connections, it is feasible to allocate logic die directly above the substrate without degradation in thermal performance due to the blocking of the memory die to the heat sink.

The above theoretical model can be extended to a 3D stack with more than two dies. Based on the thermal network, the equations for the temperature potentials and the balance of heat flow can be derived to obtain the temperatures of the various die stack. In addition, the thermal cross talk between different dies can be evaluated through the determination of the influence factors based on the temperature nodes for different dies. A more detailed discussion can be found in Ref. [36], which is not elaborated here. In the case that the die power is not uniformly distributed, a numerical solution is to be sought to derive the hot spot on the die and to optimize the design thoroughly.

3.3 Numerical modeling

3.3.1 Governing equations

In most of the cases, there exists no analytical solution for the thermal evaluation of electronic packaging. Therefore, numerical modeling based on computational fluid dynamics (CFD) techniques becomes a prevalent tool to obtain the detailed heat and fluid flows of the package. The governing equations for the laminar flow and heat transfer in electronic package in Cartesian coordinate system can be expressed as follows Ref. [1].

Continuity equation:

$$\frac{\partial \rho}{\partial t} + \nabla \cdot (\rho U) = 0 \quad (3.3.1)$$

Momentum equation:

$$\rho \frac{\partial U}{\partial t} + \rho(U \cdot \nabla)U = -\nabla p + \mu \nabla^2 U + \rho \alpha \vec{g}(T - T_{\text{ref}}) \quad (3.3.2)$$

Energy equation:

$$\frac{\partial}{\partial t}(\rho c_p T) + \nabla \cdot (\rho c_p U T) = \nabla \cdot (k \nabla T) \quad (3.3.3)$$

Here, ρ and c_p are the density and specific heat of the fluid, U the velocity vector of the liquid paraffin, ∇p the normal pressure gradient, μ the dynamic viscosity, α the thermal expansion coefficient, $\vec{g}$ the gravity vector, and T_{ref} is the reference temperature.

The objective of the CFD modeling is to provide the designer with a computer-based predictive result that enables the analysis of the heat transfer occurring within and around electronics devices, with the aim of understanding, improving, and optimizing the design of existing or new devices. Typical numerical analysis softwares include Mentor Flotherm, Ansys Icepak, Ansys Mechanical, and Comsol. The latter two have been found to be versatile in handling with coupled electro–thermal–mechanical multiphysics modeling for the electronic device and packaging, whereas the former two have been developed to aim at more flexible applications with various scales from the silicon device up to the rack system with increasing modeling complexity.

Specifically, Mentor Flotherm is one of the mostly used software in electronic cooling industry due to the user-friendliness in the building of thermal model with complex electronic devices and systems. In this solver, the conservation equations and their associated boundary conditions are discretized by subdivision of the computational domain into a set of nonoverlapping, contiguous Cartesian finite volumes, and the conservation equations are expressed in algebraic form and solved with the TDMA (tri-diagonal matrix algorithm) algorithm. These finite volumes are referred to as "grid cells" or simply as "cells", defining the spatial computational accuracy. Following the numerical procedure given by B.V. Partanka [37], the objective functions such as temperature and pressure fields can be obtained in an iterative manner.

In general, the numerical procedure is summarized as follows for a 3D software modeling the laminar flow and heat transfer:

① Build the geometrical models with thermophysical properties and meshing model.
② Initialize the fields of pressure, temperature, and velocities with boundary values.
③ Compute the coefficients for temperature field and velocity components.
④ Solve linearized algebraic equations for the value of T and velocity components (u, v, w) in each cell by performing a number of inner iterations.
⑤ Solve the continuity equations in a similar manner and make any associated adjustments to pressure and velocities.
⑥ Check for convergence and return to Step ① if not converged.

Nonetheless, because of the complexity of the physical problem, even the software running on the state-of-the-art computer may not solve the real problem thoroughly without reasonable simplification. The development of different levels of submodels or compact models is urged to simplify the parts of the model. Even a detailed model for a cross-flow heat exchanger would cause 10 million cells with at least 5–8 h of computational time on a 12-core Intel CPU.

Instead, combination of submodels in the numerical model would accelerate the numerical simulation greatly. The submodels for the package elements, such as substrate vias and TSVs etch, have been discussed in Section 3.2. In the thermal analysis and design of a practical electronic package and system, typical submodels should be developed for finned heat sink, heat pipe, vapor chamber, thermoelectric cooler (TEC), blower, etc.

No matter whatever software are used, the following modeling techniques should be taken into account in the design and development of a new package.

- Obtaining the power input and distribution: The on-die power map is mostly preferred before the numerical simulation of a newly developed package.
- Conception of cooling solution with realistic boundary conditions: The cooling solutions, such as natural convection, forced convection, and liquid cooling, should be properly conceptualized and proposed. The simulation software can be a useful tool in design optimization of the specified cooling solution.
- Obtaining necessary package material properties and dimensions: This is the step to construct a realistic package model. The power map is usually supplied by the chip makers based on the electrical analysis of power consumption in the functional blocks in the die. Sometimes the die power map is not accessible to packaging engineer; a simple uniform heating on the logic die is recommended for purpose of package analysis. Occasionally, the power split of 90%/10% on each half of the die is adopted to capture the essentially nonuniform heating in the chip. Figure 3.3.1 shows the power map for the commercial processors.
- Development of submodels to reduce the model complexity: Because of the complexity of the electronic package structure, it is neither feasible nor necessary to model all the parts in detail in most of the cases. A submodel for a certain part can reduce the computational time and efforts greatly.

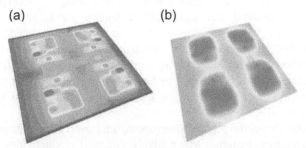

Figure 3.3.1 The nonuniform heat flux generation (a) and temperature profile (b) in a multicore processor chip.

- Consideration of possible process and assembly variance related to thermal modeling accuracy: The package and material may not perform to what is designed, subjected to the process and performance variations. One example is the TIM, which has a thermal resistance significantly different from the datasheet, varying with the operation temperature, package warpage, surface texture, and clamping pressure. An in situ measurement is important to obtain the realistic thermal properties.
- Obtaining thermal metrics: Normally the junction-to-ambient thermal resistance should be obtained. Sometimes the junction to casing thermal resistance could also be utilized to assess the thermal performance.

3.3.2 Numerical modeling of a 2.5D package

A typical 2.5D package configuration under consideration consists of a logic die and a memory die, which has been illustrated in Figure 3.2.7 in bare-die package format. Two dies with sizes of 7.6 mm × 10.9 mm and 8 mm × 8 mm are located on the silicon interposer of 18 mm × 18 mm through microbumping and underfills. The two dies represent the logic and memory chips in a typical interposer integration with different heat dissipations. The interposer wafer is built up with TSVs filled with copper, to form micro straight vias of 10 μm in diameter and 100 μm in depth, followed with a RDL process at the top side. A silicon oxide of 1 μm and corresponding barrier layer of 100 nm titanium have been grown before copper TSV to reduce the dielectric loss. The interposer of 18 mm × 18 mm × 0.1 mm in size was then assembled onto the BT (bismaleimide triazine) substrate with layered metal layers and vias and an overall size of 31 mm × 31 mm × 1 mm. For the molded packages, a molding process is conducted on the wafer level to form the epoxy molded encapsulation on the multi-die package. The wafer is then diced into the package size. After the solder ball attach process, the package is assembled on the thermal test board for electrical connections. The detailed development can be found in references such as Ref. [38].

In the package-level modeling, the TSV interposer, substrate, and interconnects are to be simulated based on the submodel development as described in Section 3.2. For the 2.5D package as shown in Figure 3.2.7, the in-line TSV array is illustrated in Figure 3.3.2 through the X-ray visualization.

In the thermal analysis, the thermal submodels for packaging parts are constructed respectively. The detailed thermophysical properties are listed in Table 3.3.1. As mentioned in the previous section, the in-plane thermal conductivity of the interposer is determined by treating the silicon oxide layer as contact resistance. The through-plane thermal conductivity of the interposer can be obtained with the parallel model. In the present case, because of the large TSV pitch to diameter ratio (100 μm: 10 μm), there is no much increase in the thermal conductivity across the interposer according to Table 3.3.2.

As mentioned above, the flip-chip microbump and underfill are simplified as solid blocks of homogeneous materials with equivalent thermal conductivities. Note that the different layers of copper traces on the substrate and printed circuits board (PCB) are modeled as separate solid blocks of the original thickness with equivalent thermal conductivity based on the copper content.

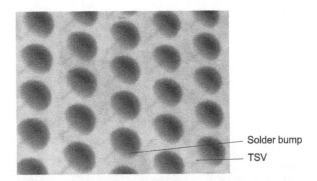

Figure 3.3.2 X-ray images of through-silicon vias (TSVs) on interposer and substrate, taken with the Nordson Dage XO 7600NT (in the absence of substrate).

Table 3.3.1 The thermophysical properties of the different parts in the package.

Package materials	Dimensions (mm)	Thermal conductivity $(W \cdot m^{-1} \cdot K^{-1})$
Thermal die	5.08 × 5.08 × 0.3	150
Si interposer at $D_v/P = 0.1$	18 × 18 × 0.1	See Table 3.3.2
Lumped microbump and underfill layer under die	5.08 × 5.08 × 0.15	$k_{//} = 0.7$, $k_\perp = 6.5$ for peripheral bump area 0.4 for center area
Bump and underfill layer under interposer	18 × 18 × 0.075	$k_{//} = 0.6$, $k_\perp = 4.9$
BT substrate with vias	31 × 31 × 1	$k_{//} = 0.2$, $k_\perp = 0.49$
Substrate copper layers 1–4	31 × 31 × 0.02	256, 320, 310, 206
Solder ball layer	31 × 31 × 0.6	$k_{//} = 0.054$, $k_\perp = 8.0$
PCB FR-4	101.6 × 114 × 1.6	$k_{//} = 0.8$, $k_\perp = 0.3$
PCB copper layers 1–4	101.6 × 104 × 0.035	20, 312, 312, 20

Table 3.3.2 The effective thermal conductivity in $W \cdot m^{-1} \cdot K^{-1}$ for through-silicon via (TSV) array in the system thermal model with SOX as insulation.

$D_v/P = 0.1$	$t_{ins} = 1.5$ μm	$t_{ins} = 1$ μm	$t_{ins} = 0.5$ μm	No oxide
$k_{eff//}$	148.3	148.3	148.5	151.2
$k_{eff\perp}$	151.0	151.3	151.6	151.8
$D_v/P = 0.5$	$t_{ins} = 1.5$ μm	$t_{ins} = 1$ μm	$t_{ins} = 0.5$ μm	No oxide
$k_{eff//}$	111.5	112.3	114.4	182.5
$k_{eff\perp}$	176.0	183.3	190.0	196.1

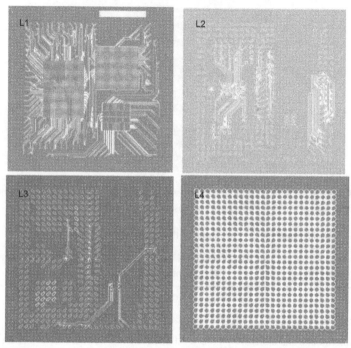

Figure 3.3.3 Schematic of the package substrate with different signal layers.

A typical substrate copper layer is illustrated in Figure 3.3.3, which has a profile of 20 μm and copper content of 65.6% for L1 as an example. The electrical design software such as Virtuoso is used to extract the exact copper content. Because it is difficult as well as unnecessary to model all the detailed routing of the copper layer, a simplified solid block with volume averaged thermal conductivity in the plane and through the plane would be sufficient to conduct the thermal analysis within given design time frame. For Layer 4 (L4) with two blocks of different metal patterns, it can be divided into two regions, and each region is modeled with their respective equivalent thermal conductivities based on the porous media theory.

The submodels for the other package parts such as solder balls and bumps are also obtained based on the Maxwell equation as given in Eq. (3.2.3) in the previous section. It should be noted that the via volume fraction is replaced by that of microbump to extract the in-plane thermal conductivity. In particular, the microbump and underfill were recalculated for the thermal die as it had peripheral pumps instead of fully populated bumps. The through-plane conductivity at the peripheral bumps was $6.5 \mathrm{W \cdot m^{-1} \cdot K^{-1}}$, whereas the center part, filled only with underfill, had a thermal conductivity the same as the underfill material. The corresponding thermal properties for the packaging parts have been listed in Table 3.3.1. The effective thermal conductivity for TSV array is given in Table 3.3.2. Although the thermal conductivity values do not deviate from the bulk silicon at small D_v/P, large discrepancies with and without the insulation layer are observed at the large D_v/P ratio of 0.5.

The microbumps/underfill layers can also be modeled as anisotropic blocks, and the effective thermal conductivities can be calculated based on the abovementioned methods. For example, the fully populated microbumps under the silicon chip, immersed in the underfill, have the bump diameter of 30 μm with a pitch of 50 μm, which leads to an in-plane thermal conductivity of $0.8 \text{W} \cdot \text{m}^{-1} \cdot \text{K}^{-1}$ based on Eq. (3.2.3). The through-plane thermal conductivity is calculated to be $8.6 \text{W} \cdot \text{m}^{-1} \cdot \text{K}^{-1}$ based on the parallel model Eq. (3.2.1). For the micropumps under the interposer, which has a large diameter of 75μm, the thermal conductivities are obtained to be $0.6 \text{W} \cdot \text{m}^{-1} \cdot \text{K}^{-1}$ in the plane and $4.4 \text{W} \cdot \text{m}^{-1} \cdot \text{K}^{-1}$ through the plane, respectively, in the similar manner. As for the substrate, which is made of BT, has four metal layers from the top to the bottom. In addition, the various vias embedded in the BT substrate have been merged with the substrate block through analytical calculations as mentioned above, which increase the bulk through-plane thermal conductivity to $0.49 \text{W} \cdot \text{m}^{-1} \cdot \text{K}^{-1}$ but has little effect on the in-plane thermal conductivity ($0.2 \text{W} \cdot \text{m}^{-1} \cdot \text{K}^{-1}$).

The numerical models for thermal design options under natural convection cooling condition are developed incorporated with the submodel for each part. The buoyancy effect is modeled as a source term in the momentum equation with the Boussinesq's approximation, whereas the other thermophysical properties are assumed to be independent on the temperature. Thermal radiation effect is considered for the natural convection conditions as a significant amount of heat is dissipated by surface thermal radiation. The emissivity for the PCB and the substrate is set to be 0.9, but the emissivity for the chip is set to be 0.1 due to the low emissivity from silicon surface. Laminar flow is assumed due to the moderate Rayleigh number ranging from 10^5 to 10^6, less than that for turbulent flow of $Ra > 10^9$ [1].

3.3.3 Package thermal performance

The corresponding thermal simulation is conducted for both $D_v/P = 0.1$ and $D_v/P = 0.5$ under natural convection condition. A detailed air flow and temperature can be obtained based on the numerical simulation by computer software. For the present package under natural convection at 2W, the flow and temperature fields are illustrated in Figure 3.3.4. The computed thermal resistances from the numerical simulation are compared with the experimental measurements in Figure 3.3.5. The experimental test was conducted at $D_v/P = 0.1$. It is seen that good agreement is found between the thermal simulation and experimental test during the heating power range up to 2.4W. Thus, thermal model is verified with the current package format.

It is reasonable to study the effect of the TSV insulation on the package thermal performance. The corresponding thermal modeling results are shown in Figure 3.3.6. To amplify the TSV effect, only cases with a high $D_v/P = 1/2$ are illustrated. As the

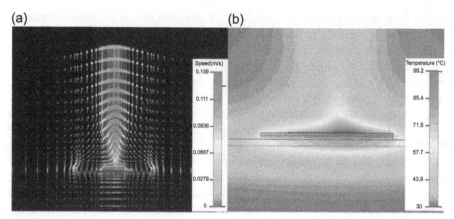

Figure 3.3.4 The computed flow and temperature profiles for a 2.5D package under natural convection.

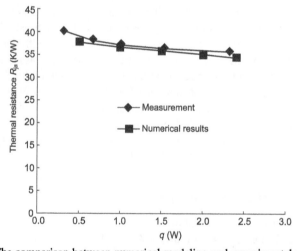

Figure 3.3.5 The comparison between numerical modeling and experimental measurement.

major thermal path for natural convection, PCBs with both 2 layer (2L) and 4 layers (4L) are examined. It is found that the consideration of SOX layer adds up to the junction-to-ambient thermal resistance by 3.6% for the 2L PCB and 7.3% for the 4L PCB in the present package configuration. This indicates that the TSV insulation layer has an unfavorable effect on the package thermal performance and may degrade the chip heat dissipation capability. Therefore, thermally aware designs such as thinning the insulation layer and avoiding densified TSVs in the proximity of high heat-generating die areas should be taken into account to minimize the chip temperature rise and thermal resistance.

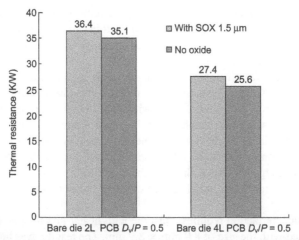

Figure 3.3.6 Effect of the through-silicon via (TSV) insulation layer on the thermal performance of the package at $D_v/P = 0.5$.

3.4 Package-level thermal enhancement

3.4.1 Thermal performance under natural convection

3D package has been viewed as the most promising packaging technology due to possible face-to-face communication. Continuous efforts have been devoted to the thermal cooling of 3D stack package [20,32−34,39], among others. A number of techniques can be used to enhance the thermal performance of the package. Thermal vias are first considered in the organic substrate and interposers made of glass or silicon whenever possible to bridge the heat transfer from chip to the ambient. More advanced cooling techniques include the use of cold plate and direct cooling with dielectric fluids. In the following, thermal enhancement of the 3D package is elaborated with various techniques.

A 3D package developed from Institute of Microelectronics, Singapore, is shown in Figure 3.4.1 [39]. It consists of a 2-die stack, each die mounted on a silicon carrier

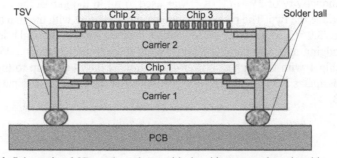

Figure 3.4.1 Schematic of 3D stack package with the chip mounted on the chip carrier.

through flip-chip bumping process. Flip-chip 1, representing an ASIC device, is mounted in carrier 1. Flip-chips 2 and 3 are mounted in carrier 2, representing memory chips. The thermal characterization for 2.5D/3D packages with TSVs is conducted under natural convection condition.

The 3D package overall dimensions are 13.5 mm × 13.5 mm × 1.4 mm, having 276 peripheral I/Os. Carrier 1 is designed to mount a flip-chip(chip 1). The chip 1 dimensions are 8.5 mm × 8.5 mm × 0.1 mm with 1000 I/Os at a bump pitch of 0.25 mm. Chip 2 and chip 3 each have overall dimensions of 4.5 mm × 9.0 mm × 0.1 mm and 80 I/Os. The two silicon carriers were connected vertically using 8 mil solder balls. Electrical connections through the silicon carrier were formed by TSV technology. Some of the advantages of using silicon as the chip carrier are low thermomechanical stress, fine line width and spacing, wafer level fabrication, and high thermal conductivity.

Thermal characterization requires a test die with heater and temperature sensor. A test die was used with the size of 8.9 mm × 8.9 mm. The test die was thinned down to 100 μm and assembled with the carrier with electrical connections. The 3D package was assembled onto a four-layer test board. The assembled board mounted horizontally inside an enclosed chamber for natural convection thermal characterization as defined in JESD-51 Standards [40,41]. The test chip on carrier 1 was supplied with 0.8W power, and the two chips on carrier 2 were supplied with 0.2W, respectively. Four such packages were characterized, and the average of maximum temperature chip 1 is compared. The maximum temperature rise of chip 1 was 24.2 °C, corresponding to a junction-to-ambient thermal resistance of 30.3K/W based on 0.8W power input.

At first, thermal enhancement techniques by adding thermal vias and adhesive thermal bridging were attempted. Thermal analysis was carried out with a heat load of 0.8 W in chip 1 and 0.1W each in chip 2 and chip 3, respectively. First, the package thermal performance was studied with 144 thermal vias made of copper in the carrier 1. Maximum temperature of chip 1 with thermal via is 77.5 °C as against 78.5 °C for the case without thermal vias. Because silicon has good thermal conductivity, the use of thermal via made of copper in the silicon carrier is not effective in reducing the package thermal resistance.

Another thermal enhancement was attempted by filling in the gap between chip 1 and carrier 2 with thermal adhesive. A thermally conductive polymer adhesive having thermal conductivity of $2W \cdot m^{-1} \cdot K^{-1}$ was used to fill in the gap of the carrier 2 and chip 2 to bridge the two. The maximum temperature of chip 1 with thermal adhesive is found to be 73.9 °C, which is 4.6 °C lower than the case without thermal bridging. The thermal bridging acts as a parallel thermal path, through which around 45% of heat from the chip 1 with the highest power dissipation of 0.8W goes up to the carrier 2 and then dissipates to the PCB. The rest heat power of the chip 1 is dissipated directly to the PCB.

3.4.2 Liquid cooling of 3D package

To achieve the design goal of 20W power dissipation, a passive cooling solution may not be available. As such, a cold plate is attached at the top to dissipate the heat through liquid cooling. With this active cooling technique, it was shown by both measurement and thermal simulation that 20W could be dissipated from chip 1 through the top cold plate cooling method.

The power dissipation of the 3D package can be further extended with direct cooling over the different layers of the package. Figure 3.4.2 shows a three-die stack, with each die flip-chip bonded onto a carrier. All the three carriers are connected with solder bumps, which are consequently assembled on the substrate and PCB. To maximize the power dissipation, a feasible way is to cool all the dies in the stack without being blocked by the top die. A connector housing is bonded to the substrate of the package, which has an inlet and outlet to connect to the remote heat exchanger to reject the heat to the ambient. In the first development, the bare die without thermal enhancement structure at the backside is conducted. Dielectric fluid FC-72 is used as the coolant. The power dissipation of 23W is demonstrated through experimental measurement when the die temperature is fixed at $T_j = 57\,°C$. Further increase the power would cause flow boiling due to die heating. In spite of the lack of modeling tool for the flow boiling in the 3D package, the experimental measurement was conducted and 40W is reached at $T_j \sim 85\,°C$.

To push the cooling limit, the package is modified by fabricating microchannels on the backside of the die. The fabricated microchannels measure 100 μm in width and 225 μm in height. A minimal assembly gap ~ 10 μm is reserved for assembly purpose

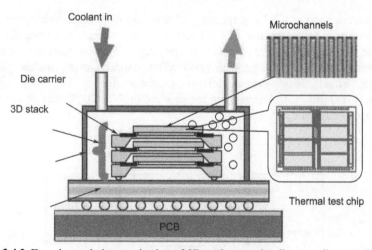

Figure 3.4.2 Experimental characterization of 3D package under direct cooling condition.

between the microchannel-fin structure and the top carrier. All the three dies with microchannel structure are assembled and integrated with the connector housing, as have been shown in Figure 3.4.3. With FC-72 working in the flow boiling regime, a power dissipation of 80W can be obtained from the 3-die stack.

In a Darpa-funded project [34], a design goal of 200W is targeted for a 2-die stack as shown in Figure 3.4.3. As demonstrated in the abovementioned study, the FC-72 has a limited cooling capacity due to its low thermal conductivity as shown in Section 3.1. Therefore, an indirect cooling design is devised with water as the highly efficient coolant. The microchannels are fabricated in the chip carrier and sealed to prevent liquid leakage. As the liquid is to be split into two streams to enter the channels of each carrier, the two carriers with channels are connected at the peripheral with hollow vias. Such hollow structure allows the liquid flow through both carriers. Experimental test shows that 200W can be rejected to the heat exchanger based on the criteria of $T_j = 85\,°C$.

Among the various efforts to elevate the power dissipation, it is important to understand the fundamental cooling limit for a single-chip assembly. Recently, IBM investigated the dielectric cooling with R1234z as the coolant in the ICECOOL project [42]. It is demonstrated that, under a new impingement cooling configuration, a power dissipation of $350W/cm^2$ is demonstrated. Such a high-performance cooling technology was operated over two months from May to June in Poughkeepsie, NY, USA, and demonstrated a reduction of 90% cooling energy compared with traditional air-cooled data centers.

3.4.3 Thermal enhancement with new materials

New materials such as CNTs, graphene, and diamond have been investigated as hot topics. In Section 3.2, the use of CNT as TSV interconnects has been discussed, and clear thermal benefits are pointed out. The graphene-based materials are used as heat spreader for the package with good effect. Jeppson et al. studied the heat spreading effect using the monolayer graphene for a hot spot chip of 390 μm × 400 μm [43]. A maximum temperature drop of 10 °C was observed at a heat flux of $460W/cm^2$.

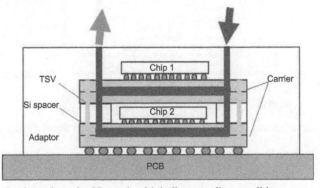

Figure 3.4.3 Configuration of a 3D stack with indirect cooling condition.

Figure 3.4.4 Fabricated lid with diamond-like coating (DLC) and the thermal tester for the assembled package with fan-heat sink.

In the research by Zhang et al. [44], the effect of diamond-like coating (DLC) on the thermal performance of the lidded package was investigated. Under the standard fan-heat sink test platform, the junction to heat sink thermal resistance is reduced from 0.308K/W to 0.275K/W, equivalent to 10.7% drop in thermal resistance. Figure 3.4.4 shows the DLC lid together with test platform.

3.5 Air cooling for electronic devices with vapor chamber configurations

Along with the thermal development in the package, the thermal design on the package is also being developed in parallel to meet various application scenarios.

Air cooling is the most prevalent cooling method used in cooling of electronic devices due to the low cost and ease of assembly with the package. In a typical electronic system, fan-cooled heat sink is applied on the high-power components such as the CPU and GPU, whereas a chipset and memory with medium power dissipation can be cooled by mounting with a heat spreader or a pin fin heat sink. With the trend in miniaturization of the package with more functionality, the heat dissipation from the package keeps increasing, which may exceed 100W for each package. This may exceed the cooling limit of the conventional air cooling with metal heat sink, which has to be extended to meet the design requirement.

One of the effective methods is the development of the vapor chamber heat sink, in which a flat chamber is evacuated, filled with working fluids, and sealed to serve as a heat spreader to replace the solid heat sink base [45–49]. Essentially, the vapor chamber is a capillary-driven planar design with a small aspect ratio, which can be viewed as flat-plate heat pipe in a disk shape. The working medium evaporates when it receives the heat at the bottom base subjected to chip heating. The vapor rises up and condenses when it contacts the top cold wall of the chamber, where the heat is released to the top and eventually exchanged to the ambient air through the finned structure. Such a two-phase heat transfer process in the vapor chamber greatly reduces the spreading resistance at the heat sink base. This renders efficient heat transfer to the fins far away from the chip so that a heat sink with a large base can be used, enabling a higher power dissipation.

Additional wick blocks between the evaporator and condenser aid in condensate return, especially when the condenser is below the evaporator in a gravity field. If the condenser is above the evaporator, there is no need to have a wick in the condenser section, as the condensate on the upper plate drips back to the evaporator. Covered over the evaporator section is a wick structure to uniformly distribute the liquid over the entire surface to prevent dry out.

The vapor chamber is an excellent candidate for use in electronic cooling applications, especially with high heat flux applications, such as desktops and server processors. Vapor chambers are preferred over a conventional heat pipe made of copper for electronic cooling with heat fluxes higher than 50 W/cm^2 as heat flow in a vapor chamber is spread out in two- or three-dimensional directions, compared with the one-dimensional heat spreading along a heat pipe. Furthermore, vapor chambers can be placed in direct contact with CPUs using TIMs, thereby reducing the overall thermal resistance of heat sinks.

The vapor chamber heat sink (VCHS) used in this study is shown in Figure 3.5.1. It consisted of a copper vapor chamber with a height of 5 mm and conventional aluminum heat sink with a base of 105 mm × 79 mm×2 mm and extruded fin thickness of 1 mm and fin height of 38 mm. The large heat sink base area was purposely selected to examine the performance limit of the air cooling technique.

An axial fan with a full flow rate 16.4 CFM and maximum pressure head 78Pa was assembled with the VCHS to provide air flow. There exist two fan configurations for the fan and heat sink assembly. In configuration (a) shown in Figure 3.5.2, the fan is located at the lateral side of the heat sink, to blow the cooling air through the heat sink. Figure 3.5.2b shows the other configuration, where the fan is located at the top of the heat sink and the air is forced to impinge onto the center of the VCHS through the fin pitches and then exit from the two ends.

Figure 3.5.1 VCHS used in the characterization.

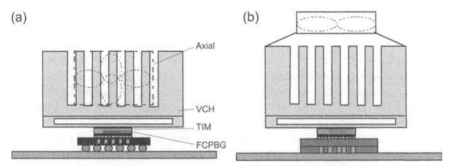

Figure 3.5.2 Two VCHS configurations with (a) side fan and (b) top fan.

Figure 3.5.3 The test configuration of VCHS, with fan assembled at the lateral side of heat sink.

It is seen that the side fan configuration has a slightly better thermal performance under the same test conditions. The VCHS was mounted onto the FC-PBGA, with four spring loaded screws at the four corners of the VCHS. Thermal grease (TM150) has been used in the experimental test. TIM1 is a type of thermal gel material and is not sensitive to the operation temperature and power input. Detailed descriptions of the TIMs are available in suppliers' datasheets.

Experimental test was also conducted for the vapor chamber heat sink assembled with flip-chip BGA package, with a chip size of 12 mm × 12 mm. The experimental test is shown in Figure 3.5.3. In the experiments, the chip temperature T_j was measured at increasing power inputs from 40W to 100W. The average chip temperature is obtained and the thermal resistance relative to the surrounding air thermal resistance can be calculated as follows:

$$R_{ji} = \frac{T_j - T_i}{q} \qquad (3.5.1)$$

Here, T_i is referred to the inlet air temperature and q is the heating power input.

The uncertainty of measured T_j was 0.2 °C and the inlet air temperature variation was estimated around 0.2 °C. The uncertainly in the heating power input was estimated to 1%. As mostly the chip temperature varied from 20 to 60 °C, the measurement uncertainty in the thermal resistance was less than 2% based on the same packaging assembly. The assembly of the heat sink to the package has been carried out carefully to minimize extra errors in the thermal resistance at the interface between the chip and the heat sink. Repeated tests have been conducted, and the deviations due to assembly were found to amount to 1%–2%.

For the VCHS characterization, the tests were first conducted for the two fan configurations shown in Figure 3.5.2. The testing was varied by adjusting the fan locations in the same assembly with FC-PBGA and with the same interface material. Figure 3.5.4 shows the chip temperature rises against the heating power inputs for the two fan configurations. It is seen that the chip temperature rises are proportional to the heating power inputs. The data were fitted into linear correlations with minor scattering within 5%. The slopes representing the thermal resistances are found to be 0.452 and 0.457 °C/W for the side and top configurations, respectively, with side configuration giving 1% better performance on average. Because the tests were conducted under the same condition except the fan location, it is assumed both configurations have the same thermal performance considering the experimental error.

Figure 3.5.4 shows the thermal resistances based on the different TIMs with varying heating power inputs. It is seen that, in most cases, the thermal resistances first decreases and then achieves stable values at high-power dissipations close to 100W, and the lowest thermal resistances are obtained at the higher power input 80–100W, which are around 0.42–0.44 °C/W. The higher thermal resistances at the lower power inputs can be due to the temperature effect on the VCHS [45]. On the other hand, the boiling limit of VCHS, which has been thought to be VCHS capacity limit [45], was not observed based on the present maximum power input of 110W or heat flux of 76W/cm^2, suggesting that the VCHS is a suitable option for power dissipation around 100W.

Figure 3.5.5 shows the computed velocity field and temperature profiles based on simulation and Fig. 3.5.6 shows the comparison between simulation and measurement.

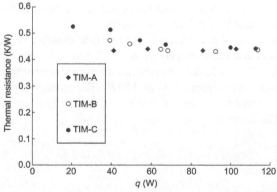

Figure 3.5.4 The junction to inlet air thermal resistances based on the different TIMs with varying heating power inputs.

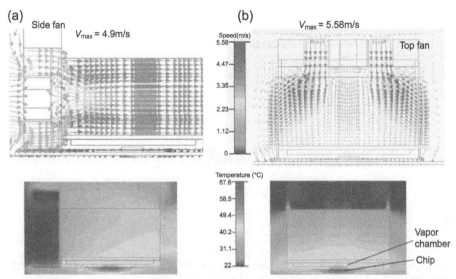

Figure 3.5.5 The computed velocity field and temperature profile: (a) the side fan configuration and (b) the top fan configuration.

A better agreement within a few percent is achieved at higher power dissipation ~100W or even higher, whereas a deviation 5%–8% is observed between simulation and measurement at relatively low power dissipation. This deviation could be due to the fact that a water-based VCHS may not be fully activated on start-up but would perform the best at higher base temperature more than 60 °C. In literature, the vapor chamber heat sink has been tested at 80W/cm^2. In the present study, the maximum power is 110W, which corresponds to the heat flux around 76W/cm^2, close to the maximum power reported in literature. Note that the vapor chamber may not work at elevated temperature more than 100 °C. Otherwise, the vapor chamber undergoes a much higher pressure and would deform permanently and affect its thermal performance. In reality, as the processor chip is less than 90 °C, the vapor chamber may maintain its stable operation much below the threshold temperature. Nonetheless, in certain application such as the cooling of power electronics, the vapor chamber must be specifically designed and fabricated to meet the application requirement (Fig. 3.5.6).

The worst disadvantage of a vapor chamber can be the negative effect of the gravity on its operation [47]. In horizontal position, the fluid is indeed distributed uniformly in-plane if the capillary pumping of the wick is sufficient. In vertical position, if the heat source is at the center of the vapor chamber, the vapor flow is divided into a gravity assisted upflow and a gravity opposed downflow. This phenomenon can involve a temperature gradient disparity between the top and the bottom of the vertical vapor chamber.

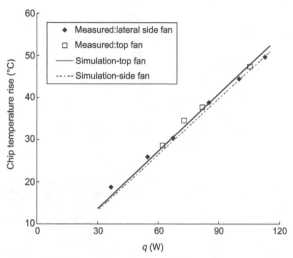

Figure 3.5.6 The computed chip temperature rise at different power levels in comparison with experimental measurements in top and lateral fan configurations.

3.6 Liquid cooling for electronic devices

Liquid cooled heat sinks of various forms have been developed in the last few decades, either in the package or around the package. Microchannels, with channel width less than 1 mm, have received the most intensive study, along with the progress in fabrication techniques. Other thermal enhancements include metal foams and honeycomb structure. Recently, impingement jet flow arrangement is also progressed to minimize the pressure drop and temperature difference on the heat sink base [34,50–54]. In practice, the microchannels can be fabricated onto the silicon as part of the package as demonstrated in Section 3.4, or they can be fabricated separately and then mounted on the package. Both in-package and on-package cooling requires an analysis of the thermal performance. In the following parts of this section, the analytical model developed for microchannel heat sink mounted on a single-chip flip-chip package is presented as an example to illustrate the thermal analysis process in liquid cooling.

3.6.1 Analytical model for finned heat sink

The analytical model developed by Zhang et al. [51] is elaborated in this section. The geometrical model is illustrated in Figure 3.6.1. An analytical technique based on fluid flow and heat transfer theory is used to perform thermal analysis and to provide design guidelines. The assumptions in the present analysis include constant thermophysical properties of the fluid and uniform laminar flow across the microchannels. All the heat is assumed to be dissipated uniformly through the top of the chip to heat sink base. Heat losses from the chip through the substrate to the ambient and through the cover plate to the surrounding air are negligibly small and ignored in the thermal

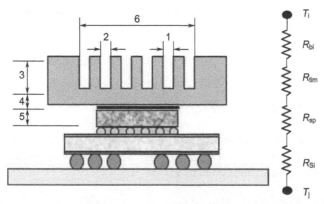

Figure 3.6.1 Cross-sectional view of the microchannel heat sink on the flip-chip ball-grid array packages and the corresponding thermal resistance network. Heat sink geometrical parameters: 1-channel width w_{ch}, 2- fin thickness t_{fin}, 3- fin height H_{fin}, 4-base thickness t_b, 5-chip thickness t_{Si}, and 6- width of fin array w (the length of fin array l is not shown).

analysis. In addition, the spreading thermal resistance occuring from the chip to the heat sink base is also examined based on the correlation by Song et al. [55,56].

The thermal resistance network is shown in Figure 3.6.1. The junction to the inlet fluid thermal resistance R_{ji} consists of four parts in series: the resistance R_{Si} due to the conduction from the active die layer near the chip bottom to the top of chip, the resistance R_{tim} through the interface material, the spreading resistance R_{sp} within the heat sink base, and the resistance R_{bi} from the top of the heat sink base through the fin array to the inlet fluid. Hence,

$$R_{ji} = R_{Si} + R_{tim} + R_{sp} + R_{bi} \quad (3.6.1)$$

Here, R_{Si} can be determined by

$$R_{Si} = t_{Si}/(k_{Si}A_{Si}) \quad (3.6.2)$$

with t_{Si}, A_{Si}, and k_{Si} the thickness, footprint, and thermal conductivity of the chip, respectively. The value of R_{tim} has been obtained based on in-house characterization and found to be 0.242 K·cm²/W with assembly deviations around 5%, independent of the flow rate. The thermal resistance R_{bi} between the heat sink base and the inlet fluid can be calculated as

$$R_{bi} = 1/(h_{eff}wl) + 1/(2\rho c_p \cdot VF) \quad (3.6.3)$$

where h_{eff} is the equivalent heat transfer coefficient from the heat sink base to the bulk fluid and can be determined with the conventional fin analysis below, ρ the density of fluid, c_p the specific heat, VF the volumetric flow rate, and w and l the width and length of the fin array, respectively (Figure 3.6.1). Note that the factor 2 is introduced in second term of the right-hand side of Eq. (3.6.3) as the average fluid temperature $T_m = (T_i + T_o)/2$ has been used for calculation of the thermal resistance from channel walls to bulk fluid [57].

The equivalent heat transfer coefficient h_{eff} can be obtained based on conventional fin analysis

$$h_{eff} = h_m[(2N\eta_f H_f + w - (N-1)t_f)]/w \qquad (3.6.4)$$

where N is the number of channels, t_f the fin thickness, η_f the fin efficiency given in Eq.(3.1.21a) in Section 3.1, and h_m the average heat transfer coefficient between fin/base surface to the bulk fluid, evaluated by

$$h_m = \frac{Nu_m \cdot k}{D_h} \qquad (3.6.5)$$

where k is thermal conductivity of the fluid, $D_h = 2w_{ch}H_f/(w_{ch} + H_f)$ the hydraulic diameter, and Nu_m the average Nusselt number. The fitted correlation from Ref. [58], which takes into account the simultaneous developing flow effect, is used to determine the heat transfer for microchannel heat sinks.

$$Nu_m = \left\{ \left[2.22(Re \cdot Pr \cdot D_h/l)^{0.33}\right]^3 + (8.31G_0 - 0.02)^3 \right\}^{1/3} \qquad (3.6.6)$$

$$G_0 = \frac{Ar^2 + 1}{(Ar + 1)^2}$$

where Pr is the Prandtl number of the fluid, $Ar = H_f/w_{ch}$ the aspect ratio, Re the Reynolds number based on the hydraulic diameter of the microchannel, and u the average flow velocity in the channel. Note that other correlations for the Nu value are available in Ref. [59]. Using the theoretically obtained average Nusselt numbers under uniform wall heat flux conditions, similar thermal performance can be obtained. It should be noted that the predictions based on Eq. (3.6.6) are applicable within a limited range of $Re \cdot Pr \cdot D_h/l$ and are not suitable for extrapolation to fully developed conditions.

It is well known that there exists spreading thermal resistance when heat conducts from a small area to a large area. Spreading thermal resistance cannot be neglected in the analysis of the thermal network. The spreading thermal resistance R_{sp} in Eq. (3.6.1) can be obtained as Ref. [55]:

$$R_{sp} = \frac{\varepsilon\tau + 0.5\sqrt{\pi}(1-\varepsilon)^{3/2}\phi}{k_b \pi a} \qquad (3.6.7)$$

$$\phi = \frac{\tanh(\lambda\tau) + \lambda/Bi}{1 + (\lambda/Bi)\tanh(\lambda\tau)} \qquad (3.6.8a)$$

$$\lambda = \pi + \frac{1}{\varepsilon\sqrt{\pi}} \qquad (3.6.8b)$$

$$Bi = \frac{1}{\pi k_b b R_{bi}} \qquad (3.6.8c)$$

with $a = \sqrt{A_{Si}/\pi}$, $b = \sqrt{wl/\pi}$, $\varepsilon = \frac{a}{b}$, and $\tau = \frac{t_b}{b}$. Here, t_b is the thickness of the heat sink base and k_b the thermal conductivity of heat sink base. With Eqs.(3.6.1) – (3.6.8), the thermal resistance R_{ji} can be determined analytically.

The pressure drop across the heat sink consists of three parts: the pressure drop across the flow channels, the pressure drop at the inlet due to the flow constriction, and at the exit due to the flow expansion. The expression for pressure drop can be written following Refs.[58,60].

$$\Delta p = \frac{\rho u^2}{2}(4 f_{app} l / D_h + K) \qquad (3.6.9)$$

$$f_{app} = \frac{1}{Re}\left\{\left[3.2(D_h \cdot Re/l)^{0.57}\right]^2 + (4.70 + 19.64 G_0)^2\right\}^{1/2} \qquad (3.6.10)$$

$$K = 0.6\sigma^2 - 2.4\sigma + 1.8 \qquad (3.6.11)$$

$$\sigma = N w_{ch} H_f / (w H_f) \qquad (3.6.12)$$

With this model, the heat dissipation from the chip to be heat sink populated with microchannels can be well described.

3.6.2 Analytical results and comparison with experimental measurements

Along the analytical model, the experimental test was conducted to verify the model. In the experimental test, a microchannel heat sink was designed and tested in assembly with a flip-chip BGA package as is shown in Figure 3.6.2. Two chip sizes of 12 mm × 12 mm and 10 mm × 10 mm were tested. Various microstructures were fabricated in house, such as the subtractive fabrication for the aluminum microchannels by wheel cutting, the copper microchannel by electrical discharge machining, the silicon channels by the deep reactive-ion etching, and the additive fabrication with the metal foam. For illustration purpose, only the aluminum channels with a design channel width of 0.2 mm are examined here. The channel height was 2 mm and the fin cross section 0.4 mm × 2 mm. The cross-sectional view of the microchannel heat sink integrated with the FCBGA is shown in Figure 3.6.1. The microchannels were fabricated by wheel cutting the aluminum block of 50 mm (length) ×24 mm (width) × 2.8 mm (height) with natural finish. It had 21 channels, with each channel found to be 0.21 mm wide and 2 mm high through optical microscope. The large

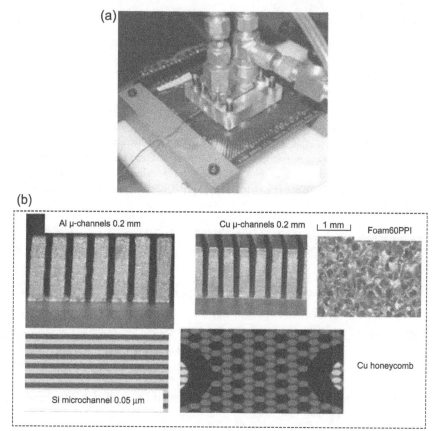

Figure 3.6.2 (a) Experimental assembly of the microchannel heat sink with the thermal test board and (b) various microstructure for thermal enhancement.

aspect ratio ~ 10 has been purposely made to enlarge the surface area for heat transfer augmentation, as compared with the low aspect ratios reported in literature [61]. Inlet and outlet plenums were also made in the aluminum block for ease of assembly and reducing the pressure loss when fluid flows into and out of the microchannels.

The analytical results are obtained on the basis of Eqs. (3.6.9)–(3.6.12) and shown for the pressure drop in Figures 3.6.3 and 3.6.4 in comparison with the experimental results for the two chip cases. For the sake of convenience, the mean fluid temperature at the flow rate of 1L/min has been used as the reference temperature for the evaluation of fluid thermophysical properties. We can see that the predicted pressure drops are somewhat higher than the experiments by around 10%–15%. The deviations could be due to the abrupt turning at the inlet and outlet plena, the possible flow bypassing over microchannels due to the use of the O-ring between the heat sink and the cover plate. In practice, such a difference is acceptable for engineering application. In spite of

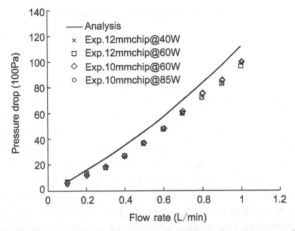

Figure 3.6.3 Comparison of analytical results with experimental measurement for pressure drop.

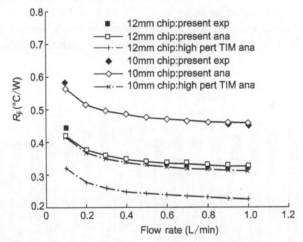

Figure 3.6.4 Comparison of analytical results with experimental measurement for thermal resistances.

the moderate difference, a similar quadratic dependence of the pressure drop on the flow rate is attained.

The comparison of R_{ji} for the analytical results and experiments at 60 W for both sizes of chips is shown in Figure 3.6.4. It is seen that the thermal resistances are predicted within 3% for all the flow range for both chip cases, except at the lowest flow rate of 0.1L/min where the deviation is close to 6%. The good agreement between the predictions and experiments also indicates that the conventional heat transfer theory can still be utilized for prediction of heat transfer at small channels of width around 0.2 mm without significant deviations from experiments.

3.6.3 Anatomy of thermal resistance elements

It is of practical interest to examine the respective thermal resistance elements on the right-hand side of Eq. (3.6.1). Such analysis has been conducted for both chip sizes and the results are illustrated in Figure 3.6.3. For both chip cases, the increase in flow rate from 0.1L/min to 1L/min leads to a reduction of 55% in R_{bi} due to the enhanced convection. Coupled with a reduction in R_{sp} of 13% for the 12 mm chip and 21% for the 10 mm chip, the overall decreases in R_{ji} is 22% and 19%, respectively.

Instead of keeping increasing the flow rate, further thermal enhancement can be achieved through other means such as thinning of the wafer thickness and, especially, the improvement of TIM performance. Considering an improved TIM with an impedance of 0.1 K·cm^2/W, the thermal resistance based on the above analysis is recalculated here for illustration. The calculated thermal resistances for the 12 and 10 mm chip cases have been shown in Figure 3.6.5. It is seen that the thermal resistances for both chips are reduced by around 30% on average due to the introduction of the advanced interface material. At a flow rate of 0.5L/min, the thermal resistance for the 12 mm chip is 0.222 °C/W and, given a temperature window of 60 °C, the estimated power dissipation goes up to 270W without further increase in system complexity and operation cost. The above example indicates that the TIM plays an important role in reducing the junction to coolant inlet thermal resistance.

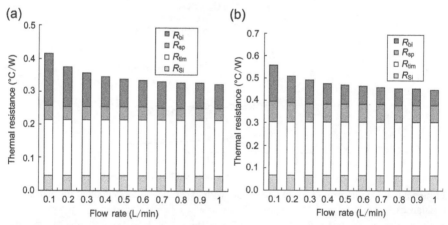

Figure 3.6.5 The breakdown of the thermal resistances for (a)12 mm chip and (b)10 mm chip.

3.7 Thermoelectric cooling

TECs have found applications in areas such as microelectronic systems, laser diodes, telecommunication, and medical devices. Recently, there is an increasing interest in the use of TEC for enhanced cooling of high-power electronic packages such as processors in both manufacturing process and test conditions. High cooling capacity TECs, in combination of air cooling or liquid cooling techniques, are being pursued to extend the conventional air cooling limits for high-power dissipating processors [62–68]. Compact in size and silent in operation, the TEC is easy to be integrated into a server or desktop computer system in comparison with the vapor compression refrigeration technology. The on-chip hot spot can also be managed with micro-TEC if properly designed [69].

In the design and development of TEC apparatus for microelectronic components, it is crucial to determine and optimize the TEC performance within the cooling system constraints. The widely used approach is the iterative method as is given in Refs. [63,64,67,70]. The iterative method offers results based on TEC pellet thermoelectric properties, which is nonetheless tedious and time-consuming for designers to use in practice. Simons and his coworkers presented an analytical solution for electronics modules based on computerized derivation [68] and analyzed the TEC-enhanced cooling performance for high-power multichip modules with available commercial TECs. Nonetheless, their solution expressions were complicated in formulation and utilized only within the same group without verification by other researchers. It is worthwhile to point out that all the abovementioned predictive methods require prior knowledge of TEC pellet geometrical dimensions and thermoelectric properties including Seebeck coefficient, thermal conductivity, and electrical resistivity. These pellet properties vary with carrier concentration and manufacturing processes and, as manufacturers' proprietary information, are mostly not available to designers. Module level properties are required to understand [71,72]. A detailed discussion on the material properties variation versus carrier concentration and composition can be found in Ref. [73].

Most of the existing optimization studies on TEC performance parameters and related thermal resistances with respect to electric current for cooling microelectronic components are still limited or presented in a roundabout manner. To address this issue, Zhang et al. developed analytical solutions for the thermoelectric cooling for electronic packaging [74,75]. The closed-form analytical solutions at both pellet and module levels including the thermal resistances at the hot side and cold side were presented for ease of usage in practical design analysis. Apart from the conventional iterative methods, their analysis approach on the TEC pellet is able to predict TEC performance metrics in a deterministic and straightforward manner given that pellet-level parameters are known. Alternatively, the module level analysis model is formulated and derived based on the TEC module parameters, which is especially useful in optimizing TEC thermal performance with supplier's datasheet in the absence of the particular pellet dimensions and thermoelectric properties.

In what follows, the analysis procedure is described based on the Zhang et al.'s work for a TEC-based cooling configuration. The high power-generating processor is exemplified for illustration. Verification of TEC analysis methodology and thermal performance across different researchers is also reported in their study.

3.7.1 Analytical solution at pellet level

The thermal analysis starts with the basic one-dimensional thermal balance equations of the TECs, which are available in previous literature such as Ref. [73]. It is noted that the temperature effect on the thermoelectric properties, the effects of ceramic plates, and joining copper traces and electrical contact resistances, which are negligibly small, are not included in the present thermal balance model.

Cooling power absorbed at the TEC cold side:

$$q_c = 2N\left(sIT_c - \frac{\rho I^2}{2Ar} - k \cdot Ar \cdot \Delta T\right) \quad (3.7.1)$$

Total heat generated at the hot side of TEC:

$$q_h = 2N\left(sIT_h + \frac{\rho I^2}{2Ar} - k \cdot Ar \cdot \Delta T\right) \quad (3.7.2)$$

TEC temperature difference ΔT from hot to cold sides:

$$\Delta T = T_h - T_c \quad (3.7.3)$$

Figure 3.7.1 illustrates the TEC cooling scenario, simulating the processor testing in SLT and burn-in test condition. Such a testing process is required for each processor product before its release to the customer. The thermal design task is to minimize the junction temperature or maximize the cooling capacity for better thermal control.

The one-dimensional thermal resistance network for both application scenarios has been shown in Figure 3.7.1. At the device side,

$$T_j = T_c + q_c R_{jc} \quad (3.7.4)$$

Thermal resistance correlation at the heat sink side is written as

$$T_h = T_a + \left[q_c + 2N\left(\frac{\rho I^2}{Ar} + sI\Delta T\right)\right] R_{hs} \quad (3.7.5)$$

It is noted that all the thermal interface resistances are already absorbed in the lumped thermal resistance R_{jc} and R_{hs}. There are six unknown quantities in Eqs. (3.7.1)–(3.7.5): T_h, T_c, ΔT, q_h, q_c, and T_j, given the known pellet properties including k, s, ρ, Ar, thermal resistances R_{jc} and R_{hs}, ambient temperature T_a, and operational current I. If any five of the six quantities are fixed, the remaining quantity is readily obtainable at a given operation current I. Among them, the cooling power q_c and device junction temperature T_j are the performance parameters of the most concern in thermal control of device. The solution expression for q_c with a given junction temperature T_j can be derived as follows.

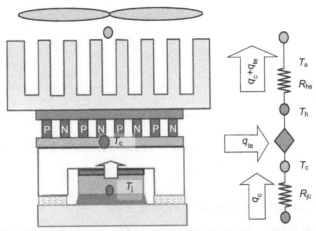

Figure 3.7.1 Schematic of the thermoelectric cooler (TEC) cooling of a processor with a heat sink at the top and the corresponding one-dimensional thermal resistance network. The arrows indicate the heat flows.

With the mathematical manipulation, one obtains the expression for the performance parameter q_c as a function of junction temperature T_j and operation current I:

$$q_c(I) = \frac{\left[(2NsI)^2 R_{hs} - 2NsI - 2NkAr\right] T_j + 2NkArT_a + \dfrac{N\rho I^2}{Ar}(4NkArR_{hs} - 2NsIR_{hs} + 1)}{-[2NkAr(R_{jc} + R_{hs}) - (2NsIR_{hs} - 1)(2NsIR_{jc} + 1)]} \quad (3.7.6)$$

Vice versa, one can obtain the expression of T_j in terms of q_c and I as follows:

$$T_j(I) = \frac{[(2NsIR_{hs} - 1)(2NsIR_{jc} + 1) - 2NkAr(R_{hs} + R_{jc})]q_c - 2NkArT_a - \dfrac{2N\rho I^2}{Ar}(2NkArR_{hs} - 2NsIR_{hs} + 1)}{(2NsI)^2 R_{hs} - 2NsI - 2NkAr}$$

(3.7.7)

We can easily confirm that the above equation holds in the simplified case with $R_{hs} = R_{jc} = 0$. The abovementioned equations (3.7.6–3.7.7) provide a basic solution approach in determining the TEC thermal performance at the pellet level. The TEC electrical power q_{te}, operational voltage V, and coefficient of performance (COP) are expressed as follows:

$$q_{te} = 2N\left(sI\Delta T + \frac{I^2\rho}{Ar}\right) \quad (3.7.8)$$

$$V = 2N\left(s\Delta T + \frac{\rho I}{Ar}\right) \quad (3.7.9)$$

COP is expressed as

$$\text{COP} = \frac{q_c}{q_{te}} \tag{3.7.10}$$

A prior knowledge of the pellet thermoelectric properties and geometry factor Ar is a prerequisite condition for the utilization of the present pellet-level analysis. This is the drawback of all the pellet-level analysis as pellet-level information is usually not available for commercial TECs, which is one of the major limiting factors in the wide usage of pellet-level analysis.

3.7.2 Analytical solution at module level

It is known that the thermoelectric properties of TEC pellet materials such as bismuth telluride vary with carrier concentration depending on manufacturers' processes. On the other hand, these details, together with the geometry factor Ar, are difficult to obtain directly from the manufacturers, who are inclined to protect their proprietary manufacturing materials and processes. To facilitate the optimization in the design and development stage, the analysis method based on TEC module parameters is presented in this section to strengthen the present analysis approach. The module parameters R_m, K_m, and S_m are given in the following equations.

$$R_m = \frac{2N\rho}{Ar}, \tag{3.7.11a}$$

$$K_m = 2NkAr, \tag{3.7.11b}$$

$$S_m = 2Ns \tag{3.7.11c}$$

Here R_m, K_m, and S_m are denoted as module electric resistance, thermal conductance, and Seebeck coefficient, respectively. Based on the module parameters, Eqs. (3.7.8) and (3.7.9) can be recast into the following form:

$$q_c(I) = \frac{(S_m^2 I^2 R_{hs} - S_m I - K_m)T_j + K_m T_a + R_m I^2 \left(K_m R_{hs} - \frac{1}{2}S_m I R_{hs} + \frac{1}{2}\right)}{-[K_m(R_{jc} + R_{hs}) - (S_m I R_{hs} - 1)(S_m I R_{jc} + 1)]} \tag{3.7.12}$$

$$T_j(I) = \frac{[(S_m I R_{hs} - 1)(S_m I R_{jc} + 1) - K_m(R_{hs} + R_{jc})]q_c - K_m T_a - R_m I^2 (K_m R_{hs} - S_m I R_{hs} + 1)}{S_m^2 I^2 R_{hs} - S_m I - K_m} \tag{3.7.13}$$

At this stage, the TEC thermal balance equations in term of the module parameters are solvable with known module parameters and operational current. The module parameters, S_m, R_m, and K_m, can be calculated with TEC specification parameters, such as ΔT_{max}, I_{max}, V_{max}, q_{max}, and T_{h0}, which are usually given in supplier's product datasheet. Here, ΔT_{max} is the maximum temperature difference obtainable between the hot and the cold ceramic plates at a given hot-side temperature T_{h0}, I_{max} is the input current that can produce the maximum ΔT_{max} across a TEC module, V_{max} is the DC voltage at the temperature difference of ΔT_{max} at $I = I_{max}$, and q_{max} is the maximum amount of heat absorbed at the TEC cold side at $I = I_{max}$ and $\Delta T = 0$. The following equations correlate the module parameters with the product specifications. Namely,

$$R_m = \frac{(T_{h0} - \Delta T_{max})V_{max}}{T_{h0}I_{max}} \quad (3.7.14a)$$

$$K_m = \frac{(T_{h0} - \Delta T_{max})V_{max}I_{max}}{2T_{h0}\Delta T_{max}} \quad (3.7.14b)$$

$$S_m = \frac{V_{max}}{T_{h0}} \quad (3.7.14c)$$

It has been shown in Eqs. (3.7.14) that the module specifications ΔT_{max}, I_{max}, V_{max}, together with T_{h0} are employed to calculate the module parameters. Based on the module level analytical solutions given in Eqs.(3.7.12)–(3.7.13) and the module parameters defined in Eqs. (3.7.14), the optimization of performance parameters T_j or q_c can be carried out even without knowing the detail of the TEC pellets.

It is noted that, different from the pellet-level analysis, the geometrical parameter Ar disappears in the module-level solution expressions. Nonetheless, a variation of Ar will change the module parameters in Eq (3.7.12) and affect T_j or q_c values accordingly. Thus, optimization of pellet geometry can also be carried out based on the module level solutions. In comparison with the pellet-level analysis, the module level analysis is more readily useful in design optimization based on necessary module parameters, which can be easily implemented through data spreadsheet such as Microsoft EXCEL.

The temperature difference between the hot side and cold side can be written as the following equations based on a given T_j or q_c.

$$\Delta T(T_j, I) = \frac{(S_m I R_{hs} - 1)T_j + (1 + S_m I R_{jc})T_a + R_m I^2 \left(\frac{1}{2}R_{hs} - \frac{1}{2}R_{jc} + S_m I R_{jc} R_{hs}\right)}{K_m(R_{jc} + R_{hs}) - (S_m I R_{hs} - 1)(S_m I R_{jc} + 1)} \quad (3.7.15)$$

$$\Delta T(q_c, I) = \frac{(1 - S_m I R_{hs})q_c - S_m I T_a + R_m I^2 \left(\frac{1}{2} - S_m I R_{hs}\right)}{S_m^2 I^2 R_{hs} - S_m I - K_m} \quad (3.7.16)$$

It is interesting to note that, if q_c and I are specified for a certain TEC, the ΔT can be obtained without being affect by thermal resistance R_{jc} at the device side.

The thermal resistances R_{hs} and R_{jc} can also be expressed as functions with respect to each other. Namely,

$$R_{hs} = \frac{(S_m I R_{jc} + K_m R_{jc} + 1)q_c + K_m T_a - (S_m I + K_m)T_j + \left(\frac{R_m I^2}{2}\right)}{[S_m^2 I^2 R_{jc} + S_m I) - K_m]q_c - (S_m I)^2 T_j + R_m I^2 \left(\frac{1}{2}S_m I - K_m\right)} \quad (3.7.17)$$

$$R_{jc} = \frac{(-S_m I R_{ha} + K_m R_{ss} + 1)q_c + K_m T_a + \left[(S_m I)^2 R_{hs} - S_m I - K_m\right]T_j + \left(R_m K_m I^2 - \frac{1}{2}S_m R_m I^3\right)R_{hs} + \left(\frac{R_m I^2}{2}\right)}{(S_m^2 I^2 R_{hs} - S_m I - K_m)}$$

(3.7.18)

Correspondingly, Eqs. (3.7.8)–(3.7.9) can be rewritten in module parameters as follows:

$$q_{te} = S_m I \Delta T + R_m I^2 \quad (3.7.19)$$

$$V = S_m \Delta T + R_m I \quad (3.7.20)$$

Alternatively, the Z factor is expressed as

$$Z = \frac{S_m^2}{R_m K_m} \quad (3.7.21)$$

The feasibility of the present analytical method, including both the pellet-level analysis and the module-level analysis, is to be demonstrated in the following section.

3.7.3 Results and discussion on the thermoelectric cooler optimization

Listed in Table 3.7.1 are the details of the TECs analyzed in this work. TEC1 is a high-power TEC, developed recently by a proprietary supplier. It has a nominal power dissipation of 330W with a packing density of 263 pairs of thermoelectric pellets in a footprint of 50 mm × 50 mm TEC2 and TEC3 are the existing TECs with traceable product specifications and thermal resistances, whereas the pellet details are also available in Refs. [64,68], respectively. The module parameters have been obtained based on the correlations given in Eqs. (3.7.14).

It would also be interesting to compare the reported TECs in the past with TEC1 to identify the advancement of TEC thermal performance. Figure 3.7.2 illustrates the acoustic image of TEC1 in planar view with the thermoelectric array through a Sonoscan Gen5 ultrasonic microscopy at 35 MHz. The TEC pellet layout can be clearly

Table 3.7.1 List of different thermoelectric coolers (TECs) and their supplied data.

Parameters	TEC1 SH330W	TEC2-Mel CP2-127-06L	TEC3-Mel127 at $T_{h0} = 25\,°C$	TEC3-Mel127 at $T_{h0} = 50\,°C$
Dimensions (mm)	50 × 50 × 4.1	62 × 62 × 4.6	39.9 × 39.9 × 3.5	
I_{max} (A)	19	14	8.5	8.45
q_{max} (W)	330	134	72	82
V_{max} (V)	32	16.4	14.4	16.1
ΔT_{max} (K)	64.5	77	67	74
T_{h0} (K)	298.15	323.15	298.15	323.15
Z (1/K)	0.00236	0.00254	0.00251	0.00238
S_m (V/K)	0.107	0.0508	0.048	0.0489
K_m (W/K)	3.69	1.14	0.71	0.70
R_m (Ω)	1.32	0.89	1.31	1.44

Figure 3.7.2 Photograph of TEC1 and corresponding C-mode acoustic image at 35 MHz.

visualized from the echoic ultrasonic image. Nonetheless, it is not possible to capture the pellet dimensions accurately due to interference from electrical traces. In view of this, a pellet-level analysis seems difficult, whereas a module-level analysis is capable of analyzing TEC performance without TEC geometrical dimensions.

3.7.3.1 Comparison with previous studies

Firstly, we compare the present analysis results with those given in the previous work by Simon et al. (2005), where a four-chip IBM module was cooled by four TECs arranged in parallel. Their analytical solutions, although not simplified, can be categorized as a pellet-level analysis and solvable in principle given the detailed pellet-level parameters. In their analysis, a cross-over point was found for either air cooling or liquid cooling techniques with and without TECs. For the no-TEC cases, the power dissipation can be obtained with the following equation:

$$q_c = \frac{T_j - T_a}{R_{jc} + R_{hs} - R_{tim,h}} \quad (3.7.22)$$

Here, the thermal interfacial resistance is denoted by $R_{tim,h}$ between the chip and heat sink. With the same pellet thermoelectric parameters and at the same current $I = I_{max}$, the pellet-level analysis based on the present solutions Eqs.(3.7.12)—(3.7.13) is conducted. The results are shown in Figure 3.7.3, which reproduces the same data for both air cooling and liquid cooling with TECs. For example, the same intersecting temperature point of $T_j = 47\,°C$ at 438W is obtained here for the air-cooled cases with and without TEC enhancement. Below the cross-over point, the TEC enhanced cooling performance outperforms the case without TECs. For purpose of comparison, the present module level analysis, based on the module parameters as obtained in Eqs. (3.7.14) and listed in Table 3.7.1, is conducted, and the results are also shown in Figure 3.7.3. It is seen that there is no significant discrepancy between the present pellet-level and module analysis results.

Another comparison is carried out with the iterative approach mostly used in the literature. Specifically, the reference by Hasan and Toh [64] is used for comparison as all the TEC properties and thermal parameters were traceable therein. In their study, the thermal performances were iteratively determined based on temperature-dependent correlations. In the present study, the same thermal resistances are used, but the module parameters are derived from suppliers' datasheet and listed in Table 3.7.1 at $T_{h0} = 25\,°C$ and $50\,°C$, respectively. Because the module parameters are also temperature-dependent, they are linearly extrapolated and tabulated at $T_{h0} = 69\,°C$. The computed T_j values versus I for all the three T_{h0} values are shown in Figure 3.7.4. It is seen that the similar level of optimal current and cooling capacity are attainable in

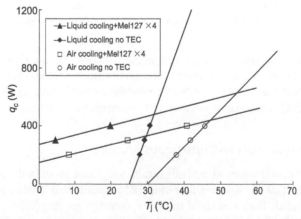

Figure 3.7.3 The thermal performance curves at Imax for TEC2 used in IBM multichip module (Simon et al.) based on the module parameters.

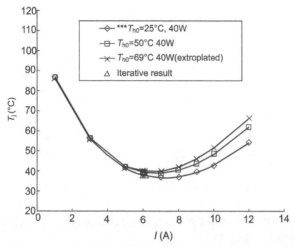

Figure 3.7.4 Comparison of the present analytical results with results with iterative result.

spite of different T_{h0}. In comparison, the case $T_{h0} = 69\,°C$ plotted in Figure 3.7.4 has the optimal current closest to that given in Ref. [64], where $T_j = 39.8\,°C$ and $T_h = 69\,°C$ at $I_{op} = 6.07$A are attained based on the iteration procedure. It is observed that a slightly better accuracy is attainable by considering the effect of the hot-side temperature in the present analysis.

3.7.3.2 Thermoelectric cooler optimization through T_j minimization

Under the test condition, a TEC can be sandwiched between the heat sink and the processor to bring down T_j in operation (Figure 3.7.1). The operational parameters are obtained as follows: $R_{jc} = 0.267$K/W is obtained based on in-house characterization for a server computer systems, $R_{hs} = 0.05$K/W as the liquid cooled heat sink condition, and $R_{hs} = 0.18$K/W as the air-cooled heat sink condition. The cooling power is set to be $q_c = 50$W, 70W, 100W, and 140W, respectively. In all these cases, $T_a = 25\,°C$ has been imposed. At a fixed processor power, the T_j minimization optimization can be conducted by setting the derivative of T_j with respect to I to be zero. Namely,

$$\left.\frac{\partial T_j(I)}{\partial I}\right|_{q_c} = 0 \tag{3.7.23}$$

Recalling the third order and second order of algebraic expressions in the nominator and denominator, respectively, with respect to I in Eq. (3.7.7), Eq. (3.7.23) results in a quartic equation with respect to I, which has two real roots and two conjugate roots. Only the positive real root is physically meaningful as the optimal operational current.

The T_j results versus current are shown in Figure 3.7.5a for TEC1 with air cooling and Figure 3.7.5b for TEC1 with enhanced liquid cooling, respectively. It is seen

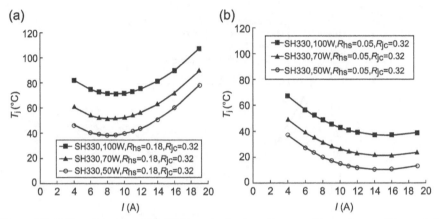

Figure 3.7.5 Plot of T_j versus I for TEC1 under fixed cooling capacities under conditions of air cooling (a) and liquid cooling (b).

that, for air cooling, the minimum T_j is obtained at optimal currents of 8–9A for the air cooling cases. Nonetheless, the larger optimal currents of 14–16A is obtained for the enhanced liquid cooling cases. Because of the high performance of the liquid cooling, the heat removal rate is much faster than the air cooling counterpart in spite of strong joule heating at larger currents. The corresponding COP is shown in Figure 3.7.6. It is seen that the COP decreases with increase in current I, which is different from the junction temperature curves with the optimal current, as displayed in Figure 3.7.5.

The temperature differences of hot to cold sides for the TEC-enhanced air cooling and liquid cooling cases are shown in Figure 3.7.7. This parameter is usually not included explicitly as a selection criterion in literature. It is seen that the temperature difference is highest at the largest power dissipation given the same electrical current. At $q_c = 140W$, the temperature difference could go up to 57°C for the liquid cooling at

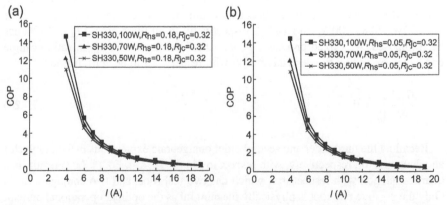

Figure 3.7.6 Plot of coefficient of performance (COP) versus I for TEC1 under fixed cooling capacities under conditions of air cooling (a) and liquid cooling (b).

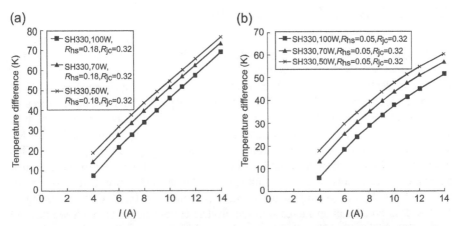

Figure 3.7.7 Plot of temperature difference versus I for TEC1 under fixed cooling capacities under conditions of air cooling (a) and liquid cooling (b).

$I = 14A$ and 47 °C for the air cooling at $I = 9A$, separately. The high temperature difference would induce unfavorably high thermomechanical stress on the pellet joints, which is detrimental to TEC operational reliability. It would be appropriate to include this parameter as one of criteria in selecting and optimizing TEC in view of reliability. In-plant SLT manufacturing test apparatus involves a liquid-cooled TEC, which shows that an operation current less than or equal to 9A would be a feasible range for prolonged TEC lifetime under dynamic processor SLT loading. Such experience could be generalized to impose an upper limit of TEC hot-side to cold-side temperature difference ~42 °C based on the present calculation under liquid cooling condition. Operation at elevated currents and larger temperature differences causes obviously earlier TEC failures and is, therefore, not recommendable in manufacturing practices.

The corresponding COP results for TEC1 under air cooling and liquid cooling conditions at different power inputs are shown in Figure 3.7.6. The COP values are calculated from Eq. (3.7.10) with q_{te} given in Eq. (3.7.19). It is worthwhile to discuss the maximization of COP herein, which has long been confused with the optimization of system cooling capacity like T_j minimization. In the present T_j minimization, the COP is an indicating instead of an objective function to maximize, taking values of 1.5–3 for air cooling and in the range of 1–0.5 for liquid cooling at the respective optimal currents. As the operational current decreases from I_{op}, the cooling power q_c decreases, and so do ΔT and TEC voltage, whereas the COP value increases. Mathematical calculation shows that an infinitely large COP is achievable at $I \sim 2A$, where $\Delta T \sim -10$ °C and TEC voltage approaches to zero. Under such circumstances, the electrical power q_{te} approaches to 0W and therefore COP would go up to infinity. However, the cooling capacity q_c drops to an unacceptably low level, which nullifies the introduction of TEC for thermal enhancement. Instead of maximizing COP, a system performance optimization shall be either to minimize T_j or to maximize q_c to fulfill the thermal performance requirement.

3.7.4 Optimizing thermal resistances

Given the thermal design target, the thermal resistance at the device side or sink side can be optimized to meet the design requirement. In this part, the effects of varying TEC thermal resistances are examined based on Eqs. (3.7.17)–(3.7.18). The plot of R_{hs} versus electrical current has been shown in Figure 3.7.8 at given $T_j = 25\,°C$, at the fixed cooling power of $q_c = 70W$ and 100W, respectively. The lidded processor with a junction to case thermal resistance of 0.267K/W is used for analysis. It is seen from Figure 3.7.8 that the maximum allowable heat sink thermal resistance R_{hs} approaches to 0.097K/W at the current of 9A and cooling power $q_c = 70W$. This means that a high performance liquid cooled heat sink shall be used to achieve such a low thermal resistance at the heat sink side with the desired performance. At a larger power of 100W, the maximum allowable R_{hs} reduces to 0.012 K/W occurring at $I = 9A$. It is noted that an increase in the electric current larger than 9A would lead to a slightly higher R_{hs}. Nonetheless, an electrical current more than 9A would cause excessive ohmic heating in the circuits, which is not recommended for actual implementation. Thus, it may not be feasible for TEC1 operated at $T_j = 25\,°C$ and $q_c = 100W$ as R_{hs} as low as 0.012K/W is difficult, if possible to achieve, based on the conventional air or liquid cooled heat sink together with the interfacial thermal resistance. When T_j is raised to a higher level of 50 °C, the thermal design constraint is greatly relieved at the heat sink side and a liquid cooled heat sink becomes applicable for both 70W and 100W power dissipations. Even an air-cooled heat sink will be sufficient for the operation at 70W in the feasible electrical current range.

Figure 3.7.9 shows the plot of R_{jc} versus I for TEC1 under fixed cooling power of 70W at both $T_j = 25\,°C$ and 50 °C. The allowable R_{jc} values at $T_j = 50\,°C$ are 0.42 K/W for air cooling and 0.74 K/W for liquid cooling at $I = 9A$, which can be easily met with conventional lidded processor packages targeted for desktop and server computers. When the performance parameter is tightened to $T_j = 25\,°C$, the maximum allowable $R_{jc} = 0.058K/W$ for air cooling and 0.42K/W for the liquid cooling at

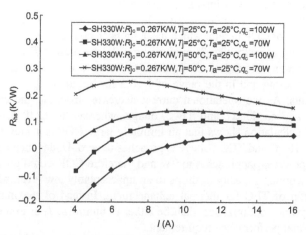

Figure 3.7.8 Plot of R_{hs} versus I for TEC1 at given T_j and q_c.

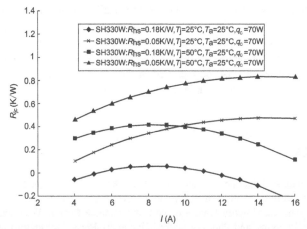

Figure 3.7.9 Plot of R_{jc} versus the electrical current for TEC1 at given T_j and q_c.

$I = 9A$. Because of the internal thermal resistance and TIM, a processor with such a low R_{jc} virtually does not exist, thus the lidded processor could only be managed with a liquid cooled heat sink. Obviously, a thermal resistance analysis based on the present module-level solution is helpful in searching for appropriate heat sinks for varying devices. It is noted that the thermal management with TEC has been used in devices such as processors and cooling. The application of the TEC temperature controller for processor burn-in aging test will be elaborated in Chapter 6. The validation of the TEC model with experimental test will also be addressed.

To summarize, a general approach in evaluating and optimizing TEC systems is presented in this section. Analytical solutions including R_{hs} and R_{jc} are presented for both the cooling power q_c and junction temperature T_j with respect to the operational current at both pellet level and module level. As against the commonly used iterative procedure, the main feature of the present analytical method lies in the fact that the TEC thermal performance can be obtained in a straightforward manner without resorting to the tedious numerical iteration. The present pellet-level solution expressions represented in Eqs.(3.7.6)—(3.7.7) are useful when the pellet parameters are all available. On the other hand, in the lack of pellet-level parameters, a module-level analysis is very helpful in TEC performance analysis. The module level analysis, based on the module parameters S_m, R_m, and K_m, provides a more convenient tool for design analysis as the pellet properties and dimensional details are often not available for a commercial TEC. It is suggested that the necessary module specifications including I_{max}, q_{max}, V_{max}, and ΔT_{max} should be reported in a thermal design involving the TEC cooling module.

It is noted that the common Be_2Te_3 material used in the commercial TECs has ZT values of 0.7—1. Progress has been made on the new thermoelectricity materials with higher ZT value than the Be_2Te_3 material. The high value of $ZT = 2.4$ for p-type superlattices of Bi_2Te_3/Sb_2Te_3 at room temperature and $ZT = 1.2$ for n-type superlattices has been reported in Refs. [76,77]. The higher the ZT value is, the higher the cooling power can be, given the same geometrical dimensions. Undoubtedly, the abovementioned TEC analysis model can be utilized to predict and improve thermal performance for electronic packaging with more advanced thermoelectric materials. The electrical contact resistance at the thermoelectric pellets, which could be significant at a smaller scale, can also be included in the analysis model for better reflection of TEC performance at microscale level.

References

[1] T.L. Bergman, A.S. Lavine, F.P. Incropera, D.P. DeWitt, Fundamentals of Heat and Mass Transfer, seventh ed., John Wiley & Sons, Inc., Hoboken, 2011.
[2] G. Chen, Nanoscale Energy Transport and Conversion: A Parallel Treatment of Electrons, Molecules, Phonons, and Photons, Oxford University Press, Inc., Oxford, 2005.
[3] K. Banerjee, A. Mohrotra, Global interconnects warming, IEEE Circuits and Devices 17 (5) (2001) 16–32.
[4] S. Im, N. Srivastava, K. Banerjee, K.E. Goodson, Scaling analysis of multilevel interconnect temperatures for high-performance ICs, IEEE Trans. on Electron Devices 52 (12) (2005) 2710–2719.
[5] Y. Shi-ming, Free convections of heat outside slender vertical cylinders and inside vertical tubes, J. Xi'an Jiaot. Univ. 14 (3) (1980) 115–131 (In Chinese).
[6] M.M. Yovanovich, Four decades of research on thermal contact, gap and joint resistance in microelectronics, IEEE Trans. Compon. Packag. Technol. 28 (2) (2005) 182–206.
[7] ASTM Standard D5470-06, Standard Test Method for Thermal Transmission Properties of Thermally Conductive Electrical Insulation Materials, ASTM International, West Conshohocken, PA, 2006.
[8] R.S. Prasher, J.C. Matayabas Jr., Thermal contact resistance of cured gel polymeric thermal interface material, in: Itherm Conference, 2004, pp. 28–35.
[9] J. Wang, Shear Modulus measurement for thermal interface materials in flip chip packages, IEEE Trans. Compon. Packag. Technol. 29 (3) (2006).
[10] H.Y. Zhang, Y.C. Mui, M. Tan, S. Ng, Thermal characterization of solid thermal interface materials (STIMs), in: EPTC Conference, 2008.
[11] I. Sauciuc, R. Yamamoto, J. Culic-Viskota, T. Yoshikawa, S. Jain, M. Yajima, N. Labanok, C. Amoah-Kusi, Carbon based thermal interface material for high performance cooling applications, in: ITHERM Conference, 2014.
[12] Y. Yang, H. Ye, et al., Heat dissipation performance of graphene enhanced electrically conductive adhesive for electronic packaging, in: 2017 IMAPS Nordic Conference on Microelectronics Packaging (NordPac), 2017, pp. 125–128.
[13] K.M.F. Shahil, A.A. Balandin, Graphene–multilayer graphene nanocomposites as highly efficient thermal interface materials, Nano Lett. 12 (2012) 861–867.
[14] K. Onda, H. Kameyama, T. Hanamoto, K. Ito, Experimental study on heat generation behavior of small lithium-ion second batteries, J. Electrochem. Soc. 150 (2003) A285–A291.
[15] J. Lau, TSV Interposer: the most cost-effective integrator for 3D IC integration, in: ASME Interpack, 2011, pp. 2011–52189.
[16] ITRS Technology Roadmap, 2013. www.itrs.org (accessed Oct. 1st 2017).

[17] Y. Yan, Y. Ji, X. Ming, 3D-TSV package technology, Electronics & Packaging 14 (2014) 1−5 (In Chinese).
[18] Y. Zhuang, D. Yu, F. Dai, G. Zhang, J. Fan, Spray coating process with polymer material for insulation in CIS-TSV wafer-level-packaging, in: Icept Conference, 2014, pp. 437−440.
[19] Di Jiang, W. Mu, S. Chen, Y. Fu, K. Jeppson, J. Liu, Vertically stacked carbon nanotube-based interconnects for through silicon via application, IEEE Electron. Device Lett. 36 (5) (2015) 499−501.
[20] B. Black, et al., Die stacking (3D) microarchitecture, Proc.IEEE/ACM Int. Symp. Microarchit. (2006) 469−479.
[21] Z. Zhang-Ming, et al., An analytical thermal model for 3D integrated circuit considering through silicon via, Acta Phys. Sin. 60 (11) (2011) 118401 (In Chinese).
[22] N. Khan, S.W. Yoon, A.G.K. Viswanath, V.P. Ganesh, Ranganathan, D. Witarsa, S. Lim, K. Vaidyanathan, Development of 3D stack package using silicon interposer for high power application, IEEE Trans. Adv. Packag. 31 (1) (2008) 44−50.
[23] J.H. Lau, T.G. Yue, Effect of TSVs (through-silicon vias) on thermal performances of 3D IC integration system-in-package (SiP), Microelectron. Reliab. 52 (2012) 2660−2669.
[24] H. Ma, D.Q. Yu, J. Wang, The development of effective model for thermal conduction analysis for 2.5D packaging using TSV interposer, Microelectron. Reliab. 54 (2014) 425−434.
[25] H. Oprins, V. Cherman, C. Torregiani, M. Stucchi, B. Vandevelde, E. Beyne, Thermal test vehicle for the validation of thermal modelling of hot spot dissipation in 3D stacked ICs, in: ESTC Conference, 2010.
[26] Y. Chen, E. Kursun, D. Motschman, C. Johnson, Y. Xie, Analysis and mitigation of lateral thermal blockage effect of through-silicon-via in 3D IC designs, in: ISLPED, 2011.
[27] H.Y. Zhang, Y.S. Wang, W.H. Zhu, T. Lin, Effective thermal conductivity model for TSVs with insulation layer as contact resistance, in: 16th International Conference on Electronic Packaging Technology, 2015, pp. 125−131.
[28] J.C. Maxwell, A Treatise on Electricity and Magnetism, third ed., Oxford University Press, Oxford, 1892.
[29] L.C. Davis, B.E. Artz, Thermal conductivity of metal-matrix composites, J. Appl. Phys. 77 (10) (1995) 4954−4960.
[30] R. Kothari, C.T. Sun, R. Dinwiddie, H. Wang, Experimental and numerical study of the effective thermal conductivity of nano composites with thermal boundary resistance, Int. J. Heat Mass Transf. 66 (2013) 823−829.
[31] H.C. Chien, J.H. Lau, Y.L. Chao, et al., Thermal evaluation and analysis of 3D IC for network system applications, in: ECTC, 2012.
[32] X.Y. Chen, K.C. Toh, T.N. Wong, J.C. Chai, D. Pinjala, O.K. Navas, Ganesh, H. Zhang, V. Kripesh, Direct liquid cooling of a stacked multi-chip module, in: ITHERM, 2004, pp. 199−206.
[33] T. Brunschwiler, B. Michel, H. Rothuizen, U. Kloter, B. Wunderle, H. Oppermann, H. Reichl, Forced convective interlayer cooling in vertically integrated packages, in: Proc. IEEE ITherm, 2008, pp. 1114−1125.
[34] N. Khan, H.Y. Li, S.P. Tan, S.W. Ho, V. Kripesh, D. Pinjala, 3-D packaging with through-silicon via (TSV) for electrical and fluidic interconnections, IEEE Trans. Compon. Packag. Manuf. Technol. 3 (2) (2013) 221.
[35] A.M. Marconnet, M.A. Panzer, K.E. Goodson, Thermal conduction phenomena in carbon nanotubes and related nanostructured materials, Rev. Mod. Phys. 85 (2013) 1296−1327.

[36] A. Jain, R.E. Jones, R. Chatterjee, P. Scott, Analytical and numerical modeling of the thermal performance of three-dimensional integrated circuits, IEEE Trans. Compon. Packag. Technol. 33 (2010) 56–63.
[37] S.V. Patankar, Numerical Heat Transfer and Fluid Flow by Patankar, Hemisphere, 1980.
[38] H.Y. Zhang, X.W. Zhang, B.L. Lau, S. Lim, L. Ding, M.B. Yu, Thermal characterization of both bare die and overmolded 2.5-D Packages on through silicon interposers, IEEE Trans. Compon. Packag. Manuf. Technol. 4 (5) (2014) 807–816.
[39] N. Khan, Seung Wook Yoon, A.G.K. Viswanath, V.P. Ganesh, R. Nagarajan, D. Witarsa, S. Lim, K. Vaidyanathan, Development of 3-D stack package using silicon interposer for high-power application, IEEE Trans. Adv. Packag. 31 (1) (2008).
[40] JEDEC 51-14, http://www.jedec.org/.
[41] JEDEC standard JESD51-2A, Integrated Circuits Thermal Test Method Environmental Conditions —Natural Convection (Still Air), http://www.jedec.org/.
[42] T.J. Chainer, M.D. Schultz, P.R. Parida, M.A. Gaynes, Improving data center energy efficiency with advanced thermal management, IEEE Trans. Compon. Packag. Manuf. Technol. 7 (8) (2017) 1228–1239.
[43] K. Jeppson, J. Bao, S. Huang, Y. Zhang, S. Sun, Y. Fu, J. Liu, Hotspot test structures for evaluating carbon nanotube microfin coolers and graphene-like heat spreaders, in: ICMTS, 2016, pp. 32–36.
[44] H. Zhang, S. Xu, P. Li, T. Lin, Thermal analysis and characterization of electronic packages with alternative lid coatings, in: Icept, 2017.
[45] Sauciuc, G. Chrysler, R. Mahajan, Spreading in the heat sink base: phase systems or solid metals?? IEEE Trans. Compon. Packag. Technol. 25 (2002) 621–628.
[46] I. Sauciuc, G. Chrysler, R. Mahajan, M. Szleper, Air-cooling extension performance limits for processor cooling applications, in: 19th IEEE Semi-therm Symposium, 2003.
[47] N. Blet, S. Lips, V. Sartre, Heat pipes for temperature homogenization: a literature review, Appl. Therm. Eng. 118 (2017) 490–509.
[48] www.thermacore.com (accessed Aug. 2017).
[49] H.Y. Zhang, D. Pinjala, P.-S. Teo, Thermal management of high power dissipation electronic packages on high density substrates: from air cooling to liquid cooling, in: EPTC, 2003.
[50] H.Y. Zhang, et al., Thermal modeling and design of liquid-cooled heat sinks for flip chip BGA packages, in: ECTC, 2003, pp. 431–437.
[51] H.Y. Zhang, D. Pinjala, Y.K. Joshi, T.N. Wong, K.C. Toh, M.K. Iyer, Fluid flow and heat transfer in liquid cooled foam heat sinks for electronic packages, IEEE Trans. Compon. Packag. Technol. 28 (2) (2005) 272.
[52] H.Y. Zhang, D. Pinjala, T.N. Wong, K.C. Toh, Y.K. Joshi, Single-phase liquid cooled microchannel heat sink for electronic packages, Appl. Therm. Eng. 25 (2005) 1472–1487.
[53] S.P. Tan, K.C. Toh, N. Khan, D. Pinjala, V. Kripesh, Development of single phase liquid cooling solution for 3-D silicon modules, IEEE Trans. Compon. Packag. Manuf. Technol. 1 (4) (2011) 536.
[54] B.L. Lau, Y. Han, H.Y. Zhang, L. Zhang, X.W. Zhang, Development of fluxless bonding using deposited gold-indium multi-layer composite for heterogeneous silicon micro-cooler stacking, in: EPTC, 2014, pp. 693–697.
[55] S. Lee, S. Song, V. Au, K.P. Moran, Constriction/spreading resistance model for electronics packaging, in: ASME/JSME Thermal Engineering Conference, 1995.
[56] S. Lee, S. Song, V. Au, K.P. Moran, Optimization and selection of heat sinks, IEEE Trans. Components, Packaging and Manufacturing Technologies Part A 18 (1995) 812–817.

[57] D. Kraus, A. Bar-Cohen, Design and Analysis of Heat Sinks, John Wiley & Sons, Inc., Hoboken, 1995.
[58] D. Copeland, Optimization of parallel plate heat sinks for forced convection, in: Proceedings of 16th IEEE SEMI-THERM Symposium, 2000, pp. 266–272.
[59] R.K. Shah, A.L. London, Laminar Flow Forced Convection in Ducts, Supplement 1 to Advances Heat Transfer, Academic Press, New York, 1978.
[60] W.M. Kays, A.L. London, Compact Heat Exchangers, third ed., McGraw-Hill, New York, 1984.
[61] W. Qu, I. Mudawar, Experimental and numerical studies of pressure drop and heat transfer in a single-phase microchannel heat sink, Int. J. Heat Mass Transf. 45 (2002) 2549–2565.
[62] J. Bierschenk, M. Gilley, Assessment of TEC thermal and reliability requirements for thermoelectrically enhanced heat sinks for CPU cooling applications, in: International Conference on Thermoelectrics, 2006, pp. 254–259.
[63] P.E. Phelan, V.A. Chiriac, T.Y.T. Lee, Current and future miniature refrigeration cooling technologies for high power microelectronics, IEEE Trans. Compon. Packag. Technol. 25 (3) (2002) 356–365.
[64] M.H. Hasan, K.C. Toh, Optimization of a thermoelectric cooler-heat sink combination for active processor cooling, in: Electronic Packaging Technology Conference, 2007, pp. 848–857.
[65] M. Ikeda, T. Nakamura, Y. Kimura, H. Noda, Thermal performance of TEC integrated heat sink and optimizing structure for low acoustic noise power consumption, in: 22nd IEEE Semi-therm Symposium, 2006.
[66] G.L. Solbrekken, K. Yazawa, Chip level refrigeration of portable electronic equipment using thermoelectric devices, in: InterPACK Conference, 2003, 35305.
[67] R. Chein, G. Huang, Thermoelectric cooler application in electronic cooling, Appl. Therm. Eng. 24 (2004) 2207–2217.
[68] R.E. Simons, M.J. Ellsworth, R.C. Chu, An assessment of module cooling enhancement with thermoelectric coolers, ASME Journal Heat Transfer 127 (2005) 76–84.
[69] S.U. Yuruker, M.C. Fish, Z. Yang, et al., System-level Pareto frontiers for on-chip thermoelectric coolers, Front. Energy 12 (1) (2018) 109–120.
[70] B.J. Huang, C.J. Chin, C.L. Duang, A design method of thermoelectric cooler, Int. J. Refrigeration 23 (2000) 208–218.
[71] S. Lineykin, S. Beb-Yaakov, User-friendly and intuitive graphical approach to the design of thermoelectric cooling systems, Int. J. Refrigeration 30 (2007) 798–804.
[72] Z. Luo, A simple method to estimate the physical characteristics of a thermoelectric cooler from vendor datasheets, Electronics Cooling 14 (3) (2008) 22–27.
[73] D.M. Rowe, CRC Handbook of Thermoelectrics, CRC Press, 1995.
[74] H.Y. Zhang, A general approach in evaluating and optimizing thermoelectric coolers, Int. J. Refrigeration 33 (2010) 1187–1196.
[75] H.Y. Zhang, et al., Analysis of thermoelectric cooler performance for high power electronic packages, Appl. Therm. Eng. 30 (2010) 561–568.
[76] Z. Tian, S. Lee, G. Chen, Heat transfer in thermoelectric materials and devices, ASME J Heat Transfer 135 (2013) 061605.
[77] C. Brian, Sales, smaller is cooler, Science 295 (2002) 1248–1249.

Stress and reliability analysis for interconnects

4.1 Fundamental of mechanical properties

Material properties can be classified as chemical properties, electrical properties, mechanical and thermomechanical properties, and so on. Among them, mechanical properties determine the mechanical response and behavior of a material when subjected to external loadings such as tension, compression, torsion, shear, bending, or combined loadings. Typical mechanical properties of materials include elastic modulus, also known as Young's modulus, Poisson's ratio, yield stress, elongation, ultimate tensile strength, hardness, toughness, etc. Mechanical properties of material usually exhibit temperature, time, and strain rate-dependent behavior. Stress—strain curve is widely used to characterize material's behavior subjected to load. Polymer materials tend to show viscoelastic properties, whereas solder materials exhibit viscoplastic properties as the stress—strain relationship changes with time. Creep is also one type of viscoplastic behaviors. Fatigue is one type of material failure phenomena when a material is subjected to repeated loading and unloading even if the maximum stress is much less than material's strength. Such mechanical properties are to be elaborated and analyzed in the subsequent sections.

4.1.1 Stress—strain relationship

The stress—strain curve is used to characterize the relationship between the stress and strain for a material. It is unique for each material and is typically described by simple uniaxial tension test. Fig. 4.1.1 shows an example of uniaxial tension test in the elastic region of material. Under uniaxial tension loading, displacement of the sample

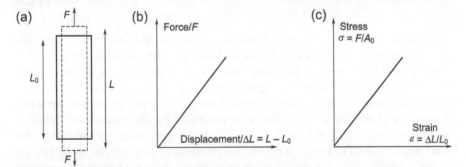

Figure 4.1.1 Typical tension test under elastic region: (a) schematic of uniaxial tension, (b) force—displacement curve, and (c) stress—strain curve.

increases linearly with the applied force, as shown in Fig. 4.1.1b. If transferring force—displacement curve to stress—strain curve as shown in Fig. 4.1.1c, strain increases lineally with the applied stress. For the uniaxial tension test, stress and strain are expressed as

$$\sigma = \frac{F}{A_0} \tag{4.1.1a}$$

$$\varepsilon = \frac{\Delta L}{L_0} \tag{4.1.1b}$$

where σ = normal stress on a plane perpendicular to the longitudinal axis of the specimen, in MPa,

F = applied load, in N,
A_0 = original cross-sectional area of the sample, in mm^2,
ε = normal strain in the longitudinal direction, in mm/mm (dimensionless),
ΔL = change in the specimen's length, i.e., $\Delta L = L - L_0$, L_0 is the original gage length of the sample, in mm.

The slope of stress—strain curve in the elastic potion is defined as elastic modulus, E, also known as Young's modulus, and expressed as

$$E = \frac{\sigma}{\varepsilon} \tag{4.1.2}$$

Young's modulus is one of the most important mechanical properties of a material, which has the same unit as stress. The stiffness is related to the Young's modulus and specimen geometry, namely,

$$stiffness = \frac{F}{\Delta L} = \frac{\sigma A_0}{\varepsilon L_0} = \frac{E A_0}{L_0} \tag{4.1.3}$$

For ductile materials, when the applied stress is higher than the yield point of the material, the stress—strain curve exhibits nonlinear behavior, as shown in Fig. 4.1.2. After the yield point, material behavior enters into nonelastic region, where it shows strain hardening effect before reaching ultimate strength of the material. Necking phenomena will happen after the ultimate strength point, and finally fracture occurs when material cannot bear further deformation. In the elastic region, stress—strain relationship follows Hooke's law as listed in Eq. (4.1.3). In the nonlinear strain hardening (also called work hardening) region, a power law relationship between the stress and the amount of plastic strain is usually used to describe the strain hardening phenomenon:

$$\sigma = K \varepsilon_p^n \tag{4.1.4}$$

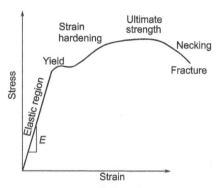

Figure 4.1.2 Typical stress–strain curve for ductile material.

where σ is the stress, K is the strength coefficient, ε_p is the plastic strain, and n is the strain hardening exponent. The value of the strain hardening exponent, n, lies between 0 and 1. A value of $n = 0$ means that a material shows a perfectly plastic behavior and a flat line follows after the yield point in stress–strain curve. A value of $n = 1$ represents a perfect elastic material. Most materials have an n value between 0.1 and 0.5.

The above discussion on stress–strain curve is for engineering stress and strain analysis, in which the stress and strain are calculated based on the original dimensions of specimen such as length and cross-sectional area. Engineering stress–strain curve is widely used and accurate in the small deformation region of the material, e.g., elastic region of the material. However, the length and cross-sectional area change in plastic region. True stress and true strain are used for accurate definition of plastic behavior of ductile materials by considering the actual (instantaneous) dimensions. Relationship between true stress-strain and engineering stress-strain is shown below:

$$\sigma' = (1+\varepsilon)\sigma \tag{4.1.5a}$$

$$\varepsilon' = \ln(1+\varepsilon) \tag{4.1.5b}$$

where σ and ε are engineering stress and strain and σ' and ε' are true stress and strain. It can be seen that true stress is larger than engineering stress, while true strain is lower than engineering strain based on above equations. Fig. 4.1.3 shows an example of comparison between engineering and true stress–strain curves. In the low strain region, both stress–strain curves have no significant difference. However, such a difference becomes significant with strain increasing. For accurate analysis, especially for plastic behavior of metals, true stress–strain relationship and material properties should be considered.

As shown in Fig. 4.1.1a, the material sample tends to contract in the direction perpendicular to the direction of tension, which is a phenomenon called Poisson effect. Conversely, if the material is compressed rather than stretched, it tends to expand in direction perpendicular to the direction of compression. Poisson's ratio is usually a

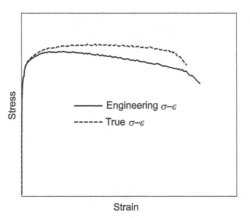

Figure 4.1.3 Comparison between engineering and true stress—strain curves.

measure of the Poisson effect, defined as a ratio of transverse strain to the longitudinal strain along the direction of stretching load, which is shown below:

$$\nu = -\frac{\varepsilon_{trans}}{\varepsilon_{axial}} \tag{4.1.6}$$

where ν is Poisson's ratio, is transverse strain (negative for axial tension [stretching], positive for axial compression), and ε_{trans} is axial strain, positive for axial tension, and negative for axial compression. Most materials have Poisson's ratio values ranging between 0 and 0.5, typically around 0.3. A perfectly incompressible material deformed elastically at small strains would have a Poisson's ratio of exactly 0.5.

Stress under tension or compression is usually called normal stress, while it is called shear stress when shear force is applied onto a sample, as shown in Fig. 4.1.4. For the shear test, shear stress and shear strain are expressed as

$$\tau_{xy} = \frac{F}{A} \tag{4.1.7a}$$

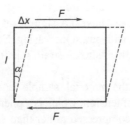

Figure 4.1.4 Typical shear test illustration.

$$\gamma_{xy} = \frac{\Delta x}{l} = \tan \alpha \tag{4.1.7b}$$

where τ_{xy} is shear stress on a transverse plane parallel to the applied force and γ_{xy} is shear strain showing shape deformation of sample. Similar to the tension test, relationship between shear stress and shear strain can also be expressed by Hooke's law as

$$\tau_{xy} = G \cdot \gamma_{xy} \tag{4.1.8}$$

where G is called shear modulus, having the same unit as Young's modulus. For an isotropic material, Young's modulus E, shear modulus G, and Poisson's ratio ν are dependent and have the following relationship:

$$E = 2G(1 + \nu) \tag{4.1.9}$$

In the most general case, the stress state at one point can be illustrated in Fig. 4.1.5. These nine stress components can be arranged in one 3×3 matrix form, giving

$$\sigma_{ij} = \begin{bmatrix} \sigma_x & \tau_{yx} & \tau_{zx} \\ \tau_{xy} & \sigma_y & \tau_{zy} \\ \tau_{xz} & \tau_{yz} & \sigma_z \end{bmatrix} \tag{4.1.10}$$

According to shear stress mutual equal law, $\tau_{xy} = \tau_{yx}, \tau_{xz} = \tau_{zx}, \tau_{yz} = \tau_{zy}$. So stress state at one point can be represented by six independent components, i.e. normal stresses of $\sigma_x, \sigma_y, \sigma_z$ and shear stresses of $\tau_{xy}, \tau_{yz}, \tau_{zx}$.

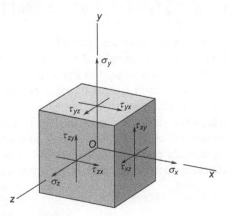

Figure 4.1.5 3D stress components.

For an isotropic elastic material, strain and stress relationship in 3D stress state as shown in Fig. 4.1.5 follows the general Hooke's law. Relationship between normal strains and stresses can be expressed as

$$\varepsilon_x = \frac{1}{E}(\sigma_x - \nu\sigma_y - \nu\sigma_z)$$

$$\varepsilon_y = \frac{1}{E}(-\nu\sigma_x + \sigma_y - \nu\sigma_z) \quad (4.1.11)$$

$$\varepsilon_z = \frac{1}{E}(-\nu\sigma_x - \nu\sigma_y + \sigma_z)$$

The relationship between shear strains and stresses are connected by the shear modulus as

$$\gamma_{xy} = \frac{\tau_{xy}}{G},\ \gamma_{yz} = \frac{\tau_{yz}}{G},\ \gamma_{zx} = \frac{\tau_{zx}}{G} \quad (4.1.12)$$

Relationship between stresses and strain above can be expressed by a matrix form as

$$\begin{bmatrix} \varepsilon_x \\ \varepsilon_y \\ \varepsilon_z \\ \gamma_{xy} \\ \gamma_{yz} \\ \gamma_{zx} \end{bmatrix} = \frac{1}{E} \begin{bmatrix} 1 & -\nu & -\nu & 0 & 0 & 0 \\ -\nu & 1 & -\nu & 0 & 0 & 0 \\ -\nu & -\nu & 1 & 0 & 0 & 0 \\ 0 & 0 & 0 & 2+2\nu & 0 & 0 \\ 0 & 0 & 0 & 0 & 2+2\nu & 0 \\ 0 & 0 & 0 & 0 & 0 & 2+2\nu \end{bmatrix} \begin{bmatrix} \sigma_x \\ \sigma_y \\ \sigma_z \\ \tau_{xy} \\ \tau_{yz} \\ \tau_{zx} \end{bmatrix} \quad (4.1.13)$$

which is called the compliance matrix of the material.

In some cases, a 3D problem can be simplified to a 2D stress state, such as the plane stress and plane strain. Plane stress is defined as a state of stress in which the normal stress, σ_z, and the shear stress, τ_{xz} and τ_{yz}, are assumed to be zero, where z is the direction perpendicular to the x-y plane. Plane stress is usually used for a thin plate case in which one dimension (thickness) is much smaller than the others, while plane strain is defined as a state of strain in which the normal strain, ε_z, and the shear strain, γ_{xz} and γ_{yz}, are assumed to be zero. In plane strain, the dimension of the structure in one direction, usually thickness z direction, is much larger compared with the other two directions.

For plane stress, the strain—stress relationship can be expressed in the matrix form as

$$\begin{bmatrix} \varepsilon_x \\ \varepsilon_y \\ \varepsilon_z \\ \gamma_{xy} \end{bmatrix} = \frac{1}{E} \begin{bmatrix} 1 & -\nu & 0 \\ -\nu & 1 & 0 \\ -\nu & -\nu & 0 \\ 0 & 0 & 2(1+\nu) \end{bmatrix} \begin{bmatrix} \sigma_x \\ \sigma_y \\ \tau_{xy} \end{bmatrix} \quad (4.1.14)$$

For plane strain, the stress—strain relationship can be expressed in the matrix form as

$$\begin{bmatrix} \sigma_x \\ \sigma_y \\ \sigma_z \\ \tau_{xy} \end{bmatrix} = \frac{E}{(1-2\nu)(1+\nu)} \begin{bmatrix} 1-\nu & \nu & 0 \\ \nu & 1-\nu & 0 \\ \nu & \nu & 0 \\ 0 & 0 & 1-2\nu/2 \end{bmatrix} \begin{bmatrix} \varepsilon_x \\ \varepsilon_y \\ \gamma_{xy} \end{bmatrix} \quad (4.1.15)$$

In the electronic products, thermal stress is a quite common stress source because temperature change occurs when electronic products work. Thermal stress is induced by thermal strain when materials are not allowed to expand and contract freely. Thermal strain is determined by the coefficient of thermal expansion (CTE) of material and temperature change, which is expressed as

$$\varepsilon_{th} = \alpha \Delta T = \alpha (T - T_0) \quad (4.1.16)$$

where T is instant temperature and T_0 is called the stress-free temperature or reference temperature.

When conducting thermomechanical analysis for advanced electronic packaging, total strain should include both mechanical strain and thermal strain.

4.1.2 Viscoplastic and viscoelastic properties

4.1.2.1 Viscoplastic Anand model of solder

Solder is one of the critical materials in electronic packaging. It exhibits creep behavior dependent on temperature and time, which is typically important at 0.4 T_m and above, with T_m the absolute melting temperature [1]. Creep deformation can be developed in any stress levels for solder alloy. Time-independent plastic strain will become significant at stress levels above $\tau/G = 10^{-3}$, which is also the critical power law breakdown condition approximately [2]. The time-independent plastic strain and time-dependent creep strain are difficult to separate in higher stress level. From the nonlinear deformation modeling point of view, the mechanism of the creep and plastic strain may be related to dislocation motion. To study the constitutive behavior using a state variable viscoplastic approach, the creep and plastic strain may be combined into inelastic strain. A suitable constitutive model for solder joints is needed to model the solder stress−strain behavior when subjected to thermomechanical loading.

A commonly used viscoplastic model for solder is Anand model [3−5]. There are two basic features in this model. First, this model needs no explicit yield condition and no loading/unloading criterion. The plastic strain is assumed to take place at all nonzero stress levels. Second, this model employs a single scalar as an internal variable, the deformation resistance s, to represent the averaged isotropic resistance to plastic flow. Anand model is broken down into a flow equation, and three evolution equations [6]:

Flow equation:

$$\dot{\varepsilon}_p = C\left[\sinh\left(\frac{\xi\sigma}{S}\right)\right]^{1/m} \exp\left(\frac{-Q}{k_B T}\right) \quad (4.1.17)$$

Evolution equations:

$$\dot{S} = \left\{h_0(|B|)^a \frac{B}{|B|}\right\}\dot{\varepsilon}_p \quad (4.1.18)$$

$$B = 1 - \frac{S}{S^*} \quad (4.1.19)$$

$$S^* = \widehat{S}\left[\frac{\dot{\varepsilon}_p}{C}\exp\left(\frac{Q}{k_B T}\right)\right]^n \quad (4.1.20)$$

where $\dot{\varepsilon}_p$ is the effective plastic strain rate, σ is effective true stress (MPa), S is the deformation resistance (MPa), Q is the activation energy, k_B is Boltzman's constant, T is the absolute temperature, C is the preexponential factor (1/s), ξ is the stress multiplier, m is the strain rate sensitivity of stress, h_0 is the hardening constant (MPa), $\widehat{S}$ is the coefficient for deformation resistance saturation value (MPa), n is the strain rate sensitivity of saturation value, and a is strain rate sensitivity of hardening. It is seen that there are nine constants (C, Q/k_B, ξ, m, $\widehat{S}$, n, h_0, a, S_0, with S_0 the initial value of S) in Anand model, which can be determined by the curve fitting the experimental data, such as creep tests and constant strain rate tensile tests. Cheng et al.[3] and Wang et al.[4] published these parameters for different lead-free and lead-based solder alloys. Anand model is a build-up model in ANSYS finite element analysis (FEA) software and has been widely used for simulation of the solder joint deformation under cyclic mechanical and thermomechanical loading with low loading frequencies.

4.1.2.2 Viscoelastic properties

There are lots of polymer materials in the advanced packaging such as molding compound, dielectrics, and underfill, which show the viscoelastic material properties. Viscoelasticity is the property of materials that exhibit both viscous and elastic characteristics when undergoing deformation. Time- and temperature-dependent viscoelastic modeling provides time effects and stress relaxation history.

Dynamic mechanical analysis (DMA) can be used to study and characterize the viscoelastic behavior of polymers. A sinusoidal stress is applied to determine the complex modulus. The temperature and/or the frequency of loading are often varied, which leads to variations in the complex modulus. The constitutive equation for an isotropic viscoelastic polymer material can be expressed as

$$\sigma_{ij} = \int_0^t 2G(t-\tau)\frac{de_{ij}}{d\tau}d\tau + \delta_{ij}\int_0^t K(t-\tau)\frac{d\varepsilon_v}{d\tau}d\tau \qquad (4.1.21)$$

where σ_{ij} are the components of the Cauchy stress, e_{ij} are deviatoric strain components, ε_v is the volumetric part of the strain, $G(t)$ and $K(t)$ are the shear and bulk modulus functions, t and τ are current and past time, and δ_{ij} are the components of the unit tensor [7].

The Maxwell model, which consists of a linear elastic Hookean spring and a linear viscous Newtonian damper in series, as shown in Fig. 4.1.6, can be used to model complicated linear viscoelastic behavior of the polymer material.

For the generalized Maxwell model, the shear and bulk modulus functions can be expressed by a Prony series [8,9]:

$$G(t) = G_\infty + \sum_{i=1}^{n} G_i e^{(-t/\tau_i)} \qquad (4.1.22)$$

$$K(t) = K_\infty + \sum_{i=1}^{n} K_i e^{(-t/\tau_i)} \qquad (4.1.23)$$

where G_∞ and G_i are the shear elastic modulus, K_∞ and K_i are the bulk elastic modulus, τ_i are the relaxation times for the various Prony series components. The time-dependent stress—strain relations are typically extended to include temperature dependence by

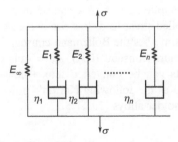

Figure 4.1.6 Generalized Maxwell model.

using the principle of time-temperature superposition and the empirical William-Landel-Ferry (WLF) shift function as

$$\text{Log}_{10}(a_T) = \text{Log}_{10}\left(\frac{t}{t_r}\right) = \frac{C_1(T-T_r)}{C_2+(T-T_r)} \tag{4.1.24}$$

where a_T is the shift factor, C_1 and C_2 are material constants, T is temperature, and T_r is the reference temperature. The experimentally characterized shear and bulk modulus as a function of time and temperature of polymer can be conducted by the dynamic mechanical analysis (DMA). Curve-fitting routines can be developed to obtain material constants of the Prony series and shift function.

4.1.3 Creep-fatigue properties

4.1.3.1 Creep model

For Sn-Ag-Cu lead-free solder, the melting point is 490K (or 217°C). Therefore, even though at room temperature of 298K, the homologous temperature ratio (in Kelvin) is more than 0.5. Hence, creep behavior in electronic solders plays a role. Typically, creep can be divided into three stages, namely primary, secondary, and tertiary stages. The secondary creep also known as the steady-state creep is widely studied as it dominates the creep rupture life of solders. In the steady-state creep stage, the creep strain rate can be generally described by Dorn's equation:

$$\dot{\varepsilon}_C = \frac{CGb}{k_B T}\left(\frac{b}{d}\right)^p \left(\frac{\sigma}{G}\right)^n D_0 \exp\left(\frac{-Q'}{k_B T}\right) \tag{4.1.25}$$

where $\dot{\varepsilon}_C$ is the steady-state creep strain rate, C is a complex constant, G is the shear modulus, b is the Burgers vector, k_B is Boltzmann's constant, T is the absolute temperature, d is the grain size, p is the grain size exponent, σ is the applied stress, D_0 is the frequency factor, n is the creep stress exponent, and Q' is the apparent activation energy with the unit eV. From the above equation, it can be seen clearly that the steady-state creep strain rate is related to not only the properties of the solder itself (shear modulus, grain size, activation energy) but also the external loading conditions, such as temperature and applied stress. A decrease in grain size or an increase in service temperature or applied stress leads to a rise in the steady-state creep strain rate and a fall in the creep lifetime. As the Boltzmann constant is equal to the ratio of the gas constant to the Avogadro constant (NA), k_B = (gas constant R)/NA = 1.38×10^{-23} J/K, some constants can be converted into a new constant, and the above equation can be simplified as the following Norton's power law constitutive model, which is widely used for solder alloys [10,11]:

$$\dot{\varepsilon}_C = C\sigma^n \exp\left(\frac{-Q}{RT}\right) \tag{4.1.26}$$

where C is a complex constant, Q is the activation energy expressed in J/mol, and n and T have the same meanings as in Eq. (4.1.25). At high stress, the simple power law creep behavior breaks down, and the creep strain rate increases more quickly with the applied stress, a phenomenon also known as the power law breakdown creep behavior. To find a wider formulation for power law creep (low and medium stresses) and power law breakdown (high stresses), double power law and hyperbolic sine function have been widely used. The double power law model can be expressed as follows [12]:

$$\dot{\varepsilon}_C = C_1 \sigma^{n_1} \exp\left(\frac{-Q_1}{RT}\right) + C_2 \sigma^{n_2} \exp\left(\frac{-Q_2}{RT}\right) \tag{4.1.27}$$

where Q_1 and Q_2 are the activation energies of grain boundary sliding and dislocation climbing, respectively. C_1 and C_2 are the fitting coefficients, and n and T have the same meanings as in Eq. (4.1.25). The first term in Eq. (4.1.27) corresponds to the creep behavior in the low stress level, at which diffusion creep along grain boundary is dominant, and the second term corresponds to the creep behavior in the high stress level, where dislocation creep is dominant. This double power law model can be realized by using the provided models for primary and secondary creep simultaneously.

The Garofalo steady-state creep behavior using hyperbolic sine function is used to model the power law breakdown region at high stress and is expressed as follows [13]:

$$\dot{\varepsilon}_C = C[\sinh(a\sigma)]^n \exp\left(\frac{-Q}{RT}\right) \tag{4.1.28}$$

where a is the multiplier of stress, and C, n, Q, and T have been defined before. In addition, the exponential creep model is also used to describe the solder creep behavior and is expressed as follows:

$$\dot{\varepsilon}_C = C \exp\left(\frac{\sigma}{b}\right) \exp\left(\frac{-Q}{RT}\right) \tag{4.1.29}$$

4.1.3.2 Elastic-plastic-creep model

It is assumed that in creep model, all inelastic strain is developed due to creep deformation, which is suitable for the condition that slow mechanical or thermomechanical loading is simulated. In viscoplastic model, plastic strain and creep strain cannot be separated distinctly, and they are unified as inelastic strain. However, elastic-plastic-creep (EPC) model is realized by combining time-independent elastic-plastic model and time-dependent creep model. Plastic strain and creep strain are separated distinctly and their summation is called total inelastic strain:

$$\varepsilon_{in} = \varepsilon_C + \varepsilon_p \tag{4.1.30}$$

where ε_{in} is the total inelastic strain, ε_C is time-dependent creep strain, which can be obtained from creep model, and ε_p is time-independent plastic strain. At high stress deformation, time-independent plastic strain can be expressed using following strain-hardening law [5]:

$$\varepsilon_p = c_p \left(\frac{\sigma}{G}\right)^{m_p} \tag{4.1.31}$$

where G is shear modulus, and c_p and m_p are material-related constants.

Plasticity theory provides a mathematical relationship that characterizes the elasto-plastic response of materials. There are three ingredients in the rate-independent plasticity theory: the yield criterion, flow rule, and the hardening rule. The yield criterion determines the stress level at which yield is initiated, and Von Mises yield criterion is widely used for metal and metal alloy. For multiaxes stress situation, the equivalent Von Mises stress and strain can be expressed as

$$\sigma_{eff} = \frac{1}{\sqrt{2}} \sqrt{(\sigma_1 - \sigma_2)^2 + (\sigma_2 - \sigma_3)^2 + (\sigma_3 - \sigma_1)^2} \tag{4.1.32}$$

$$\varepsilon_{eff} = \frac{\sqrt{2}}{3} \sqrt{(\varepsilon_1 - \varepsilon_2)^2 + (\varepsilon_2 - \varepsilon_3)^2 + (\varepsilon_3 - \varepsilon_1)^2} \tag{4.1.33}$$

The flow rule determines the direction of plastic straining, and plastic strains usually occur in a direction normal to the yield surface. The hardening rule describes the changing of the yield surface with progressive yielding, so that subsequent yielding can be established. Two hardening rules are available: work (or isotropic) hardening and kinematic hardening. In work hardening, the yield surface remains centered about its initial centerline and expands in size as the plastic strains develop. Kinematic hardening assumes that the yield surface remains constant in size and the surface translates in stress space with progressive yielding. In practical FEA simulation when electronic assembly is subjected to thermomechanical or mechanical load with low strain rate, some simple plasticity forms after yield are used: perfectly plastic behavior, bilinear kinematic or work (isotropic)-hardening plastic deformation, and multilinear kinematic or work-hardening plastic deformation.

4.1.3.3 Solder fatigue life models

Solder joints are usually considered as the weakest part in the electronic assembly. Therefore, the fatigue life of solder joints often limits the life of the electronic product. The main failure mechanisms of solder joints include fatigue, creep, and fracture. Currently, more and more lead-free solders are used in electronic products. Sn-Ag-Cu solder alloys are quite common candidate for lead-free solders. Most of the researchers investigated the effect of small volume additive as fourth element, such as Ni, Fe, Al, Zn, In, Co, and rear-earth elements [14–19], on Sn-Ag-Cu solder properties such as strength, creep, and fatigue behavior.

There are two approaches to develop a fatigue life prediction model for solder. One approach is developed by conducting displacement-controlled isothermal mechanical fatigue test using bulk solder material. The other is developed by using actual reliability test data and finite element simulation to estimate the fatigue driving parameter. For the first method, the bulk solder material is tested under displacement-controlled isothermal fatigue test condition. Inelastic strain energy density, inelastic strain, and total strain in one cycle can be determined from stress—strain hysteresis loop. A fatigue failure criterion is usually set using 50% load drop based on load cell measurement. Then fatigue life model can be developed using Coffin-Manson strain-based model and/or Morrow's energy-based model. For the second method, to develop a life prediction model for solder joints, four steps are required. First, a constitutive equation and appropriate material properties need to be defined or chosen. Second, the constitutive equation is translated into FEA modeling to simulate stress—strain behavior of solder joints according to actual reliability test loading. Third, the volume-averaged numerical results such as strain or strain energy density from FEA simulation can be used as a damage parameter in fatigue life prediction model. Fourth, actual thermal cycling (TC) test data (or Weibull failure distribution data) from different test conditions and different package types were used to correlate the FEA damage parameter with the mean time to failure (MTTF) cycles to failure in the fatigue model. The coefficient and exponent in fatigue model can be determined by combining numerical result and actual reliability test data based on the Weibull cumulative failure distribution.

Lee et al. [20] divided fatigue models into five categories, namely, (a) stress-based, (b) plastic strain—based, (c) creep strain—based, (d) energy-based, and (e) damage accumulated based. Sometimes, plastic strain—based and creep strain—based models can be combined into inelastic strain—based model.

4.1.3.3.1 Stress-based fatigue models

This stress-based fatigue model, expressed by the stress cycles or S-N curve, is used for high cycle fatigue failure assessment ($>10^5$ cycles) at low stress amplitude levels. This approach applies when plastic deformation is small or negligible. S-N curve data are usually presented on a lg—lg plot. The S-N curve expression is given as follows:

$$\sigma_{ar} = \sigma'_f (2N_f)^b \qquad (4.1.34)$$

where σ_{ar} is stress amplitude for the zero mean stress case, σ'_f is fatigue strength coefficient, b is fatigue strength exponent, and $2N_f$ is reversals of failure (1 reversal = $\frac{1}{2}$ cycle). When nonzero mean stress effect is considered for the fatigue life, Goodman criterion is generally used and expressed as follows:

$$\frac{\sigma_a}{\sigma_{ar}} + \frac{\sigma_m}{\sigma'_f} = 1 \qquad (4.1.35)$$

where σ_a is stress amplitude for the case of nonzero mean stress, i.e., an equivalent completely reversed stress, and σ_m is the mean stress. Substituting into Eq. (4.1.34), we get

$$\sigma_a = \left(1 - \frac{\sigma_m}{\sigma'_f}\right)\sigma'_f(2N_f)^b = (\sigma'_f - \sigma_m)(2N_f)^b \qquad (4.1.36)$$

For a condition of variable amplitude loading, the Palmgren-Miner rule, also called the cumulative damage theory, states that fatigue failure is expected when summation of life fractions, N_i/N_{fi}, reaches unity. Namely,

$$\frac{N_1}{N_{f1}} + \frac{N_2}{N_{f2}} + \frac{N_3}{N_{f3}} + \cdots = \sum \frac{N_i}{N_{fi}} = 1 \qquad (4.1.37)$$

where N_i is a number of cycles when a certain stress amplitude, σ_{ai}, is applied, N_{fi} is the number of cycles to failure from the S-N curve at σ_{ai}. In practice, failure is often assumed to occur at a more conservative value such as 0.7 or lower for electronic equipment. High cycle fatigue is relevant to electronic assembly subjected to vibration loading. The cumulative damage index (CDI) approach has been used to predict the fatigue life of solder joints subjected to vibration loading [21].

4.1.3.3.2 Strain-based fatigue models

When the electronic products are subjected to TC, low cycle fatigue ($<10^4$ cycles) failure occurs and the fatigue life can be predicted using strain-based approach. Strain-induced fatigue can be further divided into three groups, namely plastic strain, creep strain, or total strain. Plastic strain deformation focuses on the time-independent plastic effects, while creep strain accounts for the time-dependent creep effects. Total strain considers all strain deformation due to not only inelastic deformation but also elastic deformation.

4.1.3.3.2.1 Plastic strain fatigue model

Plastic strain fatigue model applies to fatigue situations where cyclic plastic yielding of ductile materials leads to short fatigue lives and hence low cycle fatigue behavior. The Coffin-Manson low cycle fatigue model is perhaps the most widely used model currently in solder joint fatigue analysis and is given by Ref. [22]:

$$\varepsilon_p = \varepsilon'_f (2N_f)^c \qquad (4.1.38)$$

where ε_p is plastic strain amplitude, N_f is total number of cycles to failure, ε'_f is fatigue ductility coefficient, which is approximately equal to the true fracture ductility ε_f, and c is fatigue ductility exponent (varying between -0.5 and -0.7). Experimental data are required to determine the material constants. To account for the effect of frequency, a frequency-modified Coffin-Manson model was given by Shi et al. [23]:

$$\left[N_f f^{(n-1)}\right]^m \varepsilon_p = C \qquad (4.1.39)$$

where f is frequency and n is a frequency exponent that can be determined from the relationship between fatigue life and frequency.

Solomon's low cycle fatigue model relating the plastic shear strain to fatigue life is expressed by Ref. [24]:

$$\Delta \gamma_p N_p^a = \theta \qquad (4.1.40)$$

where $\Delta \gamma_p$ the plastic shear strain range, N_p is the number of cycles to failure, θ is the inverse of the fatigue ductility coefficient, and a is a material constant. As this model does not consider creep effect, it is limited in practical use for solder joints subjected to TC load. For multiaxes stress situation, the equivalent Von Mises strain as shown in Eq. (4.1.32) is used to output the state of strain in the solder joint. Then fatigue life prediction models with shear strain component require conversion from equivalent strain. The relationship between equivalent strain and shear strain can be expressed as follows:

$$\gamma = \sqrt{3}\varepsilon_{eq} \qquad (4.1.41)$$

4.1.3.3.2.2 Creep strain fatigue model Syed [25] proposed creep strain fatigue models for Sn-Pb and Sn-Ag-Cu solder by partitioning creep strain into two parts corresponding to transient creep and steady-state creep stages:

For Sn-Pb: $N_f = (0.02\varepsilon_{cr,i} + 0.063\varepsilon_{cr,ii})^{-1}$ \qquad (4.1.42a)

For Sn-Ag-Cu: $N_f = (0.106\varepsilon_{cr,i} + 0.045\varepsilon_{cr,ii})^{-1}$ \qquad (4.1.42b)

Syed [25] also presented creep strain fatigue model for Sn-Ag-Cu using total accumulated creep strain by considering two different creep constitutive models of power law and hyperbolic sine creep model:

Power law: $N_f = (0.0468\varepsilon_{cr})^{-1}$ \qquad (4.1.43a)

Hyperbolic sine: $N_f = (0.0513\varepsilon_{cr})^{-1}$ \qquad (4.1.43b)

Schubert et al. [26] proposed creep strain fatigue model for Sn-Pb and Sn-Ag-Cu using total accumulated creep strain by experiments and simulations:

For Sn-Pb: $\quad N_f = 0.69 \varepsilon_{cr}^{(-1.80)}$ (4.1.44a)

For Sn-Ag-Cu: $\quad N_f = 4.5 \varepsilon_{cr}^{(-1.295)}$ (4.1.44b)

where ε_{cr} from Eqs. (4.1.42) to (4.1.44) is the accumulated creep strain per cycle calculated by volume-averaging technique [27]:

$$\Delta \varepsilon_{cr} = \frac{\sum_{i}^{n} \varepsilon_{cr2,i} \cdot v_{2,i}}{\sum_{i}^{n} v_{2,i}} - \frac{\sum_{i}^{n} \varepsilon_{cr1,i} \cdot v_{1,i}}{\sum_{i}^{n} v_{1,i}} \quad (4.1.45)$$

where $\varepsilon_{cr2,i}$ and $\varepsilon_{cr1,i}$ are the total accumulated creep strain in one element at the end point and start point of the ith cycle, respectively, $v_{2,i}$ and $v_{1,i}$ are the element volume at the end point and start point of the ith cycle, respectively, and n is the amount of selected elements to calculate averaged creep strain.

Creep strain fatigue model provides a more comprehensive approach because it can consider dwell time and strain amplitude in TC load. However, one limitation of creep strain fatigue model is the absence of plastic strain effects. Plastic strain effects can be neglected only if the strain rate is low enough, thus resulting in a constant stress situation and the strain is indeed time dependent.

4.1.3.3.2.3 Total strain fatigue model In this model, total strain including elastic, plastic, and creep strain is considered as a damage parameter. Sometimes, just inelastic strain including plastic and creep strain is used as fatigue damage parameter. These two cases here are all called total strain fatigue models.

Conffin-Manson Eq. (4.1.38) is often combined with the stress-based approach to account for elastic deformation as well. The resulting equation is known as the total strain fatigue model:

$$\varepsilon_a = \varepsilon_{ea} + \varepsilon_{pa} = \frac{\sigma_f'}{E}(2N_f)^b + \varepsilon_f'(2N_f)^c \quad (4.1.46)$$

where the explanation of all the material constants can be referred to Eqs. (4.1.34) and (4.1.38).

Shi et al. [28] developed a temperature- and frequency-dependent fatigue life prediction model using the total strain range in the frequency-modified low cycle fatigue relationship:

$$N_f f^{(n-1)} = \left(\frac{C}{\Delta \varepsilon_t}\right)^{\frac{1}{m}} \quad (4.1.47)$$

It is shown that fatigue life of solder actually follows a different trend when frequencies change over at 10^{-3} Hz, which is expressed as follows [28]:

$$f^{(n-1)} = \begin{cases} f^{n_1-1} & 10^{-3} \leqslant f \leqslant 1\text{Hz} \\ \left(\dfrac{f}{10^{-3}}\right)^{n_2-1} (10^{-3})^{n_1-1} & 10^{-4} \leqslant f \leqslant 10^{-3}\text{Hz} \end{cases} \qquad (4.1.48)$$

where the parameters of n_1, n_2, m, and C can be determined by polynomial expression fitting from temperature-dependent experimental data.

Practical applications at high temperature often involve both creep and fatigue, and these phenomena may act together in a synergistic manner. One simple approach is to sum the life fractions due to both creep and fatigue. By applying Miner's linear superposition principal, both plastic and creep strain can be accounted for in strain-based fatigue model [13]:

$$\frac{1}{N_\text{f}} = \frac{1}{N_\text{p}} + \frac{1}{N_\text{c}} \qquad (4.1.49)$$

where N_p refers to the number of cycles to failure due to plastic fatigue and is obtained directly from the plastic strain fatigue model and N_c refers to the number of cycles to failure due to the creep fatigue and is obtained from the creep strain fatigue model. However, this approach is an approximate method as the physical processes of creep and fatigue can have an interaction effect.

4.1.3.3.3 Energy-based fatigue models

The energy-based fatigue model is based on calculating the overall stress–strain hysteresis energy of solder joints. The Morrow energy-based fatigue model is widely used for solder joint fatigue life prediction:

$$N_\text{f}^m W_\text{p} = C \qquad (4.1.50)$$

where m is fatigue exponent and C is material ductility coefficient. W_p is plastic strain energy density (unit: MPa). Typically, the hysteresis loop is observed to stabilize after a few cycles. The plastic strain energy density is determined from the area within the stable hysteresis loop. However, it should be noted that the fatigue life strongly depends on the test frequency at any given strain energy density value. Solomon and Tolksdorf [29] adopted a different approach by introducing a frequency-modified energy model, which incorporates both the frequency-modified strain energy density and the frequency-modified fatigue life, as shown below:

$$\left[N_\text{f} f^{(n-1)}\right]^m \frac{W_\text{p}}{f^n} = C \qquad (4.1.51)$$

where f is load frequency, n is a frequency exponent like in Eq. (4.1.47), and n is another frequency exponent that is determined from the relationship between strain energy density and frequency.

A modified energy-based fatigue model for eutectic Sn-Pb solder was proposed by Shi et al. [23] and is given below:

$$\left[N_f f^{(n-1)}\right]^m \frac{W_p}{2\sigma_f} = C \tag{4.1.52}$$

where σ_f is called flow stress, which is the averaged values of yield point and the highest point in the stress-strain curve in a hysteresis loop. The averaged material constants for eutectic Sn-Pb were $m = 0.70$, $C = 1.69$, $(1-n) = 0.1$ for $f > 10^{-3}$ Hz, or $(1-n) = 0.59$ for $f < 10^{-3}$ Hz.

Syed [25] proposed energy-based fatigue models for Sn-Ag-Cu solder using creep strain energy as fatigue damage parameter based on simulation and test data from different BGA specimens:

$$N_f = (0.0019 W_{cr})^{-1} \tag{4.1.53}$$

Schubert et al. [26] proposed creep strain energy-based fatigue model for Sn-Pb and Sn-Ag-Cu using total accumulated creep strain energy by simulations and experiments using plastic ball grid array (PBGA) and flip-chip-on-board (FCOB) with/without underfill as samples:

For Sn-Pb: $$N_f = 210 W_{cr}^{(-1.20)} \tag{4.1.54a}$$

For Sn-Ag-Cu: $$N_f = 345 W_{cr}^{(-1.02)} \tag{4.1.54b}$$

where W_{cr} in Eq. (4.1.54) is the accumulated creep strain energy density per cycle calculated by the volume-averaging technique.

Darveaux [5] proposed a solder life prediction model for 62Sn-36Pb-2Ag solder based on the plastic strain energy density accumulated per cycle. Based on Darveaux's model, the fatigue life consists of two parts, one for crack initiation and the other for crack propagation. Crack initiation and crack growth are given as follows:

$$N_0 = m_1 \Delta W_{ave}^{m_2} \tag{4.1.55}$$

$$\frac{da}{dN} = m_3 \Delta W_{ave}^{m_4} \tag{4.1.56}$$

where m_1, m_2, m_3, and m_4 are constants for all the various modeling methodologies employed and a is the crack length. The plastic work per unit volume, or plastic strain energy density, ΔW_{ave} is averaged across the elements along the solder joint surface where the crack initiates and propagates.

Based on the above equations, the fatigue life can be predicted as

$$N_f = N_0 + \frac{D}{da/dN} \qquad (4.1.57)$$

where D is the solder joint diameter at interface between solder ball and substrate or chip side. This method is sensitive to the mesh size and model form. To obtain accurate results, the elements near the interface used to calculate averaged plastic work must be fine enough.

4.1.3.3.4 Fracture mechanics approach

Many failures in electronic products exhibit the crack fracture of the interface between solder and substrate or chip. Many researchers [30,31] studied this phenomenon by observing microstructures of solder joints using scanning electron microscope (SEM).

At high stress-strain amplitude, initiation life may be relatively short compared with propagation life. Propagation life is determined in terms of fatigue crack growth rate of da/dN, which is related to stress intensity factor range, ΔK_I. The Paris law, a power law relationship, is frequently used to determine da/dN:

$$\frac{da}{dN} = C(\Delta K_I)^m \qquad (4.1.58)$$

where C and m are constants that are influenced by material properties, temperature, environment, cyclic load frequency, waveform, and load ratio, $\Delta K_I = F\Delta S\sqrt{\pi a}$, and F is a dimensionless function of geometry. ΔS means stress variation in one cycle. Therefore, fatigue life can be estimated for constant amplitude loading by integrating above equation:

$$N_{if} = \int_{a_i}^{a_f} \frac{da}{C(\Delta K_I)^m} = \int_{a_i}^{a_f} \frac{da}{C(F\Delta S\sqrt{\pi a})^m} = \frac{a_f^{1-m/2} - a_i^{1-m/2}}{C(F\Delta S\sqrt{\pi})^m \left(1 - \frac{m}{2}\right)}, (m \neq 2)$$

(4.1.59)

where N_{if} is fatigue life of crack propagating from a_i of initial crack length to a_f of final crack length.

4.2 Reliability test and analysis methods

The reliability of a packaged microelectronic system is related to the probability that will be operational within acceptable limits for a given period of time [32]. Electronic assemblies are subjected to life cycle testing by TC or thermal shock (TS), cyclic bend, repeated drop impact, and sinusoidal and random vibration loading. Failure mechanisms for electronic solders subjected to these different loadings need to be investigated. For microelectronic packages, surface mount technology and soldering are used for assembly. The solder joint in the electronic assembly is often the weakest link and solder joint reliability becomes even more important with further minimization of electronic assemblies. The solder joint is particularly prone to fatigue failure due to power and temperature cycling loading.

Reliability tests of electronic assemblies are very important concerns to ensure long-term reliability of electronic products. In service, electronic products failures can occur within a year but often much longer. Therefore, it is not practical to test at service load level, and it is uneconomical with respect to test time and cost. Accelerated life testing (ALT) methods are employed for the purpose of accelerating reliability test results, where stresses are applied to a product in excess of the specified operating limits to precipitate latent flaws to the point of detection via testing.

Reliability tests aim at revealing and understanding the physics of failure. Another objective of reliability tests is to accumulate representative failure statistics. Accelerated tests use elevated stress levels and/or higher stress cycle frequency to precipitate failures over a much shorter time. Stress loading includes constant stress, step stress, progressive stress, and random stress [33]. Typical accelerating stresses are temperature, mechanical load, TC or TS, cyclic bend, drop impact, and vibration. Such tests can facilitate physics of failure for reliability tests that cost effectively, shorten the product process, and improve long-term product reliability. Accelerated tests can be performed at any level, such as part level, component level, board-level, equipment level, and system level. The degree of stress acceleration is usually determined by an acceleration factor (AF). This factor is defined as the ratio of lifetime under normal service conditions to the lifetime under accelerated conditions.

In this section, TC, cyclic bending, vibration test, and drop impact test will be discussed with examples to present testing method and reliability analysis for advanced electronic packaging. Thermal cycling loading (high strain, low cycle fatigue) induces viscoplastic deformation in the solder joints. Vibration loading (low stress, high cycle fatigue) primarily induces elastic or elastic-plastic deformation in the solder joints. Drop impact test is used to investigate impact reliability of electronic assemblies for portable electronic products.

4.2.1 Thermal cycling test and analysis

When accelerated thermal cycling (ATC) tests are carried out, AF given by Norris and Landzberg [34] has been used to relate test conditions to field conditions for solder:

$$\text{AF} = \left(\frac{\Delta T_t}{\Delta T_f}\right)^{1.9} \left(\frac{f_f}{f_t}\right)^{1/3} \exp\left[1414\left(\frac{1}{T_{\max.f}} - \frac{1}{T_{\max.t}}\right)\right] \quad (4.2.1)$$

where ΔT_t and ΔT_f are temperature ranges at test and field conditions; f_t and f_f are the cyclic frequencies at test and field conditions; and $T_{\max.t}$ and $T_{\max.f}$ are the highest temperature in K at test and field conditions, respectively. When the accelerated factor,

AF, is determined and reliability test data are known, the life distribution, reliability function, and failure rate for field condition can be calculated using best-fit Weibull life distribution model considering linear acceleration case in which the Weibull plots of the field condition and testing condition have the same Weibull slope:

$$F_f(x_f) = 1 - \exp\left(- (x_f/\mathrm{AF}\cdot\theta_t)^{\beta_t}\right) \qquad (4.2.2a)$$

$$R_f(x_f) = \exp\left(- (x_f/\mathrm{AF}\cdot\theta_t)^{\beta_t}\right) \qquad (4.2.2b)$$

$$h_f(x_f) = (\beta_t / \mathrm{AF} \cdot \theta_t)(x_f/\mathrm{AF}\cdot\theta_t)^{\beta_t-1} \qquad (4.2.2c)$$

where x_f, F_f, R_f, and h_f are the time to failure, life distribution, reliability function, and failure rate at field condition, respectively, β_t and θ_t are the shape parameter and characteristic life at testing condition, respectively, and AF represents the accelerated factor.

The two common ATC tests used for solder joint reliability used in industry are TC [35] and TS [36] tests. Cyclic stress—strain will deform the solder material when subjected to TC or TS test due to CTE mismatch in different electronic materials, thus inducing thermal fatigue failure in solder joints. TC has a ramp rate varying from 8 to 33°C/min, whereas TS condition has a ramp rate between 33 and 55°C/min. Dwell time (soak time) is included at the cycle extremes to allow for creep and stress relaxation process to take place. TC testing condition is in air environment, while TS testing condition is in liquid environment to provide fast temperature change. Different temperature ranges are shown in Table 4.2.1 for different service environments for electronic products. In ATC test, such temperature range will be considered.

Surface mounted components, such as PBGA, are commonly used in electronic assemblies. In PBGA, the solder joints are very important parts for the integrity of electronic products as they are prone to fatigue failure at test or field condition. Lead-free solders are rapidly replacing lead-based solders as the EU legislations on

Table 4.2.1 Thermal environments for electronic products.

Use condition	Temperature range(°C)
Consumer electronics	0–60
Telecommunications	−40–85
Commercial aircraft	−55–95
Military aircraft	−55–125
Space	−40–85
Automotive passenger	−55–65
Automotive under the hood	−55–150

RoHS (Reduction of Hazardous Substances) and WEEE (waste electrical and electronic equipment) took effective in July 2006. The lead-free solder alloy recommended by the iNEMI, Sn-3.9Ag-0.6Cu, is gaining widely spread use with a narrow range Sn-(3—4)Ag-(0.5—1) Cu. In this section, thermal fatigue test from −40 to 125°C and FEA simulation of PBGA package with Sn-3.8Ag-0.7Cu solder were investigated as an example. Weibull distribution analysis was conducted for test results, and the MTTF was used for validation of FEA simulation result.

4.2.1.1 Thermal cycling test and failure analysis

Fig. 4.2.1 shows a PBGA specimen comprising 316 solder joints with 0.76 mm solder ball diameter, 1.27 mm solder joint pitch, and 0.58 mm solder ball stand-off height. The die size is 7.6 mm × 7.6 mm × 0.4 mm. The size of BT substrate is 227 mm × 0.28 mm. The PBGA component is mounted on the 1.63 mm thick PCB using Sn-3.8Ag-0.7Cu lead-free solder. Electroless nickel immersion gold (ENIG) surface finish is used for both PCB and package substrate copper (Cu) pads. The thickness of over mold is 0.7 mm. Thermal cycling test with temperature ranging from −40 to 125°C was conducted for the PBGA specimen in a three-zone thermal chamber. As shown in Fig. 4.2.2, the TC profile has a 60 min cycle time including 15 min ramp up and ramp down and 15 min dwell time at extreme low and high temperature. One daisy-chain loop connecting all the solder joints for each PBGA package was designed to monitor resistance change using a data logger. The initial resistance of the daisy-chain loop is about 10Ω. Solder joint failure criterion is defined as having resistance values more than 300Ω, which is one of the most used failure criteria in defining the solder joint failure of the electronic assemblies under accelerated reliability tests. Total of 24 samples were tested in the thermal chamber. The test was stopped at 3500 TC cycles. The total of 16 failure data was obtained and shown in Fig. 4.2.3. The PBGA package used in this work has a good TC reliability with the first failure life more than 1000 cycles. The calculated MTTF is 2742 cycles and the characteristic life (63.2% failure) is 3081 cycles based on the Weibull distribution model analysis. The characteristic life will be used as a reference for comparison with numerical life prediction. Failure analysis was performed for the PBGA package to ascertain the failure site and mode through cross-sectioning and scanning electron microscope (SEM).

Figure 4.2.1 Plastic ball grid array (PBGA) specimen and solder joint layout.

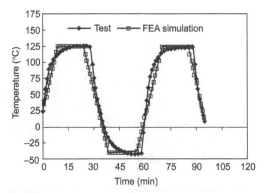

Figure 4.2.2 Thermal cycling profiles from actual test and finite element analysis (FEA) simulation.

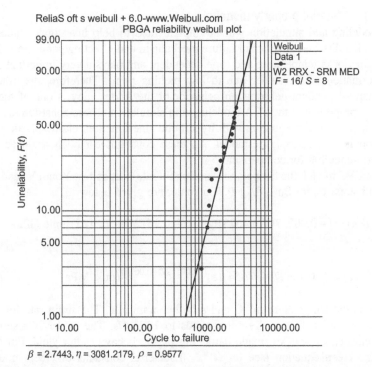

$\beta = 2.7443, \eta = 3081.2179, \rho = 0.9577$

Figure 4.2.3 Weibull plot of thermal cycling fatigue life for plastic ball grid array (PBGA) package.

The solder joint under silicon die corner was found to be total fatigue cracking failure. Fig. 4.2.4 shows the SEM image of typical solder joint failure site. The observed failure mode was solder joint fatigue cracks initiating from the solder joint corner and propagating through the bulk solder close to the solder/IMC interface at the component side. The IMC was identified as Cu-Ni-Sn ternary IMC as shown in Fig. 4.2.4.

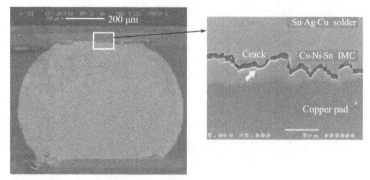

Figure 4.2.4 SEM images showing solder joint failure mode and location.

4.2.1.2 Thermal cycling fatigue life analysis

4.2.1.2.1 Material property model

FEA modeling and simulation will be implemented to help investigate solder joint fatigue reliability and understand failure mechanism and affecting parameters. Solder creep behavior is temperature and time dependent and plays a very important role in the deformation of the solder due to low melting point. Therefore, the time- and temperature-dependent deformation behavior of the solder alloy is one of the most important properties in the FEA modeling and simulation. These constitutive models include a single creep equation with hyperbolic sine function, elastic-plastic model, EPC models, and viscoplastic model. All these constitutive models are used in TC reliability modeling for comparison.

In the EPC model, the temperature- and strain rate–dependent Young's modulus E and yield stress σ_y for Sn-3.8Ag-0.7Cu solder are given below [37]:

$$E(T,\dot{\varepsilon}) = (-0.0005T + 6.4625)\lg\dot{\varepsilon} + (-0.2512T + 71.123)(\text{unit: GPa}) \quad (4.2.3)$$

$$\sigma_y(T,\dot{\varepsilon}) = (-0.1362T + 67.54)(\dot{\varepsilon})^{(0.000559T+0.0675)}(\text{unit: MPa}) \quad (4.2.4)$$

where T is the temperature in °C and $\dot{\varepsilon}$ is the strain rate. The strain rate for solder deformation under TC usually corresponds to 10^{-4} 1/s. The solder is assumed to exhibit elastic, bilinear kinematic hardening plastic behavior after yield. For hyperbolic sine creep equation (see Eq. (4.1.28)), material constants of creep model for Sn-Ag-Cu solder was presented in Ref. [38]. Nine constants in the viscoplastic Anand model (see Eqs.(4.1.17)–(4.1.20)), A, Q/k_B, ξ, m, $\widehat{S}$, n, h_0, a, and S_0, for Sn-Ag-Cu can be found in Ref. [39]. Such material properties can be used in FEA modeling to simulate stress−strain behavior of solder joints under TC loading condition for life prediction. Fatigue damage parameters such as inelastic strain energy density W_{in} can be expressed as follows:

$$W_{in} + W_p + W_c \quad (4.2.5)$$

where subscripts in, p, and c represent inelastic, plastic, and creep, respectively. Some researchers only considered solder creep behavior without considering the plastic aspect in the FEA simulation of TC. Under such a condition, all creep strain energy density contributes to a whole inelastic strain energy density.

There are two ways to develop a fatigue model for the solder joints. One way is to develop a fatigue model by conducting displacement-controlled isothermal mechanical fatigue test using bulk solder material or solder joint specimen. The other way is to develop a fatigue model by combining actual thermal fatigue reliability test and FEA simulation results. For the first method, energy-based life models are given in Eq. (4.1.50). The fatigue ductility coefficient, C, and the fatigue exponent, m, for Sn-3.8Ag-0.7Cu can be obtained from the energy–life curves. In a TC test, temperature changed from −40 to 125°C with a frequency of a cycle per hour. The isothermal tension test condition is required similar to the TC test. The extreme high temperature and strain rate in the TC test have a significant effect on solder joint fatigue failure. According to above conditions, the isothermal test condition was selected as at 125°C with a frequency of 0.001 Hz. The material constants n and A for Sn-3.8 Ag-0.7Cu solder are 0.897 and 311.7 MPa, respectively, obtained from the isothermal fatigue tension tests [40].

For the second method, the coefficient and exponent in fatigue model can be determined by combining the FEA simulation results and actual test data. Schubert et al. [26] proposed the creep strain energy-based fatigue life models (Eq.(4.1.54b)) for Sn-Ag-Cu solder with high Ag content based on FEA simulation results (averaging the accumulated creep strain and energy density for interface layer of solder at package side) and actual reliability test data from different types of assemblies such as FCOB and PBGA under different TC conditions. Syed [25] proposed fatigue models (Eq.(4.1.53))for Sn-Ag-Cu solder with high Ag content using FEA simulation results and actual reliability test data from different BGA assemblies such as PBGA, Ceramic BGA, and fleXBGA under different TC conditions. Different fatigue life models mentioned above will be used to predict PBGA solder joint fatigue life to investigate the effect of different life models on the solder joint fatigue life prediction.

4.2.1.2.2 Finite element analysis modeling and life prediction
4.2.1.2.2.1 Finite element analysis model Fig. 4.2.5 shows different FEA models for PBGA package including quarter, octant, slice, and 2D models. Quarter model was obtained by cutting from two symmetric centerline planes. Octant model was obtained by cutting from diagonal and centerline planes based on geometric symmetry. Slice models were extracted from a diagonal line and two different slice models were considered, one was obtained by cutting plane through package diagonal, thus all solder joints along diagonal are in the shape of half solder ball (S2). The other slice model contains all entire solder joints along package diagonal (S1). Because of the complexity of octant and quarter models, submodeling method was used in these two models. A 2D model is chosen from package diagonal with a diagonal pitch distance ($\sqrt{2}$ times normal pitch) between adjacent solder joints. For the 2D model, 2D strain, 2D stress with unit thickness, 2D stress with thickness equivalent to solder joint

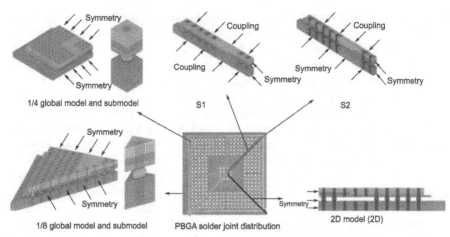

Figure 4.2.5 2D and 3D finite element analysis (FEA) models of plastic ball grid array (PBGA) package.

pitch, and 2D symmetric models were investigated. The proper symmetric and/or coupled boundary conditions were applied to different FE models. All FEA models that are commonly encountered in the simulation of TC for electronic assemblies were simulated. Errors arising from different numerical models were identified qualitatively and quantitatively through result comparison by considering the quarter model as a reference model because the quarter model was the most accurate model compared with the other models.

In the FEA simulation of PBGA package under TC, the elements required for solder joints occupy a majority of all elements because finer meshing is needed for solder joints. Therefore, it is desirable to reduce elements size for solder joints. According to a beam theory, a ball-shape solder joint can be replaced by a cuboid-shape solder joint with the equivalent tensile and shear stiffness because tension compression and shear loads are the major loading type when electronic package subjects to TC. Fig. 4.2.6 shows the meshed solder joints in an equivalent global quarter model.

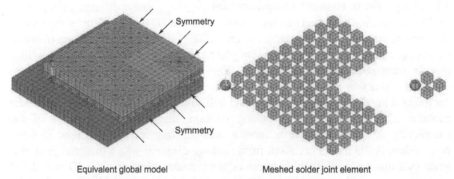

Figure 4.2.6 Meshed ball-shape and cuboid-shape solder joints in an equivalent global model.

The critical solder joints, for example, solder joints under die corner and package corner, were modeled as a real shape solder joint and others were modeled as the equivalent cuboid-shape solder joints. The submodel used for the effective solder joint model is the same as that for the quarter model as shown in Fig. 4.2.5. This effective global model reduces elements size significantly compared with global model with all meshed ball-shape solder joints.

To investigate the effect of element size on the simulation results, the mesh sensitivity study was carried out. The quarter global FEA model has the same element density, while the submodel has different element densities. Fig. 4.2.7 shows the different FEA models of solder ball used in the submodel containing coarse, medium, fine, and very fine element mesh. Usually, the top interface layer of component/solder is a critical area in terms of failure site. The thickness of the interface layer elements is 25 µm for all FEA models, which was used for BGA solder joints. Different mesh densities along the radial direction and across the thickness of the 25 µm interface layer were implemented as shown in Fig. 4.2.7. The effect of the mesh density across the interface thickness on the simulation results was also studied by comparing a fine model (Fig. 4.2.7c) with a very fine model (Fig. 4.2.7d). The fine FEA model has one layer elements across the 25 µm interface thickness, while the very fine FE model has 2-layer elements across the same interface thickness. The medium and coarse models have grid size 2 times and 3 times as the fine model along the radial direction, respectively. Anand constitutive model was used for the solder in the mesh sensitivity study.

Tables 4.2.2 and 4.2.3 list the material properties used in the FEA simulation for the PBGA package [37,41]. Solder joint material properties are critical in the FEA simulation. The elastic-plastic, EPC, elastic-creep, and viscoplastic Anand models were simulated for the Sn-Ag-Cu solder for comparison. Temperature-dependent material properties were used for Cu and silicon die [41]. Fig. 4.2.2 shows the TC profile used in the FEA simulation.

4.2.1.2.2.2 Finite element analysis simulation results and discussion The accumulated strain energy density in the top interface layer elements based on volume-averaging method was compared in Fig. 4.2.8. The volume-averaged method was used for plastic strain energy density (W) calculation [27]:

$$\Delta W_{ave} = \frac{\sum_{i}^{n} W_{2,i} \cdot v_{2,i}}{\sum_{0}^{n} v_{2,i}} - \frac{\sum_{i}^{n} W_{1,i} \cdot v_{1,i}}{\sum_{i}^{n} v_{1,i}} \qquad (4.2.6)$$

where $W_{2,i}$ and $W_{1,i}$ are the total accumulated strain energy density in one element at the end point and the starting point of the ith thermal cycle, respectively, $v_{2,i}$ and $v_{1,i}$ are the element volume at the end point and start point of the ith thermal cycle, respectively, and n is the number of selected elements to calculate averaged strain energy density. The element size effect on the simulation result is not significant because the volume-averaging method was used. The difference of the simulation

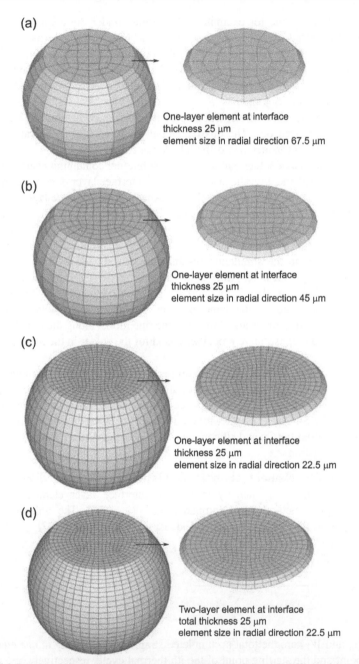

Figure 4.2.7 Solder joint finite element analysis (FEA) models with different mesh density: (a) coarse mesh, (b) medium mesh, (c) fine mesh, and (d) very fine mesh.

Table 4.2.2 Material properties used in finite element analysis (FEA) simulation for plastic ball grid array (PBGA) package.

Materials	Young's Modulus (GPa)	Poisson's ratio	CTE(ppm/°C)
Solder	Table 4.2.3	0.35	24.5
Copper	155.17	0.34	Table 4.2.3
FR4 PCB	x,z:20;y:9.8	x,z:0.28;y:0.11	x,z:18;y:50
BT substrate	x,z:26;y:11	x,z:0.39;y:0.11	x,z:15;y:52
Die	Table 4.2.3	0.278	Table 4.2.3
Adhesive	7.38	0.3	52
Molding compound	16	0.25	15

Table 4.2.3 Temperature-dependent material properties.

Temperature(°C)	−40	25	50	125
Solder modulus (GPa)	54.43	41.73	36.84	22.19
Copper CTE (ppm/°C)	15.3	16.4	16.7	17.3
Die modulus (GPa)	192.1	191	190.6	190
Die CTE (ppm/°C)	1.5	2.6	2.8	3.1

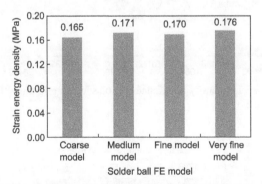

Figure 4.2.8 Effect of mesh density on strain energy density result of top interface layer.

results between the fine model and very fine model is less than 4%, which means that the volume-averaged result over the interface layer is not sensitive to the element size across the thickness of the interface. The medium model has a closer result compared with the fine model. Therefore, the medium model would be used in this chapter for parametric studies because this model needs lower computing sources compared with the fine model.

From FEA simulation results, all FEA models predict the same critical solder joint location, which is the solder joint under the die corner as shown in Fig. 4.2.9. In the FE simulation, three thermal cycles were simulated because it is enough to obtain the converged strain or strain energy density results. Fig. 4.2.10 shows the accumulated plastic strain energy density per cycle averaged over elements of an interface layer on solder/component and solder/PCB based on the quarter FEA model simulation results based on Eq. (4.2.6). The accumulated plastic strain energy density converges at the third thermal cycle for both component side and PCB side interface elements. Therefore, simulating three thermal cycles is enough to obtain a stable plastic strain energy density accumulation per cycle. The accumulated plastic strain energy density is larger on the solder/component interface layer than on the solder/PCB interface layer such that solder joint fatigue failure occurs on the component side, which was verified by failure analysis in TC test results.

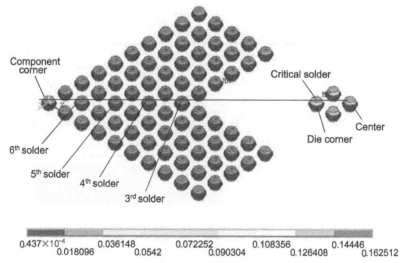

Figure 4.2.9 Strain energy density accumulated in plastic ball grid array (PBGA) solder joint after one thermal cycle.

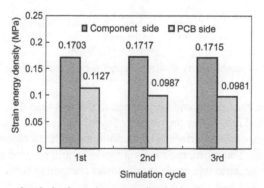

Figure 4.2.10 Accumulated plastic strain energy density in three thermal cycles.

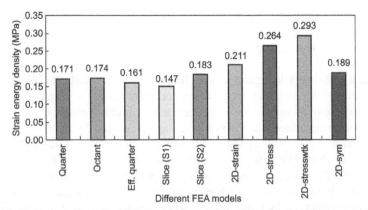

Figure 4.2.11 Accumulated plastic strain energy densities per cycle for different finite element analysis (FEA) models.

Fig. 4.2.11 shows volume-averaged plastic strain energy densities accumulation on component/solder interface elements for different FEA models. Firstly, the 2D stress models result in the highest plastic strain energy density, while the slice models S1 result in the lowest plastic strain energy density due to more assumptions about geometry and boundary conditions used in these models compared with real condition. Secondly, the 2D strain model results in the higher plastic strain energy density than the 2D axisymmetric model. Finally, octant, equivalent quarter, S2 slice, and 2D axisymmetric FE models can give reasonable results compared with the reference model (quarter FE model). Solder fatigue life predictions based on different FEA models were carried out using the energy-based fatigue model. For simplicity, the normalized fatigue life, or fatigue life ratio as shown in Fig. 4.2.12, was calculated considering fatigue life prediction for quarter reference model as a unity. It can be seen that all 2D models underestimate the solder fatigue life, especially for 2D stress

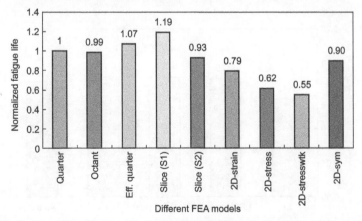

Figure 4.2.12 Normalized fatigue life predicted by different finite element analysis (FEA) models.

model. Octant, equivalent quarter, S2 slice, and 2D symmetric models result in accurate fatigue life predictions with error less than 10% compared with quarter reference model. These findings are applicable for the large PBGA solder joints.

4.2.1.2.2.3 Parametric study using finite element analysis (FEA) simulation

Fig. 4.2.13 shows a comparison of volume-averaged strain energy density results for different constitutive models. The strain energy density values obtained from the EPC model, creep model, and Anand model are consistent. In the EPC model, total inelastic strain energy was separated into two parts, one for creep and the other for plastic deformation. When the EPC model was used for solder, the time-dependent creep effect (EPC creep) is dominant compared with time-independent plastic effect. In the Anand model, plastic and creep strain energy density cannot be separated and they were together as inelastic strain energy density. In the creep model, time-dependent creep behavior was developed. The EP model does not consider the time-dependent creep deformation but only the irreversible plastic deformation. Therefore, the energy density resulting from the EP model is lower than those from the other constitutive models. Therefore, the EP model is not a suitable constitutive model for solder joints subjected to TC.

Volume-averaging technique can be used to determine strain energy density to minimize the effect of mesh sensitivity and stress concentration on fatigue life prediction. As such, selecting suitable elements for volume-averaged strain energy density is very important as it affects fatigue life prediction significantly. Fig. 4.2.14 shows different volumes from solder/component interface elements in the quarter FEA model containing whole layer elements with 540 μm diameter, two outer rings elements with width of 1/3 radius of the interface, i.e., 90 μm ring width, the outermost ring elements with width of 1/6 radius of the interface, i.e., 45 μm ring width, and segment (1/4 ring) of the outmost ring. Fig. 4.2.15 shows the predicted fatigue life of solder joints based on different averaging volumes. Fatigue life prediction based on the segment elements results in a low life compared with test result due to stress concentration effect on

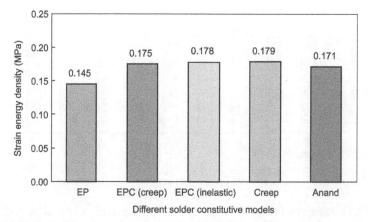

Figure 4.2.13 Inelastic strain energy density from different solder constitutive models.

Stress and reliability analysis for interconnects

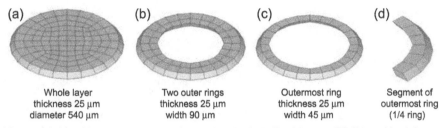

(a)	(b)	(c)	(d)
Whole layer thickness 25 μm diameter 540 μm	Two outer rings thickness 25 μm width 90 μm	Outermost ring thickness 25 μm width 45 μm	Segment of outermost ring (1/4 ring)

Figure 4.2.14 Different averaging volumes for plastic ball grid array (PBGA) solder joint.

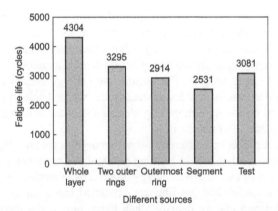

Figure 4.2.15 Comparison of solder joint fatigue life based on different averaging volumes.

such small volume. Fatigue life based on the outermost ring elements has a good agreement with test result. During a TC test, a crack always initiates at the outermost area of solder/component interface due to stress concentration near this area. Then, the crack propagates along the interface so that the stress concentration area will move toward the center of the interface. The majority of the solder joint reliability will be consumed before the crack propagates to the center of the interface because the effective interface area used for supporting thermal-induced loading becomes smaller and smaller. Therefore, averaging the whole solder/component interface layer volume will underestimate strain energy density and overestimate solder fatigue life accordingly.

To explain why volume-averaging method based on the outermost ring elements can result in accurate fatigue life prediction, a nonuniformity factor is introduced, which is defined as a ratio of the maximum plastic strain energy density to the minimum result among the selected elements. Fig. 4.2.16 shows the plastic strain energy density contours for different averaging volumes. The nonuniformity factors of plastic strain energy density are 10.2, 5.1, and 2.9 for the whole layer, two outer rings, and outermost ring elements, respectively. The peripheral area of the interface has the higher stress level under TC loading compared with the center area of the interface. The resulted strain energy density will reduce from the periphery to the center of the interface. For the result of the whole interface layer, the nonuniformity factor is the ratio of the maximum result at the edge of the interface to the minimum result at

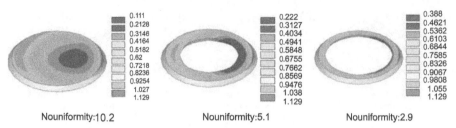

Figure 4.2.16 Plastic strain energy density contour for different averaging volumes.

the center of the interface, so the nonuniformity factor is larger. When using a ring as an averaged-volume, the maximum result is the same as that of the whole layer volume, but the minimum result of the ring is larger than that of the whole layer. Therefore, the nonuniformity factor for the ring volume becomes smaller compared with that based on the whole layer volume. The plastic strain energy density averaged on the whole interface layer results in a lower magnitude, which leads to a higher fatigue life. Volume based on the outermost ring elements is a desirable averaging volume for solder joint fatigue life prediction.

To investigate the effect of different fatigue models on solder fatigue life, fatigue life prediction was carried out by substituting accumulated strain energy density results as shown in Fig. 4.2.13 into different energy-based fatigue models such as Schubert's model and Syed's model. Fatigue life was predicted based on the outermost ring volume-averaged strain energy density when using our model developed by bulk solder specimen. Fig. 4.2.17 shows a comparison of fatigue life predictions using the energy-based fatigue life models. For a given fatigue life model, different solder constitutive models lead to similar fatigue life prediction. Syed's model was developed by using the experimental results of different types of BGA packages and FEA simulation results where the fatigue damage parameter was averaged based on the whole solder/component interface elements. Schubert's model has a similar method as Syed's

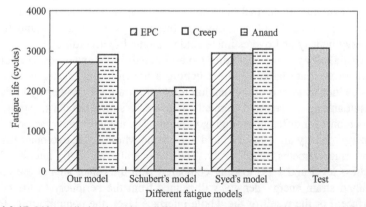

Figure 4.2.17 Life prediction based on different solder constitutive models and fatigue life models.

model except that different types of packages such as PBGA and FCOB with and without underfill were used in Schubert's model. Three different fatigue models result in an acceptable fatigue life prediction compared with experimental result. It should be noted that the whole layer element averaging method is reasonable in fatigue life prediction when using Schubert's model and Syed's model, while the outermost ring averaging method is desirable when using our fatigue model.

In the thermomechanical analysis, the initial stress-free condition (also called reference temperature) is important. Solder material is in a relatively stress-free state when solder joint is formed at soldering temperature. When the electronic assemblies are cooled down to room temperature, stresses have been induced in the assemblies due to CTE mismatch among different materials. Many researchers assumed room temperature or some other arbitrary temperatures as a reference temperature. This assumption is based on the belief that any residual stresses in solder will relax due to the creep characteristics of the solder [42]. Through stress relaxation tests for lead-free solders, Mavooti et al. [43] observed that the relaxation time (time required for stress drop to 1/e of the initial stress) decreased with increasing temperature and increasing strain. At 80°C, the stress relaxed completely to zero in less than 1 day for lead-free solder. At 25°C, most of stress (more than 50%) relaxed rapidly in less than 120s, and then the stress relaxation was insignificant over a period of 1 day. Some researchers [43,44] used 25°C as a reference temperature in the FEA simulation for investigating BGA solder joint fatigue life subjected to TC loading and they achieved a good agreement between life prediction and experimental results. Pyland et al. [45] used underfill cure temperature, 150°C, as a stress-free temperature in the FEA simulation for an underfilled BGA package to predict a consistent fatigue life compared with reliability test results. Fan et al. [46] recommended high dwell temperature, T_{max}, e. g., 125°C, for a TC test from −40 to 125°C, as a stress-free condition because the packages always relax to the lowest stress during the high-temperature dwell period after several cycles. Through their simulation results, it was found that the stabilized strain energy density almost converged to the same value for different stress-free conditions such as T_{max} and room temperature. In the study for fine pitch BGA packages [47], stress-free conditions affected the location and history of the stress—strain hysteresis loop, while the area of the hysteresis loop (strain energy density per cycle) was similar for different stress-free condition of 125 and 25°C. It seems that the selection of the stress-free temperature in the FEA simulation has been a debate for different packages and loading conditions. Here, three stress-free temperatures at 25°C, 125°C, and 150°C were simulated to investigate the effect of the stress-free temperature on the simulation results for the PBGA solder joints. Different solder material constitutive models, including Anand model, creep model, and EPC model, were also considered in the study of stress-free condition. Energy-based fatigue model and the outermost ring elements for averaging volume were used in solder joint fatigue life prediction. Fig. 4.2.18 shows a comparison of fatigue life prediction for different stress-free conditions when using different solder constitutive models. Solder joint fatigue life prediction decreases with increasing reference temperature for all solder constitutive models. However, fatigue life prediction is more sensitive to the reference temperature when using the Anand model for solder material than that when using the creep and EPC

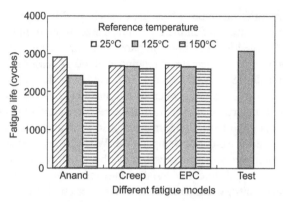

Figure 4.2.18 Effect of stress-free temperature on solder joint fatigue life prediction.

models. Selecting 25°C, 125°C, or 150°C as a reference temperature results in a reasonable fatigue life with error less than 25% compared with the test result. Generally, package warpage measured using shadow moiré and solder joint deformation measurement by digital image correlation [48] are two direct ways to benchmark the simulation results.

4.2.1.2.2.4 Effect of IMC layer on life prediction The microstructure change and its influence on solder joint reliability have received much attention recently due to the miniaturization trend of electronic packages. Chen et al. [49] reported the effect of microstructure change on the solder joint reliability under TC. The failure modes are dependent on the localized microstructural changes in the stress concentration regions and the crack propagation paths follow the networks of grain boundaries by recrystallization. Li et al. [50] developed a new method to predict the onset and progress of recrystallization in solder interconnection under TC test by combining the finite element simulation and Monte Carlo (MC) method. The stored energy distribution from FEA simulation was mapped onto the MC model to predict the onset and progress of recrystallization. However, the predicted fatigue life of solder joint is sensitive to the selection of the retained stored energy fraction.

The Cu-Ni-Sn ternary IMCs form between the Sn–Ag–Cu solder joints and Cu/Ni(Au) pad during reflow and thermal aging [51]. The reliability of the electronic assembly was affected by the IMC layer during TC [52]. The thickness of the IMC layer increases with aging time when the PBGA package subjects to TC. The evolution in the IMC morphology and microstructure were investigated using an isothermal aging test, and results were shown in Fig. 4.2.19 with aging time of 0 h, 120 h, 260 h, and 500 h, respectively. The microstructure of interfacial IMC changing with aging time can be roughly divided into several steps, from initial needle-type, transitional scallops + needles to finally planer type. Then, the IMC layer becomes smoother and smoother. The thickness of the IMC layer is around 3 to 6 µm after 2000 thermal cycles from −40 to 125°C [51]. However, it cannot be realized in the finite element modeling to model the IMC growth dynamically. Therefore, different constant IMC

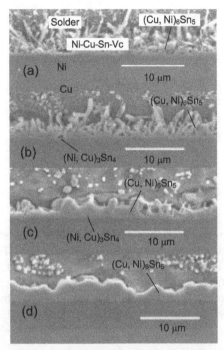

Figure 4.2.19 Microstructure change of the IMC layer at solder/Ni(Au) interface under different aging times: (a) 0 h, (b) 120 h, (c) 260 h, and (d) 500 h.

thicknesses, e. g., 1 μm, 2 μm, 3 μm, 4 μm, and 6 μm, were simulated separately in the FEA models to investigate the effect of the IMC thickness on the solder joint fatigue life, thus it can mimic the IMC growth influence approximately. In a real condition, the IMC surface is not flat with random roughness during growing. For convenience, a flat IMC layer was modeled. The IMC layers with the same thickness were modeled at both package side and PCB board side, as shown in Fig. 4.2.20. The solder joint stand-off height was the same for both FEA models with and without IMC layer. The elastic modulus and CTE of the Cu-Ni-Sn IMC are 160 GPa [53] and 15.3 ppm/°C [54], respectively.

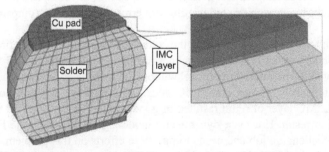

Figure 4.2.20 Finite element analysis (FEA) model including thin IMC layers in solder joint.

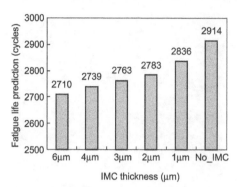

Figure 4.2.21 Effect of IMC layer thickness on solder fatigue life prediction.

Fig. 4.2.21 shows a comparison of the solder joint fatigue life when modeling different IMC layer thickness. Thicker IMC layer results in a lower solder joint fatigue life, which is consistent with the results given by Yao et al. [55]. The thicker IMC layer makes solder joint stiffer, which induces higher strain energy in solder ball and accelerates the solder joint fatigue failure accordingly. From a numerical simulation point of view, simulating a 6 μm IMC layer in the FEA model reduces the solder joint fatigue life prediction about 8% compared with the case without simulating the IMC layer.

4.2.1.3 Summary on thermal cycling reliability test and analysis

Thermal cycling reliability test and FEA simulation were carried out for a PBGA package with the Sn-Ag-Cu lead-free solder joints as an example. Failure analysis shows solder fatigue failure with fatigue crack close to solder/component interface, which is consistent with FEA simulation results. In the FEA simulation, 3D quarter and octant models lead to more accurate results than 2D model and 3D slice model. Solder constitutive models, such as the EPC model, elastic-creep model, and viscoplastic model, lead to the similar strain energy density and life prediction. Elastic-plastic model is not suitable for solder material in the thermomechanical FEA simulation. Fatigue damage parameter based on the outermost ring volume-averaging method is reasonable and proper in solder fatigue life prediction when using fatigue model developed by the bulk solder fatigue tests. The predicted fatigue life of solders decreases with increasing reference temperature. Selecting 25°C, 125°C, or 150°C as a reference temperature results in a reasonable fatigue life with error less than 20% compared with the test result.

The simulation results showed that the solder joint fatigue life prediction incorporating the modeling of IMC layer thickness results in much shorter TC fatigue life compared with the base reference model without the IMC layer present. The reduction is more significant with increasing IMC thickness. The finite element model results with a thick IMC layer of 6 μm resulted in good correlation with the TC test fatigue life prediction result. However, dynamically modeling IMC layer in the FEA model is still one challenging job and needs to put more efforts on this problem.

4.2.2 Mechanical bending test and analysis

Bend loading is encountered for hand-phone board-level keypad loading test. Three-point and four-point bend tests with cyclic or monotonic loading may be used. Cyclic bend test can evaluate the fatigue reliability of electronic assembly, while monotonic bend test is intended to characterize the fracture strength and overstress of solder joints. Three-point bend test is useful for generating reliability data for multiple packages tested under different load levels (bending moments) in a single test, while four-point bend test is suitable for testing large sample size of packages at similar loading condition (same bending moments) between the inner load span regions.

Shetty et al. [56,57] reported cyclic three-point bend tests from zero curvature to maximum curvature for chip-scale packages (CSP) mounted on both sides of PCB. Before cyclic bend test, overstress monotonic three-point bend tests were conducted to determine the overstress limits for CSP assembly. Fatigue life was developed based on bending strain and failure data. However, bending strain of PCB board measured during bend test is not a useful fatigue damage parameter for determining solder fatigue failure because bending strain can be affected by many factors such as PCB thickness and material properties, load profile, support span and load span, package type, etc. Fatigue damage parameters for solder fatigue failure should be used and the solder stress (S-N) or strain (ε-N) or energy (W_p-N) fatigue life approach needs to be developed and is reported in this chapter.

Mercado et al. [58] introduced four-point bend fixture with two-sided bend test from positive to negative deflections and investigated the effect of loading frequency with 1 Hz, 3 Hz, and 5 Hz and deflection levels with 1.5 mm, 2 mm, 3 mm, and 4 mm on fatigue reliability of handheld components. Correlation between four-point and three-point bend was also developed. Geng et al. [59] studied the effect of loading rate on PBGA solder joint reliability using four-point bend test with three different loading speeds of 2.54 mm/min, 25.4 mm/min, and 254 mm/min. It was shown that solder joint fatigue failure is dependent on bending loading rate. At high loading rate, solder joint fails at less board deflection. Therefore, traditional quasistatic bending test is not sufficient to quantify solder joint failure under shock load.

All the bend tests from existing literature were conducted at room temperature (25°C). In this section, VQFN (very-thin quad-flat-no-lead package) assembly with Sn-Ag-Cu lead-free solder was chosen as an example and tested under three-point and four-point cyclic bending load at room temperature (25°C) and high temperature (125°C). The correlation between three-point bend and four-point bend was developed and verified by test results and FEA simulation results. Temperature effect on bending fatigue life was evaluated, and accelerated factor was presented. In this study, FEA modeling and simulation were also performed for different testing conditions to investigate the stress—strain behavior of solder joints. Fatigue models were developed for Sn-Ag-Cu lead-free solder subjected to cyclic bending load at 25 and 125°C based on FEA results and bend test data using strain energy density as a fatigue damage parameter.

4.2.2.1 Isothermal cyclic bend test and analysis

4.2.2.1.1 Test vehicle and experimental procedure

VQFN assembly with Sn-Ag-Cu solder was selected for cyclic bend test in this study. The specimen size as shown in Fig. 4.2.22 follows JEDEC standards [60,61]. Two different board surface finishes of Ni/Au and OSP (organic solderability preservative) were investigated. For specimen with OSP finish, two types of test boards were used: one type of test board has 5 components and the other one has 15 components. For specimen with Ni/Au finish, one type of test board with 5 components was used. The structure of VQFN component is shown in Fig. 4.2.23. VQFN package with geometry size of 7 mm × 7 mm × 0.8 mm was mounted on the FR-4 PCB board of 132 mm × 77 mm × 1 mm in size by 48 Sn-Ag-Cu solder joints with 0.1 mm joint thickness. One daisy-chain loop containing all 48 solder joints was designed for each component to take in situ resistance measurement.

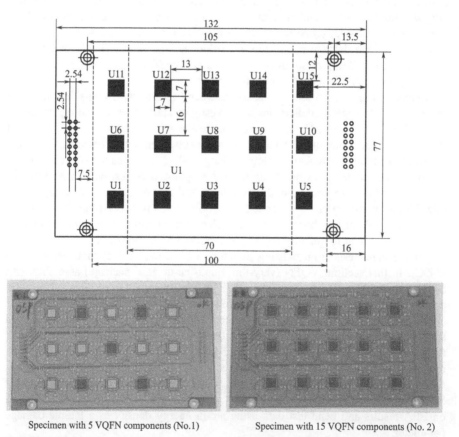

Specimen with 5 VQFN components (No.1)   Specimen with 15 VQFN components (No. 2)

Figure 4.2.22 Board size and layout of very-thin quad-flat-no-lead package (VQFN) specimens.

Stress and reliability analysis for interconnects   171

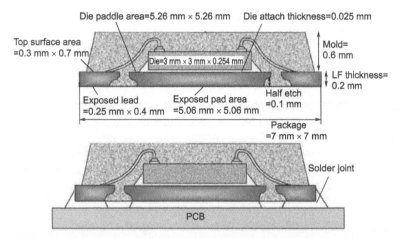

Figure 4.2.23 Structure of very-thin quad-flat-no-lead package (VQFN)

Cyclic bend test of specimen was conducted by an Instron tensile tester with a thermal chamber. The failure event was detected when resistance of the daisy-chain loop is greater than 300Ω lasting for 0.1 s or longer. Four-point and three-point bend tests were conducted to make correlation between them. Fig. 4.2.24 shows the fixture for four-point bend and three-point bend. Anvil radius and length of fixture satisfy requirement of bend test given by IPC/JEDEC-9702 standard [62]. The support span of 100 mm and load span of 70 mm as shown in Fig. 4.2.22 were used for four-point bend test. For three-point bend test, support span of 100 mm was used. Constant bending moment can be obtained between load head in four-point bend test so that all the samples within the load span are subjected to the same stress level (bending

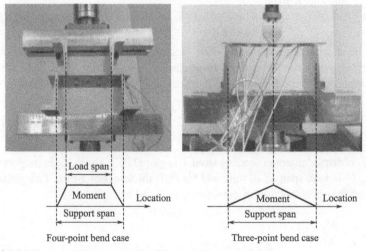

Figure 4.2.24 Four- and three-point bending test fixtures and setup.

moment or curvature), which significantly increases the sample size suitable for Weibull analysis. Therefore, in four-point bend test, all five components are used for test board No. 1, and nine components along three center rows for test board No. 2 are used for reliability analysis.

4.2.2.1.2 Correlation between three-point bend and four-point bend test

Correlation between three-point bend and four-point bend was performed considering components in four-point bend test subjected to the same bending moment value as components in the center, right, and left rows in three-point bend test. Based on classic beam theory, the maximum deflection, D_3, for three-point bend with centered loading condition can be expressed as

$$D_3 = \frac{Fl^3}{48EI} \tag{4.2.7}$$

where F is the load applied on specimen, l is support span (100 mm in this study), and E and I are Young's modulus and moment of inertia of specimen, respectively. For centered loading case, the maximum deflection and maximum moment occur at centered loading point. The maximum moment can be obtained by

$$M_C = M_{max} = \frac{Fl}{4} = \frac{12EID_3}{l^2} \tag{4.2.8}$$

where M_C is moment applied on components of the center row in three-point bend test. The moment value of M_R applied on components of the right row can be obtained considering specimen layout as shown in Fig. 4.2.22:

$$M_R = \frac{3}{5}M_C = \frac{36EID_3}{5l^2} \tag{4.2.9}$$

Pure bend condition occurs between loading anvils in four-point bend test. The following equation can be derived from classic beam theory when any effects due to the packages and the Poisson's ratio effect of a plate in bending are ignored [62]:

$$D_4 = \frac{\varepsilon(L_S - L_L)(L_S + 2L_L)}{6t} \tag{4.2.10}$$

where D_4 is displacement of loading anvil, ε is global PCB strain, L_S is support span of 100 mm, L_L is load span of 70 mm, and t is PCB thickness of 1 mm. The global strain of PCB surface is a function of curvature expressed as

$$\varepsilon = \frac{y}{\rho} = \frac{My}{EI} \tag{4.2.11}$$

where y is distance of PCB surface to middle plane, equating to half thickness of PCB board. Combining above two equations and substituting known value, the relationship between moment and displacement of loading anvil can be obtained as follows:

$$M_4 = \frac{D_4 EI}{600} \qquad (4.2.12)$$

The same moment value leads to the same global stress-strain level. Therefore, to use four-point bend test to simulate stress level of the center row components in three-point bend test, let Eq. (4.2.8) equal to Eq. (4.2.12), obtain

$$D_{4C} = 0.72 D_3 \qquad (4.2.13)$$

where D_{4C} and D_3 are displacements of loading anvil in four-point bend and three-point bend tests, respectively. Similarly, let Eq. (4.2.9) equal to Eq. (4.2.12), the relationship of displacement in four-point bend and three-point bend can be obtained to use four-point bend to simulate the same stress level of right/left row components in three-point bend:

$$D_{4R} = 0.432 D_3 \qquad (4.2.14)$$

4.2.2.1.3 Three-point cyclic bend test

VQFN assembly with OSP finish was used in three-point bend test under displacement control. Displacement from 1 to 5 mm with a frequency of 1 Hz was applied to specimen center with chip facing down condition as shown in Fig. 4.2.24, which gives a positive curvature to positive curvature fatigue test. Three components (U3, U8, and U13) located at the center row of specimen as shown in Fig. 4.2.22 have the same stress level. Three components at left row (U2, U7, and U12) and three components at right row (U4, U9, and U14) have the same stress level due to symmetry. Table 4.2.4 shows the failure data for three-point bend test including averaged fatigue life and life range. The fatigue life of components at the center row is less than that at the right and left rows dramatically due to high stress level. It is one major disadvantage of three-point bend test that failure is dependent of component location, which reduces the number of components subjected to the same stress level. Moreover, the limited

Table 4.2.4 Fatigue life (cycles) correlation between three-point and four-point cyclic bend tests.

Bend type	Conditions	Averaged life	Life range
Three-point	Center row	512	422–563
Four-point	0.72–3.6mm	444	317–678
Three-point	Right/left row	2111	787–3033
Four-point	0.432–2.16 mm	2249	1363–3426

sample size and large board-to-board variation will result in large variation in failure data for three-point bend condition.

4.2.2.1.4 Four-point cyclic bend test for correlation

In the four-point bend test, failures of components between loading anvils are independent of component location due to uniform stress-strain region, which increases the sample size for reliability analysis. Correlation between three-point bend and four-point bend was performed considering components in four-point bend test subjected to the same bending moment value as components of the center, right, and left rows in three-point bend test.

VQFN assemblies with OSP surface finish were selected for four-point bend tests to conduct comparison between four-point bend and three-point bend. Displacement range from 0.72 to 3.6 mm and from 0.432 to 2.16 mm was used in four-point bend tests to simulate stress level of components at the center row and components at the right/left rows in three-point bend test with displacement range from 1 to 5 mm according to Eq. (4.2.13) and Eq. (4.2.14), respectively. Three uniaxial strain gages as shown in Fig. 4.2.25 were mounted on the test board according to requirement of standard [62]. Strain gage mounted on opposite side of PCB as component and positioned at package center was used to measure the local stiffening effect of VQFN package. Strain gage mounted on opposite side of PCB as component and positioned at package corner was used to measure the maximum PCB strain. Strain gage mounted on the same side of PCB as component and centered between package edge and anvil centerline was used to measure the global PCB strain. The sensing direction of all strain gages was aligned with the longitudinal board direction, which is coincident with PCB principal strain angle. The strain measurement was used to validate FEA simulation results.

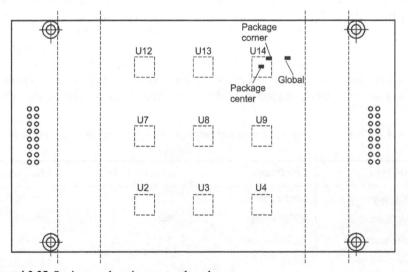

Figure 4.2.25 Strain gage location on test board.

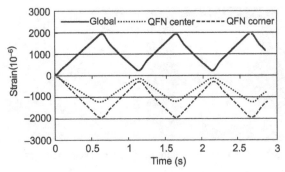

Figure 4.2.26 Strain data of different positions during four-point cyclic bend test.

Fig. 4.2.26 shows the strain values measured by three strain gages during four-point cyclic bend test at 25°C with displacement from 0.432 to 2.16 mm. The maximum strains are $2.000 \times 10^{-3}\varepsilon$, $1.960 \times 10^{-3}\varepsilon$, and $1.280 \times 10^{-3}\varepsilon$ for package corner, global, and package center strain gages, respectively. Therefore, the local stiffening effect of VQFN package on global PCB strain is obvious. The package corner strain is slightly more than global PCB strain.

Fig. 4.2.27 shows the typical resistance change of daisy-chain loop with cycles during four-point bend test with displacement from 0.72 to 3.6 mm. The package failure is defined when resistance of daisy chain is more than 300Ω.

The fatigue comparison between three-point bend and four-point bend is listed in Table 4.2.4. The fatigue lives of package in four-point bend tests with displacement from 0.72 to 3.6 mm and from 0.432 to 2.16 mm are consistent with those of components at the center row and right/left rows in three-point bend test with displacement from 1 to 5 mm, which verifies the good correlation between three-point bend and four-point bend tests. Failure mode analysis was carried out through cross section and optical microscopy after bend test. The typical solder fatigue failure was identified as shown in Fig. 4.2.28.

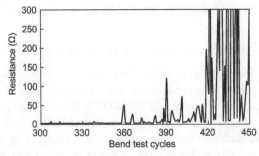

Figure 4.2.27 Typical in situ resistance measurement during cyclic bend test (four-point bend: 0.72–3.6 mm).

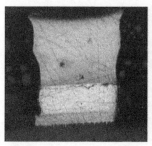 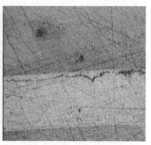

Figure 4.2.28 Typical solder fatigue failure mode under cyclic bend test.

4.2.2.1.5 Four-point cyclic bend test at 25 and 125°C

Different displacement ranges were used in four-point bend tests to develop bending fatigue life model of VQFN solder. Two different surface finishes containing Ni/Au and OSP were selected for VQFN assembly to investigate finish effect on bending fatigue life. All cyclic bend tests reported in limited literatures were conducted at room temperature of 25°C. In this study, bend test was also conducted at high temperature of 125°C to investigate AF of fatigue life at high-temperature condition compared with room temperature condition. All four-point bend test conditions are listed in Table 4.2.5. For convenience, one term was given for each test condition, for example, Ni/Au-25_3.6 indicates that specimen with Ni/Au finish was tested at room temperature of 25°C with the maximum displacement of 3.6 mm. The minimum displacement was selected as one-fifth of the maximum displacement for all displacement ranges used in four-point cyclic bend tests.

Figs. 4.2.29 and 4.2.30 show the Weibull plots of cycle to failure for VQFN assembly with OSP finish under four-point cyclic bent tests conducted at 25 and 125°C, respectively. It can be seen that failure data satisfy the Weibull distribution very well. The cycle to failure increases significantly with decreasing displacement range for both bend tests at 25 and 125°C. However, Weibull slopes for all test conditions have a similar value. These phenomena are also fit for failure data of VQFN assembly with Ni/Au finish.

Fig. 4.2.31 shows comparison of failure data for OSP finish case tested at 25 and 125°C. It was shown that high temperature (125°C) accelerates the failure of VQFN.

4.2.2.2 Finite element analysis for cyclic bend test

FEA modeling for bend fatigue tests were conducted to investigate the stress—strain behavior of solder joints. Based on bend test data and FEA simulation results, fatigue life prediction models were proposed for VQFN solder fatigue with OSP finish subjected to cyclic bend at 25 and 125°C. Submodeling technique was used for board-level FEA simulation to save computing source. Firstly, two-level submodeling and one-level submodeling methods were compared to investigate the effect of mesh density on FEA simulation results.

Table 4.2.5 Summary for four-point cyclic bend testing data.

Test conditions	Slope	Characteristic life	MTTF	Averaged life	Life range
Ni/Au-25_3.6	1.76	447	398	388	158–691
Ni/Au-25_2.16	2.62	2375	2110	2089	1103–3034
OSP-25_3.6	4.04	485	440	444	317–678
OSP-25_2.16	4.27	2458	2237	2249	1363–3426
OSP-25_1.5	2.31	14,097	12,489	12,434	6735–19331
OSP-25_1.25	4.16	69,058	62,739	62,914	43553–84640
Ni/Au-125_2.16	2.47	590	523	520	272–854
Ni/Au-125_1.5	3.99	1509	1368	1374	984–1920
Ni/Au-125_1.0	5.24	31,932	29,400	29,526	21372–36102
OSP-125_2.16	3.38	699	628	631	330–1098
OSP-125_1.5	2.93	1968	1756	1778	1076–3062
OSP-125_1.25	3.3	9185	8239	8331	5131–13895
OSP-125_1.0	3.15	32,949	29,488	29,455	14880–46175

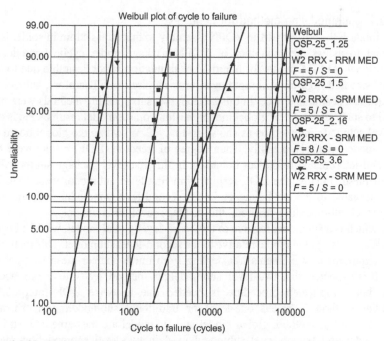

Figure 4.2.29 Weibull plot of cycle to failure for four-point bend test at 25°C.

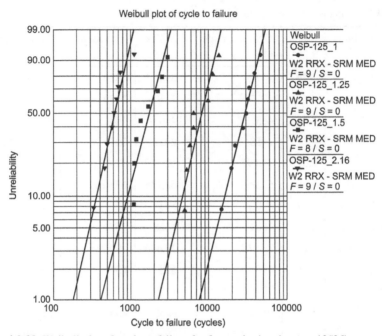

Figure 4.2.30 Weibull plot of cycle to failure for four-point bend test at 125°C.

4.2.2.2.1 Submodeling method for bend test

Solder joints is the critical part in VQFN assembly so that stress-strain in solder is very important to assess the reliability of VQFN package under cyclic bend loading. Conventional 3D FEA model for board-level simulation is not desirable due to larger element size and high-level computing source requirement. Therefore, submodeling technique is a desirable choice. In this section, two submodeling methods were implemented to study the effect of global model meshing on submodel results. One is called two-level submodeling method as shown in Fig. 4.2.32, where the global board-level quarter model was solved firstly. Then degree of freedom (DOF) results were transferred and interpolated to the cut boundary in the first-level package submodel with medium mesh. Finally, the DOF results from first-level submodel were transferred and interpolated to the cut boundary in the second-level submodel with fine mesh. In two-level submodeling method, the package-level FEA model is a transitional model, which is a submodel relative to the board-level global model, while it becomes the package-level global model relative to the second-level submodel. This method can satisfy requirement of elements transferring from much coarser mesh to much finer mesh. It is expected that two-level submodeling method leads to more accurate results, but complicated procedure is needed due to twice DOF interpolations. The other method is called one-level or traditional submodeling as shown in Fig. 4.2.33. In this method, global model and submodel are the same as board-level global model and second-level submodel used in two-level submodeling method, respectively. The DOF results from global model were transferred to final submodel

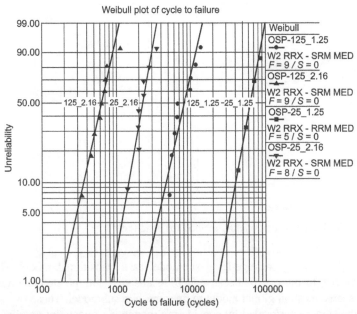

Figure 4.2.31 Comparison of cycle to failure between 25 and 125°C test condition assembly compared with room temperature (25°C) in four-point bend test with same displacement range because the solder fatigue resistance would be reduced and simultaneously solder joint deformation becomes larger at 125°C compared with 25°C. AF of cycle to failure due to high temperature effect (125°C) is around 4–7 times compared with room temperature case for cyclic bend test. All bending fatigue life data were summarized in Table 4.2.5. The cycle to failure increases significantly with decreasing displacement range for both bend tests at 25 and 125°C. The VQFN assemblies with OSP finish have slightly longer fatigue life than VQFN assemblies with Ni/Au finish for all test conditions. The MTTF is almost the same as the averaged fatigue life. Weibull slope is larger when life range is narrow.

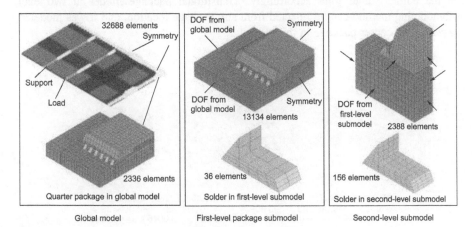

Figure 4.2.32 Two-level submodeling method for bend test.

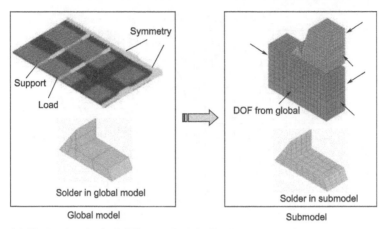

Figure 4.2.33 One-level submodeling method for bend test.

directly without using package-level submodel as a transitional model. However, it may give rise to more error because DOF results were interpolated and transferred from too coarse mesh in global model to finer mesh in submodel. These two submodeling methods were performed by considering monotonic four-point bending load to find a desirable method for following more simulation.

The displacement range from 0 to 6 mm was added on loading position of global model in 1 s using 6 load steps in FEA simulation. Viscoplastic Anand model was used for solder material [45]. Material properties refer to Table 4.2.2. Fig. 4.2.34 shows the global strain contour with direction normal to load and support anvil at the displacement of 6 mm. It is shown that all packages are subjected to the same strain level in four-point bending test. Strain value close to package corner is more than global strain due to localized stiffening effect of package. The outmost corner solder joints of VQFN package were prone to failure so that the submodel was created based on the corner solder joint accordingly. Transitional package model in two-level submodeling method can be selected from any package location, and it was modeled based on the center package in this study.

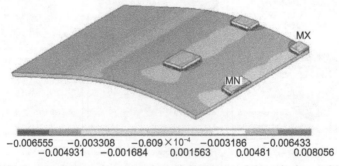

Figure 4.2.34 Longitudinal strain contour at a displacement of 6 mm in four-point bend test.

Figs. 4.2.35 and 4.2.36 show the volume-averaged Von Mises strain and plastic strain energy density for different models considering the whole solder joint as an averaging volume, respectively. It can be seen that one-level submodel and two-level submodel result in almost the same Von Mises strain and plastic strain energy density. Therefore, one-level submodeling method is an effective and sufficient FEA model for modeling and simulation of VQFN assembly subjected to bending load. The important advantage of one-level submodeling exists in saving time and computing source compared with two-level submodeling and providing more accurate results compared with coarse global model, especially for simulating cyclic bend test with more load steps.

4.2.2.2.2 Finite element model modeling for three-point cyclic bend test

One-level global-local modeling approach discussed above was used to simulate three-point bend test. Quarter model was used to model board-level assembly due to symmetry of geometry and load. According to test conditions, board-level global model was constrained at both symmetry planes and fixed in the out-of-plane direction at

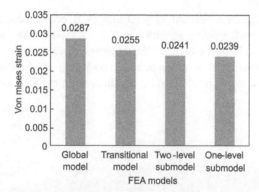

Figure 4.2.35 Comparison of solder joint strain among different finite element model (FEA) models.

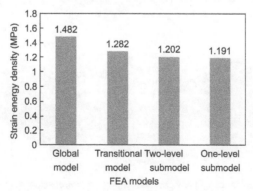

Figure 4.2.36 Comparison of solder joint plastic strain energy density among different finite element model (FEA) models.

the support edge. In three-point cyclic bend test, a displacement range from 1 to 5 mm was applied at the centerline of specimen and repeated with a frequency of 1 Hz without dwell time. For convenience, only four load steps with equal displacement increment were specified for both loading and unloading periods to facilitate the DOF transfer between the global model and submodel.

Fig. 4.2.37 shows strain contour with direction normal to support/load edge and energy density contour of solder joints at different components. The global strain distribution shows very strong location dependent and package localized stiffening effect. The global strain increases with joint location close to loading edge at the centerline of specimen, which verifies the fact that packages at different locations are subjected to different stress levels in three-point bend test. The stress levels for packages at the same row show similar value. Therefore, it is assumed that all packages in the same row are subjected to the same stress level and only one submodel is created for packages at the same row to simplify the simulation without loss of generalization. It can be seen from strain energy density contour that the critical solder joint is located at package corner. For package located at the left row, the critical solder joint locates at package corner and closes to loading edge. The submodel is created based on such a critical solder joint.

The submodel contains one solder joint and PCB board, copper pad, and mold compound in one pitch volume as shown in Fig.4.2.33.The solder joint was modeled with Anand's viscoplastic behavior. Displacement results from global model was interpolated and transferred to corresponding nodes on cut boundary in submodel step by step. The accurate results can be obtained from detailed submodel with fine

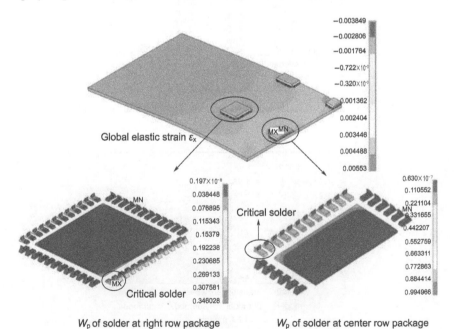

W_p of solder at right row package W_p of solder at center row package

Figure 4.2.37 Global simulation results for three-point bend test.

mesh based on the assumption that the influence of the submodel on the global model can be ignored. Usually, strain energy density is used as damage control parameter for solder joints. For Anand's model, unified plastic strain energy density can be obtained. Fig. 4.2.38 shows the solder joint mesh in submodel and plastic strain energy contour for elements at the top and bottom interfaces. The maximum energy density locates at the corner node of bottom interface or top interface due to stress concentration effect, which is exactly a crack initiation site. To reduce stress concentration effect, volume-averaged approach was used to calculate strain energy density based on elements at the top or bottom interfaces.

Generally, mechanical loading is much faster than TC loading with cycle time from several minutes to hours. Therefore, it takes only a few simulation cycles, for example, three cycles [63], to reach results convergence for thermomechanical loading. However, for mechanical loading with frequency around 1 Hz, it always takes more simulation cycles to obtain the converged results because the creep of solder material cannot fully develop during short cycle period. Usually, 10–15 cycles are needed to obtain converged results for bending load with 1 Hz frequency [56]. In this study, 15 cycles were simulated for both global and local models.

Figs. 4.2.39 and 4.2.40 show the accumulated strain energy density per cycle averaged on elements at the top or bottom interface from submodel results for the center row and left row, respectively. Strain energy density is larger for center row component case than for left row component case, which is consistent with test results where center row components fail faster than right/left row components. At high stress level for center row component, strain energy density at the bottom interface converges faster than at the top interface due to larger plastic deformation. At low stress level for left row component, strain energy density convergence is similar for bottom and top layers. After 10 cycles, volume-averaged strain energy densities are similar for bottom and top interfaces. FEA simulation by Shetty and Reinikainen [57] showed that strain energy density appeared to be nearly same on the top and bottom BGA solder interfaces.

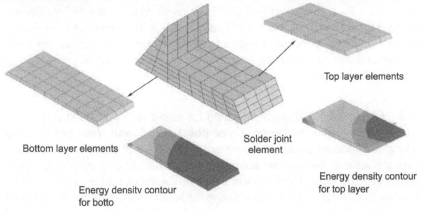

Figure 4.2.38 Solder joint mesh and energy density contour at bottom and top interfaces.

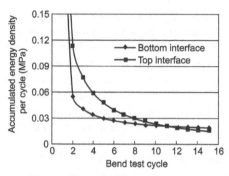

Figure 4.2.39 Convergence history of accumulated strain energy density per cycle for the critical solder joint located at center row component.

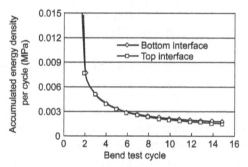

Figure 4.2.40 Convergence history of accumulated strain energy density per cycle for the critical solder joint located at left row component.

To explain strain energy convergence further, relative strain energy density change was calculated based on energy density difference between two consecutive cycles and plotted in Fig. 4.2.41. It was verified that FEA result convergence for mechanical loading is slower than that for thermomechanical loading. It is shown that 15 simulation cycles are needed to obtain change of energy density accumulation per cycle between two consecutive cycles less than 5%, which is considered as a convergence criterion in FEA simulation.

4.2.2.2.3 Finite element analysis modeling for four-point cyclic bend test
4.2.2.2.3.1 Correlation between three-point bend and four-point bend test The global-local modeling procedure and FEA models in four-point bend simulation are the same as those as in three-point bend simulation except displacement loading position and value. Displacement load was applied on the centerline of specimen in three-point bend while it was added at the loading span position in four-point bend. Four-point bend with displacement loading from 0.72 to 3.6 mm was simulated to correlate FEA results between four-point bend and center row components in three-point bend. Fig. 4.2.42 shows comparison of energy density accumulation per cycle

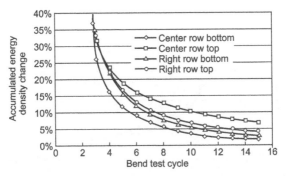

Figure 4.2.41 Change of accumulated strain energy density per cycle from finite element analysis (FEA) simulation for three-point bend test.

between three-point bend and four-point bend. It can be seen that energy density accumulation values appear to be similar for four-point bend and three-point bend after 10 bend test simulation cycles, which is consistent with four-point bend and three-point bend test results.

4.2.2.2.3.2 Finite element analysis modeling and result discussion on bend test at 25°C In four-point bend FEA simulation, it was shown that components between loading span are subjected to almost the same stress level, so global board-level simulation can be conducted just considering one center component. Global board-level quarter model with full VQFN components as shown in Fig. 4.2.33 has 32,688 elements, while global quarter model just with the center VQFN component (named U8 in Fig. 4.2.22) has 8195 elements. The simulation time used by global model with full VQFN is 10 times that used by global model with one VQFN. Fig. 4.2.43 shows the strain energy density accumulations of bottom interface layer from submodel results with DOF transferred from different global models. It was shown that energy density accumulations are nearly the same for different cases.

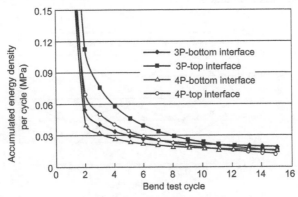

Figure 4.2.42 Finite element analysis (FEA) modeling results correlation between four-point and three-point bend.

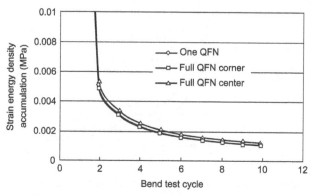

Figure 4.2.43 Comparison of strain energy density of bottom layer for different cases.

Different components on specimen result in almost the same energy density value, which verifies the assumption of different components subjected to the same stress level under four-point bend test. The stiffening effect of one component on another component can be neglected. Therefore, global model with one center VQFN is equivalent to global model with full VQFNs in terms of simulation results. Therefore, the global model with one center VQFN would be used in following FEA simulation to save solving time without loss of accuracy.

Fig. 4.2.44 shows the strain results from FEA simulation for four-point bend with displacement from 0.432 to 2.16 mm at 25°C. FEA simulation results are consistent with experimental results as shown in Fig. 4.2.26, which validates the modeling methodology and results.

Fig. 4.2.45 shows the plastic strain energy density accumulation convergence history of bottom interface layer for different four-point bending loads at 25°C. It was shown that all energy density accumulations reach a convergence after 15 simulation cycles with 5% difference between two consecutive cycles. The energy density values converge faster for larger displacement, which are consistent with results from three-point bend FEA simulation.

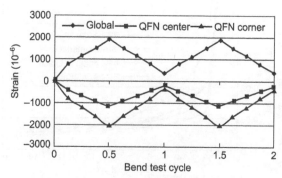

Figure 4.2.44 Strain simulation results of PCB board with displacement from 0.432 to 2.16 mm.

Stress and reliability analysis for interconnects

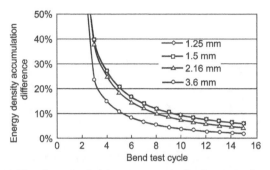

Figure 4.2.45 Accumulated energy density convergence for four-point bend test simulation at 25°C.

Table 4.2.6 lists energy density accumulations at 15th simulation cycle for different displacement loads. It can be seen that volume-averaged energy density accumulation value on bottom interface layer is larger than that on top interface layer. The energy density increases significantly with increasing displacement load. The relationship between displacement and energy density accumulation can be expressed by a power law equation as shown in Fig. 4.2.46. Therefore, once one displacement load and its corresponding energy density accumulation are known, energy density accumulation can be estimated for another displacement load using following equation:

$$\frac{W_{kn}}{W_{un}} = 0.773 \left(\frac{D_{kn}}{D}\right)^{4.581} \quad (4.2.15)$$

where D_{kn} and W_{kn} are known displacement and corresponding energy density accumulation, respectively. Unknown W_{un} can be obtained when D is known.

4.2.2.2.3.3 Finite element analysis modeling and result discussion on bend test at 125°C
FEA modeling and simulation for four-point bend at high temperature (125°C) were also conducted. The geometry of FEA model for high temperature bend test is the same as that for room temperature bend test. The difference is that some

Table 4.2.6 Summary of results for four-point bend test finite element analysis (FEA) simulations at 25°C.

Maximum displacement	1.25 mm	1.5 mm	2.16 mm	3.6 mm
Energy density at bottom layer (MPa)	1.27×10^{-4}	2.10×10^{-4}	8.08×10^{-4}	1.56×10^{-2}
Energy density at top layer (MPa)	8.53×10^{-5}	1.76×10^{-4}	7.75×10^{-4}	1.25×10^{-2}
Energy density factor (bottom/top)	1.49	1.19	1.04	1.25
Relative energy density (bottom)	1	1.65	6.34	122.64
Relative displacement	1	1.20	1.73	2.88

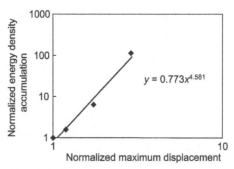

Figure 4.2.46 Relationship of displacement and accumulated energy density for bend test at 25°C.

material properties used in high temperature bend test simulation are temperature-dependent like those used in TC simulation (Table 4.2.3). Before applying mechanical bending load, temperature ramp up from 25 to 125°C was simulated with time period of 10 min, which was consistent with physical bend test condition. Then cyclic bending load was simulated with frequency of 1 Hz and at dwell temperature of 125°C. Viscoplastic Anand model was used for solder material. Fifteen bending cycles were simulated to obtain converged results. It was found that accumulated energy density on top interface layer is slightly higher than that on bottom interface layer due to high residual stress effect close to top solder interface after temperature ramp up to high dwell temperature, while higher volume-averaged energy density accumulation occurs on bottom interface layer at room temperature bend test. Fig. 4.2.47 shows the energy density accumulation convergence history based on top layer volume averaging for different four-point bend tests at 125°C. It is similar to room temperature bend test simulation that convergence of 5% difference between two consecutive cycles can be obtained after 15 simulation cycles for bend test at high temperature.

Table 4.2.7 lists converged energy density accumulation results for different displacement loads at 125°C. It can be seen that volume-averaged energy density accumulation value on top layer is nearly twice that on bottom layer. The relationship

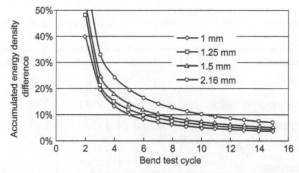

Figure 4.2.47 Energy density convergence for four-point bend test simulation at 125°C.

Table 4.2.7 Summary of results for four-point bend test finite element analysis (FEA) simulations at 125°C.

Maximum displacement	1 mm	1.25 mm	1.5 mm	2.16 mm
Energy density at bottom layer (MPa)	4.45×10^{-4}	8.31×10^{-4}	1.41×10^{-3}	3.52×10^{-3}
Energy density at top layer (MPa)	8.25×10^{-4}	1.60×10^{-3}	2.80×10^{-3}	6.45×10^{-3}
Energy density factor (top/bottom)	1.85	1.92	1.99	1.83
Relative energy density (top)	1	1.94	3.40	7.82
Relative displacement	1	1.25	1.5	2.16

between displacement and energy density accumulation as shown in Fig. 4.2.48 can be fitted by a power law curve. Energy density accumulation in four-point bend at 125°C can be calculated by following equation (method: refer to Eq. (4.2.15) for bend test at 25°C):

$$\frac{W_{kn}}{W_{un}} = 1.054 \left(\frac{D_{kn}}{D}\right)^{2.669} \tag{4.2.16}$$

Fig. 4.2.49 shows the comparison between FEA simulation results for bending test at 25 and 125°C with the same displacement load of 2.16 mm. Accumulated energy density per cycle at high temperature is more than that at room temperature significantly, which indicates that high temperature accelerates bending fatigue of solder joints. The simulation results are consistent with experimental data.

4.2.2.3 Fatigue model development for cyclic bend test

Some literatures showed fatigue life curves with fatigue life as a function of board strain, or even board deflection and reaction load. These fatigue models cannot be extended when board or package parameters change. In fact, the solder fatigue model

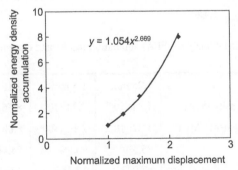

Figure 4.2.48 Relationship of displacement and accumulated energy density for bend test at 125°C.

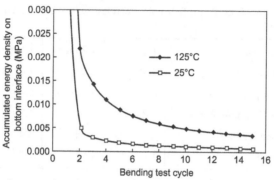

Figure 4.2.49 Result comparison between bending test simulation at 25°C and at 125°C.

should correlate fatigue life to damage parameter extracted from solder material. Skipor and Leicht [64] developed a bending fatigue model for PBGA Sn/Pb solder joints using inelastic strain as a damage parameter. Mercado et al. [58] proposed a total energy-based bending fatigue model for CSP with eutectic Sn/Pb solder. All bending fatigue models were developed at room temperature from literatures. Based on limited literature, bending fatigue model for Sn-Ag-Cu solder was not documented. In this section, bending fatigue models were presented for VQFN package with Sn-Ag-Cu solder at both test conditions of 25 and 125°C.

Strain energy density was selected as a damage parameter because it is a comprehensive factor considering both stress and strain. The strain energy density results shown in above section are all plastic strain energy density from Anand's model. The total energy density accumulation on the bottom interface layer was also calculated and listed in Table 4.2.8 for FEA simulation of bend test at 25°C. It can be seen that plastic energy density is dominant. For high temperature case, the ratio of plastic energy density to total energy density is more than 99% from FEA simulation results.

Fig. 4.2.50 shows the relationship between MTTF from experimental data and plastic strain energy density from FEA simulation for bend test at 25°C. The power law function can fit data well. Bending fatigue life model at 25°C can be established by a power law function below:

$$N_f = 6.025(\Delta W_P)^{-0.946} \tag{4.2.17}$$

Table 4.2.8 Finite element analysis (FEA) simulation results and experimental data for four-point bend at 25°C.

Displacement	1.25 mm	1.5 mm	2.16 mm	3.6 mm
Total strain energy density (MPa)	1.34×10^{-4}	2.24×10^{-4}	8.56×10^{-4}	1.56×10^{-2}
Plastic strain energy density (MPa)	1.27×10^{-4}	2.10×10^{-4}	8.08×10^{-4}	1.55×10^{-2}
Ratio (plastic/total)	95.3%	93.8%	94.4%	99.7%
MTTF (cycles)	67,073	12,489	2237	440

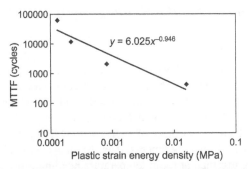

Figure 4.2.50 Bending fatigue model of Sn-Ag-Cu solder for bend test at 25°C.

$$N_f = 5.725(\Delta W_T)^{-0.959} \tag{4.2.18}$$

where ΔW_P and ΔW_T are plastic strain energy density and total strain energy density, respectively. The power law exponents are -0.946 and -0.959 for plastic strain energy-based and total strain energy-based fatigue models, respectively. These exponents are quite close due to plastic deformation dominant.

Fig. 4.2.51 shows the power law relationship between MTTF and plastic strain energy density for high temperature bend test. As plastic strain energy density is as high as 99% total strain energy density, plastic strain energy-based fatigue model is very close to total strain energy-based fatigue model. Therefore, bending fatigue life model for bend test at 125°C was established according to Eq.(4.1.50) as below:

$$N_f = 0.032(\Delta W_P)^{-1.922} \tag{4.2.19}$$

where ΔW_P is plastic strain energy density. It was found that the power law exponent of bending fatigue model at 125°C is about twice that of bending fatigue model at 25°C, which indicates that AF of cycle to failure is larger at high temperature bend test than at room temperature bend test.

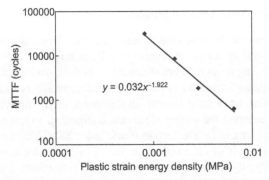

Figure 4.2.51 Bending fatigue model of Sn-Ag-Cu solder for bend test at 125°C.

4.2.2.4 Summary on bend test and analysis

Cyclic four-point and three-point bend tests were performed for reliability assessment of VQFN assembly with Sn–Ag–Cu solder joints. The correlation between three-point bend and four-point bend was demonstrated and validated by experimental and simulation results. Four-point bend tests were conducted at 25°C and at 125°C for VQFN with Ni/Au or OSP surface finish, respectively. Failure data satisfy Weibull distribution very well. Cycle to failure increases significantly with decreasing bending displacement range. Fatigue resistance of VQFN with OSP finish is slightly better than that of Ni/Au finish case from experimental data. AF of cycle to failure is larger at high temperature bending test than at room temperature bending test.

FEA modeling and simulation were performed for different testing conditions. Before cyclic bend simulation, monotonic bending load was simulated using one-level and two-level submodeling methods. FEA simulation results show that one-level submodeling method is an effective and sufficient method. Fifteen simulation cycles are needed for cyclic bending test to obtain converged results. Simulation results show the location-dependent global strain for three-point bend and location-independent global strain for four-point bend. Strain results from FEA simulation are consistent with experimental data, and localized stiffening effect due to package is also observed in four-point bend. Volume-averaged strain energy density accumulation on the bottom interface is slightly higher than that at the top interface for bend test simulation at 25°C, while it is higher at the top interface than the bottom interface for bend test simulation at 125°C. The relationship between displacement and energy density accumulation fits power law curve for both bend tests at 25 and 125°C Accumulated strain energy density per cycle at 125°C is more than that at 25°C significantly, which indicates that higher temperature accelerates bending fatigue of solder joints.

Energy-based fatigue models were developed for Sn–Ag–Cu solder subjected to cyclic bending load at 25 and 125°C. It was found that the power law exponent of bending fatigue model at 125°C is about twice the exponent of bending fatigue model at 25°C, which indicates that AF of cycle to failure is larger for bending fatigue at 125°C than bending fatigue at 25°C.

4.2.3 Vibration test and analysis

Electronic equipment can be subjected to many different forms of vibration over wide frequency ranges and acceleration levels [65]. Vibration, in a broad sense, is considered as an oscillating motion where structure or body moves back and forth. The simplest form of periodic motion is harmonic motion, which is usually represented by a continuous sine wave on a plot of displacement versus time, and this type of vibration is often selected for testing electronic equipment. Vibration-induced stress can usually lead to fatigue failure for electronic assemblies. There are three steps in vibration fatigue analysis of electronic assembly. Firstly, modal analysis of the PCB assembly; secondly, dynamic analysis of the PCB assembly; and thirdly, stress analysis of the individual soldered component.

For vibration analysis, vibration mode and natural frequency of the vibrating body must be determined first through experimental and/or numerical methods. The first harmonic mode often has the greatest displacement amplitude and usually the greatest displacement induced stresses. Vibration-induced stresses usually lead to fatigue failure for electronic assemblies. For surface mount microelectronic components, an approximation of modal analysis for PCB assembly can be made by assuming PCB as a bare unpopulated thin plate because the increase in stiffness of PCB due to the mounting of the components is approximately offset by the increase in total mass of the populated PCB [66]. However, this approximation may lead to errors in natural frequency prediction for different packages such as FCOB and PBGA assemblies [67,68]. When the component has small profile, the approximation of PCB assembly as a bare PCB can provide satisfactory modal analysis results because the stiffness and mass contribution of small component to PCB assembly is not significant.

The second step in vibration fatigue life calculation is dynamic analysis where the response of the PCB is calculated under vibrating load. The first few mode shapes are used as they represent the largest deflections and curvatures of the PCB and will account for the majority of fatigue damage. The dynamic analysis provides information about the PCB deflection at each component location and for each mode shape. The vibration analyses by Basaran and Chandaroy [69,70] and Zhao et al. [71] showed that solder joints respond elastically mainly at room temperature vibration loading. So, vibration fatigue failure of solder joints is often assessed for reliability using a high cycle fatigue life model, which is represented by an S-N (stress-life) curve. Then FEA modeling and simulation are used to determine the stress of the solder joint subjected to vibration load. Because of the complication of the geometry of electronic assembly, global-local or submodeling method [72,73] has been used for the board-level vibration test simulation. In this section, the constant g-level and block g-level vibration tests for FCOB assembly were conducted. Four different FEA models were investigated for modal analysis to calculate the first-order natural frequency of the FCOB assembly. A quasi-static analysis approach was conducted for the FCOB assembly to evaluate the stress—strain behavior of the solder joints. A harmonic analysis was also conducted to investigate the dynamic response of the FCOB assembly subjected to vibration load. Fatigue life prediction results from quasi-static analysis and harmonic analysis approaches were compared with experimental results.

4.2.3.1 Vibration test approach

4.2.3.1.1 Test vehicle and setup

FCOB assembly was selected as test vehicle in vibration fatigue tests to analyze the dynamic response of FCOB assembly and solder joint fatigue subjected to sinusoidal vibration loading. The specimen and chip number are shown in Fig. 4.2.52. Six larger flip-chip modules with 8.5 mm × 8.5 mm × 0.65 mm silicon die numbered from 1 to 6 and six smaller flip-chip modules with 3.5 mm × 3.5 mm × 0.65 mm silicon die numbered from 7 to 12 were mounted on the FR4 PCB with 185 mm × 150 mm × 1.13 mm in size. The solder joints were eutectic 63Sn/37 Pb solder with a diameter of 0.16 mm, ball pitch of 0.35 mm, and standoff height of 0.1 mm. Capillary underfill

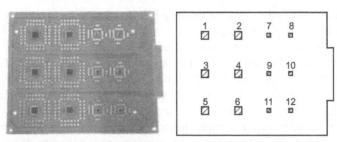

Figure 4.2.52 Flip-chip-on-board (FCOB) specimen and chip numbering.

is used between silicon die and PCB board through underfilling process. FCOB specimens were tested at different acceleration levels to assess the solder joint reliability subjected to constant g-level (g is the gravitational acceleration) or varying g-level (also called block g-level) vibration loads of 3-g, 5-g, and 10-g, respectively.

Vibration tests were conducted by using an electrodynamic shaker and vibration control test facility. Fig. 4.2.53 shows the vibration test setup. Two accelerometers were used to determine the dynamic response of the FCOB specimen with one on fixture and the other on the PCB board; thus, the vibration transmissibility of the PCB can be determined. The FCOB specimen was clamped along two longer edges

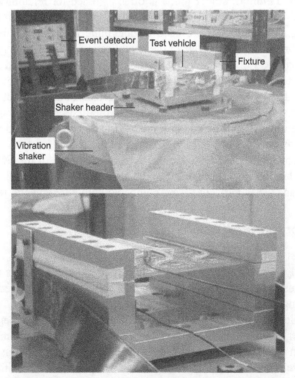

Figure 4.2.53 Vibration test setup and accelerometers mounted on fixture and specimen.

using an aluminum fixture, which was bolted to a shaker header. The daisy-chain loop was designed for each flip-chip module to detect solder joint failure by measuring the resistance change of the loop. Daisy-chain loops were monitored simultaneously by an event detector during vibration test. Any resistance change exceeding a preset threshold with minimum duration of 0.1 μs can be detected by the event detector. When a crack is initiated in the solder joint during the vibration test, the resistance will increase with increasing fatigue damage. Different failure criteria have been used by different researchers. For example, fatigue failure was defined as 10% increase in resistance for PBGA package under dynamic vibration test by Kim et al. [74]. The failure criterion was set as 20% increase in daisy-chain resistance reported by Stepniak [75] for flip-chip joint and for PBGA package with eutectic solder under sinusoidal vibration test with clamped condition by Chen et al. [76]. Wong et al. [77] used 100% increase in resistance as failure criteria for PBGA package under sinusoidal vibration test. Perkins et al. [78] used a resistance value over 100Ω as the failure criteria for ceramic column grid array (CCGA) under linear sweep vibration test. The failure criteria recommended by IPC standard [79] is defined as an increase of the daisy-chain resistance by at least 20% of the initial value for six or more consecutive occurrences. In this study, 50% increase in daisy-chain resistance was defined as the failure criterion, which meets the requirements by the IPC standard.

4.2.3.1.2 Frequency-scanning test

The frequency-scanning test was conducted with sweep frequency from 20 to 1000 Hz to determine the fundamental resonance frequency of the specimen. The transmissibility of the specimen can be estimated using the following equation:

$$\text{TR} = \frac{G_{\text{out}}}{G_{\text{in}}} \quad (4.2.20)$$

where G_{out} is the maximum acceleration measured at the center of the specimen, and G_{in} is the acceleration amplitude of the sinusoidal excitation, which is measured at the fixture. The fundamental resonance frequency of 195 Hz was determined by scanning test with 0.5-g level input.

The screening test was conducted at 0.5, 1, 2, 3, 4, 5, 6, 7, 8, 10, 12, 14, 16, 18, and 20-g levels with the same frequency range from 20 to 1000 Hz. According to Meirovitch [80], for a linear vibration system, its transmissibility should remain constant regardless of any input change. However, the varying transmissibility, TR, at different acceleration input shows nonlinear relationship as shown in Fig. 4.2.54. Therefore, the FCOB assembly under the out-of-plane sinusoidal vibration is not a linear system, similar to the results reported by Yang et al. [81]. The transmissibility is also location and test frequency dependent.

For a vibration system subjected to a harmonic excitation, its displacement response at any location can be expressed as

$$D = D_{\text{max}} \sin(\Omega t + \phi) \quad (4.2.21)$$

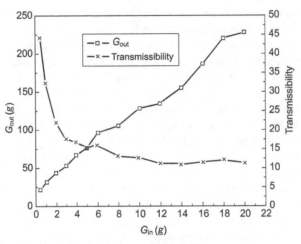

Figure 4.2.54 Nonlinear relationship between G_{in} and G_{out}.

where D_{max} is the maximum displacement and Ω is the natural frequency in radians per second, which can be changed into natural frequency in Hertz by

$$\Omega = 2\pi f \tag{4.2.22}$$

Differentiating Eq. (4.2.21) twice with respect to time, the acceleration can be obtained as

$$\ddot{D} = -D_{max}\Omega^2 \sin(\Omega t + \phi) \tag{4.2.23}$$

Hence, maximum acceleration of $\ddot{D}_{max}$ can be expressed as

$$\ddot{D}_{max} = D_{max}\Omega^2 \tag{4.2.24}$$

And the maximum displacement can be determined as

$$D_{max} = \frac{\ddot{D}_{max}}{\Omega^2} = \frac{9.8 G_{in} TR}{4\pi^2 f^2} = \frac{0.248 G_{in} TR}{f^2} \tag{4.2.25}$$

where G_{in} has a unit of the gravity acceleration, and D_{max} in meter. According to Eq. (4.2.25), the maximum displacements for different g-level tests can be estimated.

4.2.3.1.3 Sweep vibration reliability test

The fundamental natural frequency was obtained from frequency-scanning test, and then reliability test can be conducted based on the known natural frequency. In reliability vibration test, the sweep frequency range was 10% around natural frequency of 195 Hz, that is, from 175 to 215 Hz. The sine sweep vibration swept from low frequency to high frequency and then from high frequency to low frequency and repeated until a preset time was reached. For the varying g-level vibration test, 3-g, 5-g, and 10-g level vibration tests were conducted in turn for the same specimen and 5 h was used for each acceleration level. Constant g-level vibration test was conducted at 3-g for 200 h, 5-g for 100 h, and 10-g for 36 h on different FCOB specimens, respectively. The cycles to failure were calculated based on duration time recorded by the event detector and the averaged frequency of the sweep sinusoidal excitation. No failure results were observed for the small chip components numbered 7–12. The test results of the large components (chips 3 and 4) satisfy two-parameter Weibull distribution well as shown in Fig. 4.2.55. Fig. 4.2.55 also shows the Weibull plot of vibration test results for PBGA assembly from Yang's work [81], where the PCB has the same length and width as that used in this study while the clamped-clamped boundary condition at two shorter edges was used.

Table 4.2.9 lists the Weibull parameters together with the MTTF and first-time-to-failure (FTTF) for the FCOB and PBGA assemblies. It can be seen that the fatigue life of flip-chip solder joint reduces rapidly with increasing acceleration input. The Weibull slope is similar for PBGA and FCOB assemblies. However, the fatigue life of PBGA solder joint is much less than that of flip-chip solder joints even under low

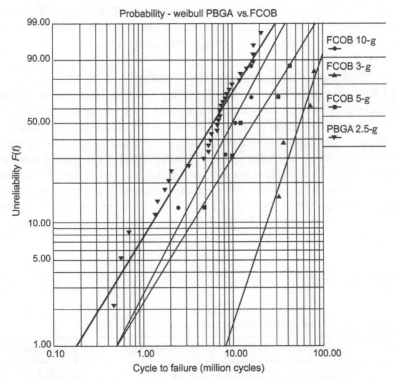

Figure 4.2.55 Weibull plots of test results for flip-chip-on-board (FCOB) and plastic ball grid array (PBGA) assembly.

Table 4.2.9 Vibration test fatigue data of flip-chip-on-board (FCOB) and plastic ball grid array (PBGA) packages.

G_{in}	$\eta\,(10^6)$	β	MTTF(10^6)	FTTF(10^6)
FCOB 3-g	65.30	2.22	57.84	32.30
FCOB 5-g	23.28	1.20	21.91	4.93
FCOB 10-g	12.89	1.41	11.73	2.47
PBGA 2.5-g	8.31	1.20	7.82	0.46

g-level vibration test. This is because components in FCOB assembly have less mass and dimension than those in PBGA assembly. Furthermore, underfill is used in the FCOB assembly, which can increase the fatigue life of flip-chip solder joints when subjected to vibration load. The failure analysis from SEM image showed the solder joint fatigue cracking failure mode.

The high-speed camera with recording rate up to 4500 frames/s was used to capture the dynamic response of the specimen. Displacement was measured at the center of the PCB free edge. The relative displacement of free edge can be calculated using a mark at the fixture as a reference point. The maximum theoretical displacement can be calculated by Eq. (4.2.24). Table 4.2.10 gives comparison of the displacement results between high-speed camera and theoretical data. It can be seen that high-speed camera can provide accurate result without contact effect on specimen.

4.2.3.1.4 Cumulative damage index fatigue analysis

For the electronic assembly subjected to varying g-level vibration tests, the solder fatigue damage due to each acceleration level (stress level) can be superimposed using a linear superposition method of Miner's law [82] if solder joint stress is in the elastic stage. Miner's cumulative fatigue damage ratio is calculated by summing up the ratio of the actual number of fatigue cycles (n) accumulated in a specific element, in different environments, divided by the number of fatigue cycles (N) required to produce a fatigue failure in the same specific element in the same environment. When the ratios are added together, a sum of 1.0 or greater means that the fatigue life has been used up and should fail. CDI using Miner's cumulative damage law can be given by

$$\text{CDI} = D_{\text{total}} = \sum_{i=1}^{n} \frac{n_i}{N_i} \qquad (4.2.26)$$

Table 4.2.10 Displacement results comparison.

G_{in} (g)	Eq. (4.2.24) (mm)	Camera (mm)	Difference (%)
3	1.07	0.96	−10.3
10	2.18	2.36	8.3

Eq. (4.2.26) was used to assess the reliability of solder joints subjected to varying loading conditions if the solder fatigue life under constant loading condition and duration under different loading condition were known.

The Miner's CDI is widely used to estimate component life under different loading conditions. Perkins et al. [83] applied Miner's law for life prediction of CCGA package subjected to sinusoidal sweep vibration test, and the MTTF predicted using linear Miner's law was validated by experimental data considering CDI as unity (1.0). Chen et al. [76] applied Miner's law for PBGA package subjected to sinusoidal vibration loads, and the calculated CDIs were happened to be roughly equal to 1 (0.9–1.07). Wong et al. [77] calculated CDI values for PBGA package under different random vibration conditions and observed that the variation of CDI is significant from 0.25 to 3.39. When the solder stress is in the elastic range, the Miner's law can be used to predict the solder fatigue life in the high cycle fatigue life regime. For higher g-levels or plastic deformation loading, the Miner's rule can significantly underestimate the cumulative damage result as reported by Basaran et al. [84] for a package subjected coupled TC and dynamic vibration loads.

In this study, 3-g, 5-g, and 10-g input vibration reliability tests were conducted for the FCOB assemblies. From finite element simulation results under these three test conditions, no plastic deformation was observed in the solder material. Hence, the linear Miner's law can be used to predict the solder fatigue life. For the varying g-level test, chips 3 and 4 were selected for the CDI analysis. The values of n_i were obtained from the test result as shown in Fig. 4.2.56 for chip 3 and chip 4, while N_i are MTTF in Table 4.2.9. Substituting the data as shown in Fig. 4.2.56 into Eq. (4.2.26),

$$\text{CDI}_{\text{chip3}} = \frac{n_1}{N_1} + \frac{n_2}{N_2} + \frac{n_3}{N_3} = \frac{3.42}{57.84} + \frac{3.42}{21.91} + \frac{2.76}{11.73} = 0.451$$

(4.2.27)

$$\text{CDI}_{\text{chip4}} = \frac{n_1}{N_1} + \frac{n_2}{N_2} + \frac{n_3}{N_3} = \frac{6.84}{57.84} + \frac{6.84}{21.91} + \frac{3.65}{11.73} = 0.742$$

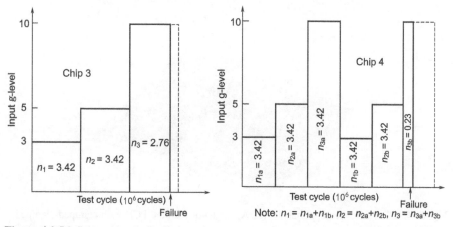

Figure 4.2.56 Schematic of vibration cycles at varying g-level tests for chip 3 and chip 4.

When the CDI value reached 0.451 and 0.742 for chip 3 and chip 4, respectively, failure should occur for FCOB solder joint. Therefore, the CDI of unity is nonconservative for FCOB solder joint when subjected to varying g-level vibration test. Based on the vibration test data in this work, a safety factor of 2–3, which relates to the reciprocal of the CDI value, is recommended for FCOB solder joint fatigue failure evaluation when subjected to the varying sinusoidal vibration test.

4.2.3.2 Finite element analysis for vibration test

4.2.3.2.1 Modal analysis of flip-chip-on-board assembly

The PCB was modeled with shell elements in the modal analysis. The clamped-clamped boundary condition was applied according to test condition. Firstly, the bare PCB without mounting flip-chip components was simulated to study the effect of different variables such as material properties and element size. The FEA model of the bare PCB is shown in Fig. 4.2.57. Table 4.2.11 shows the natural frequency results of the first three modes for the bare PCB model. It can be seen that the element size and considering PCB with orthotropic material properties have slight effect on the natural frequency. In the subsequent FEA modeling, the 5 mm × 5 mm element size and PCB with orthotropic material properties were used. Fig. 4.2.58 shows the first six mode shapes for the bare PCB. It is obvious that the mode shape becomes more complicated when the order of mode increases. Generally, the first order mode has the most significant effect on the reliability of the electronic assembly.

To obtain accurate modal analysis result for FCOB assembly, four FEA models for modal analyses were implemented. In model 1, the flip-chip components were modeled as distributed masses on the PCB. In model 2, the flip-chip components were modeled as concentrated masses at the center locations of the corresponding components on the PCB. In model 3, the components were modeled as a solid part of flip-chip package without considering the effects of the solder joint and underfill. In model 4, the integrated circuit (IC) chips were modeled with shell elements and the solder joints were modeled as the effective two-node beam elements with equivalent stiffness

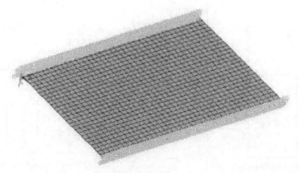

Figure 4.2.57 Finite element analysis (FEA) model of the bare PCB with clamped-clamped boundary condition.

Table 4.2.11 Natural frequencies of the bare PCB from modal simulation results.

No.	Modulus, E (GPa)	Poisson's ratio, ν	f_1 (Hz)	f_2 (Hz)	f_3 (Hz)	Element size (mm × mm)
1	22	0.28	208.75	232.86	324.0	2.5 × 2.5
2	22	0.28	208.61	232.51	323.17	5 × 5
3	22	0.28	208.59	232.44	322.9	7.5 × 7.5
4	22	0.28	208.57	232.34	322.51	10 × 10
5	22	0.28	208.5	232.07	321.48	15 × 15
6	$E_x = 22$ $E_y = 22$ $E_z = 10$	$\nu_{xy} = 0.28$ $\nu_{yz} = 0.11$ $\nu_{zx} = 0.11$	208.51	232.36	322.88	5 × 5

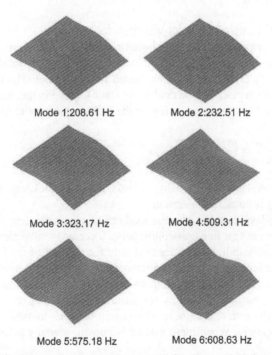

Mode 1: 208.61 Hz   Mode 2: 232.51 Hz
Mode 3: 323.17 Hz   Mode 4: 509.31 Hz
Mode 5: 575.18 Hz   Mode 6: 608.63 Hz

Figure 4.2.58 Mode shape and corresponding natural frequency of the bare PCB.

[85]. The natural frequency of the bare PCB with the clamped-clamped boundary condition also can be obtained from the following equation [65]:

$$f_n = \frac{3.55}{L^2}\sqrt{\frac{D}{\rho}} \qquad (4.2.28)$$

where $D = \frac{EH^3}{12(1-\nu^2)}$, $\rho = \frac{Mass}{Area}$, $E = 22$ GPa, (Young's modulus), $\nu = 0.28$ (Poisson's ratio), $H = 1.13$ mm, (thickness of the PCB), $L = 140$ mm (free edge length of the PCB), and for the shell model $\rho = 2.147$ kg/m^2 = 2.147×10^{-9} N·s^2/mm^3.

Table 4.2.12 shows the first-order natural frequencies of the PCB obtained from different FEA models. The frequency obtained from the bare PCB modal analysis has a good agreement with theoretical result. The result of modal analysis using global-local-beam model agrees with the test result. Therefore, model 4 is an optimal simplified method among four FEA models for the modal analysis of FCOB assembly. From modal simulation results, approximation of FCOB assembly modal response can be made by assuming the FCOB assembly as a bare unpopulated PCB to obtain the approximate natural frequency result quickly. However, the same approximation may introduce more error for the PCB with large PBGA package [68] because the contribution of mass and stiffness of the PBGA package for the PCB assembly cannot be ignored due to the PBGA package with larger volume and mass compared with small flip-chip package.

4.2.3.2.2 Quasi-static analysis for vibration fatigue

4.2.3.2.2.1 Determination of quasi-static load A quasi-static analysis method was developed to evaluate the stress–strain behavior of solder joints, which can be used for fatigue life prediction by stress-based fatigue model. The dynamic loading due to vibration was replaced by an equivalent static loading in this method. According to Newton's second law, pressure acting on the PCB or component can be obtained by

$$p = \frac{F}{A} = \frac{mG_{out}}{A} = \frac{\rho v G_{out}}{A} = \rho t G_{out} g \qquad (4.2.29)$$

where m, ρ, v, A, and t are mass, density, volume, area, and thickness of PCB assembly, respectively. G_{out} is output acceleration.

For constant g-level test, the pressure load can be obtained when transmissibility is known. It is observed that transmissibility is not a constant value along transverse locations. The transmissibility was measured only for the half of the PCB due to symmetry. Eight locations uniformly distributed along the PCB transverse direction as shown in Fig. 4.2.59 were selected to measure transmissibility of the PCB. Fig. 4.2.60 shows the transmissibility for different locations at natural frequency for 10-g input vibration test, and a linear relationship between transmissibility and location is observed clearly. When input g-level is known, output g at different locations

Table 4.2.12 Comparison of the natural frequency for flip-chip-on-board (FCOB) assembly from different sources.

Methods	Bare PCB	Eq. (4.2.28)	Model 1	Model 2	Model 3	Model 4	Test
Frequency (Hz)	208.5	209.4	206.8	206.8	209.6	201.9	195

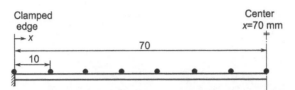

Figure 4.2.59 Schematic of location on the PCB board for transmissibility measurement.

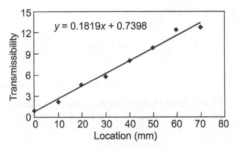

Figure 4.2.60 Transmissibility versus location at natural frequency from vibration test (10-g input).

can be obtained based on transmissibility in Fig. 4.2.60, and the pressure loading can be calculated by Eq. (4.2.29).

4.2.3.2.2.2 Determination of stress amplitude A submodeling method was implemented for quasi-static simulation. Fig. 4.2.61 shows the schematic for submodeling. For the whole global model, only two large chips 3 and 4 located at PCB transverse direction were modeled. Elastic-plastic material model for solder was used. Table 4.2.13 lists material properties used in the FEA simulation. No plastic deformation occurs for solder material from simulation result, so the Miner's law can be implemented for solder fatigue life analysis. The stress—strain distribution for two chips was almost the same from simulation results, so only chip 3 was used for further analysis. The corner solder joint has the maximum stress, which means the first failure will occur at the corner solder joint. Thus, submodel was created based on the corner solder joint. For multiaxes stress situation, the equivalent Von Mises stress is commonly used:

$$\sigma_{\text{eff}} = \frac{1}{\sqrt{2}} \sqrt{(\sigma_1 - \sigma_2)^2 + (\sigma_2 - \sigma_3)^2 + (\sigma_3 - \sigma_1)^2} \quad (4.2.30)$$

To minimize the effect of stress concentration, the volume-weight method was used to calculate the Von Mises stress at the die/solder interface layer as shown in Fig. 4.2.61. The volume-weight averaged Von Mises stresses at natural frequency were calculated for die/solder interface layer for different g-level tests as shown in Fig. 4.2.62. It can be seen that the stress amplitude is proportional to G_{out} of the PCB center for different g-level vibration tests.

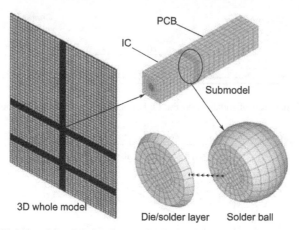

Figure 4.2.61 Global and local finite element analysis (FEA) models for flip-chip-on-board (FCOB) assembly.

Table 4.2.13 Material properties used in finite element analysis (FEA) simulation for flip-chip-on-board (FCOB) assembly.

Materials	Solder joint	FR4 PCB	Silicon die	Underfill	Copper
Young modulus (GPa)	30.67	22(x, y), 9.8(z)	131	5.6	120
Shear modulus (GPa)	–	3.5(xy), 2.5(xz, yz)	–	–	–
Poisson's ratio	0.35	0.11(xy), 0.28(xz, yz)	0.3	0.3	0.35
ρ (kg/m^3)	8500	1900	2300	1700	8900

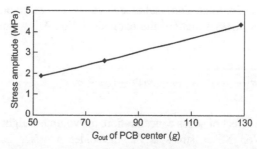

Figure 4.2.62 Plot of G_{out} versus Von Mises stress amplitude of solder joint at natural frequency.

The frequency range from 175 to 215 Hz was applied in vibration reliability test. Different transmissibility of the PCB center occurs at different frequencies as shown in Fig. 4.2.63. Therefore, specimen was subjected to different effective pressures during test, thus different stress amplitudes will be applied on the solder joint during vibration test with frequencies from 175 to 215 Hz. The stress amplitude can be determined using linear relationship as shown in Fig. 4.2.62 when the G_{out} or transmissibility of the PCB center is known. The stress amplitude (σ_a) can be calculated for three different g-level tests (10-g, 5-g, and 3-g) using the following equations:

$$\sigma_a = \frac{4.35 \times G_{out}}{128.5} = \frac{4.35 \times 10 \times T}{128.5} \text{ for } 10\text{-}g \text{ (in MPa)}$$

$$\sigma_a = \frac{2.63 \times G_{out}}{77.25} = \frac{2.63 \times 5 \times T}{77.25} \text{ for } 5\text{-}g \text{ (in MPa)} \quad (4.2.31)$$

$$\sigma_a = \frac{1.87 \times G_{out}}{53.22} = \frac{1.87 \times 3 \times T}{53.22} \text{ for } 3\text{-}g \text{ (in MPa)}$$

The G_{out} (in g) at the center of the PCB are 128.50, 77.25, and 53.22 for vibration test with G_{in} of 10-g, 5-g, and 3-g, respectively. The corresponding critical solder joint stress amplitude at natural frequency when subjected to 10-g, 5-g, and 3-g vibration tests are 4.35, 2.63, and 1.87 MPa (refer to Fig. 4.2.62), respectively.

4.2.3.2.2.3 Vibration fatigue life prediction The stress-based high cycle fatigue model listed in Eq. (4.1.34) was used to predict flip-chip solder joint fatigue life. For eutectic Sn/Pb solder, the material constant σ'_f and b can be determined by curve fitting as shown in Fig. 4.2.64 using experimental data by Yao et al. [86], and they are 177.15 MPa and −0.2427, respectively. The fatigue life can be predicted by Eq. (4.1.34) when stress amplitude is known. From the above analysis, different stress amplitude will be applied on solder joints during vibration test with frequencies changing from 175 to 215 Hz, which causes different cycles to failure at different frequencies. Miner's law of Eq. (4.2.26) should be used and solder fatigue life under vibration test can be predicted by combining Eqs. (4.2.26) and (4.1.34). The procedure is given below for calculating solder joint fatigue life.

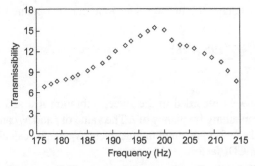

Figure 4.2.63 Plot of transmissibility versus frequency from 5-g input vibration test.

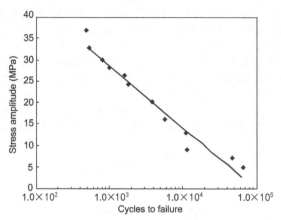

Figure 4.2.64 Stress-based fatigue life curve of eutectic Sn/Pb solder.

Assuming the failure time is t (in second), the fatigue cycle, N_f, can be determined using the mean frequency or natural frequency of 195 Hz:

$$N_f = 195t \qquad (4.2.32)$$

In total, 28 different frequencies (increment steps) were recorded in the sweep vibration test with frequency sweep from 175 to 215 Hz. Because the sweep velocity is constant, the same period of time ($t/28$) is consumed at each frequency. Then the actual cycle (n_i) for each frequency (f) can be determined:

$$n_i = (t/28) \times f = N_f \times f / (195 \times 28) \qquad (4.2.33)$$

The CDI in Eq. (4.2.26) can be expressed by

$$\text{CDI} = \sum_{i=1}^{28} \frac{n_i}{N_i} = \sum_{i=1}^{28} \frac{N_f \times f/(195 \times 28)}{N_i} = N_f \cdot \sum_{i=1}^{28} \frac{f/(195 \times 28)}{N_i} \qquad (4.2.34)$$

So, the vibration fatigue life of the solder joint can be obtained by

$$N_f = \text{CDI} \bigg/ \sum_{i=1}^{28} \frac{f/(195 \times 28)}{N_i} \qquad (4.2.35)$$

where f is the frequency recorded in the sweep vibration test and N_i is the cycle to failure for the corresponding frequency of f. The value of f and N_i can be obtained from previous analysis. Therefore, the total fatigue cycle, N_f, can be calculated by using Eq. (4.2.35) for a given CDI factor.

Different CDI values were used to predict the flip-chip solder joint fatigue life to examine suitable CDI values leading to better correlation between fatigue life prediction and experimental data. The factor of normalized fatigue life is defined as the ratio of fatigue life predicted by quasi-static simulation result to experimental data. Different factors considering four CDI values of 1, 0.7, 0.5, and 0.33 were calculated for chip 3, as shown in Fig. 4.2.65. Fatigue life prediction shows good agreement with the test data for 5 g-level vibration test (factor near 1). For input 3 g-level vibration test, lower CDI value, such as 0.5 or 0.33, provides a reasonable fatigue life. For input 10 g-level vibration test, the quasi-static analysis method results in a conservative fatigue life compared to test result.

4.2.3.2.3 Harmonic response analysis

Harmonic response analysis is a technique used to determine the steady-state response of a linear structure to loads that vary sinusoidally (*harmonically*) with time. In this section, harmonic response analysis for FCOB assembly was conducted for solder joint fatigue life prediction to compare with quasi-static analysis result. Submodeling technique used in quasi-static analysis mentioned above was implemented in harmonic analysis. The FEA model as shown in Fig. 4.2.61 and the material properties as listed in Table 4.2.13 are the same for quasi-static and harmonic simulation. For the global model, the displacement amplitude applied onto the clamped-clamped boundary can be obtained by Eq. (4.2.25) with considering the transmissibility TR as a unity. The displacement results from the global model were transferred to the cut boundary of submodel. Damping in some form should be specified in harmonic analysis; otherwise, the response will be infinity at the resonant frequencies. In this study, a constant damping ratio was used. The trial and error of damping ratios from 1% to 10% was conducted to find suitable damping ratio value. A suitable damping ratio value was determined by comparing displacement amplitude between simulation result and test result measured by a high-speed camera, and then the damping ratio of 2% was determined

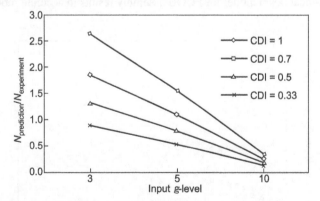

Figure 4.2.65 Comparison of fatigue life prediction with experimental life considering different cumulative damage index (CDI) values.

accordingly. The frequency range used in harmonic simulation was the same as that used in the vibration test mentioned above. Volume-averaged Von Mises stresses based on elements at die/solder interface layer of the corner solder joint were calculated from submodel result for three different input g-level tests as shown in Fig. 4.2.66. Stress increases with frequency before the fundamental natural frequency while stress decreases with frequency after the fundamental natural frequency. Fatigue life prediction of FCOB solder joints based on harmonic simulation results was carried out by the similar procedure used in the previous section for quasi-static analysis. Table 4.2.14 lists comparisons of predicted fatigue life between quasi-static and harmonic simulation results by considering CDI of 0.5. It can be seen that both harmonic and quasi-static analysis provide consistent fatigue life prediction results. It is very important and sometimes difficult for harmonic analysis to select a desired damping ratio. Therefore, quasi-static analysis can be used to calculate the stress amplitude for fatigue life prediction of solder joints when subjected to vibration load without damping ratio requirement, especially for low input g-level vibration test.

4.2.3.3 Summary on vibration test and analysis

FCOB assemblies were tested under out-of-plane sinusoidal vibration load with constant input g and varying input g excitation and with clamped-clamped boundary condition at longer PCB edges. The FCOB assembly shows nonlinear dynamic behavior under vibration load and the transmissibility value decreases with input g-level. The transmissibility is location and frequency dependent. The fatigue life of FCOB assembly reduces rapidly in vibration test when acceleration input increases. Based on CDI fatigue analysis, a CDI of unity (1.0) leads to nonconservative results by a factor of 2–3. Hence, it is recommended that a CDI value of 0.5 should be used for the FCOB solder joint when subjected to vibration tests with varying g-level loading.

Considering the PCB as a bare unpopulated thin plate for the FCOB assembly or using global-local beam model for FCOB assembly results in accurate modal analysis

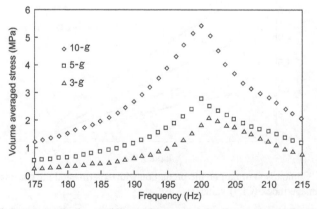

Figure 4.2.66 Stress amplitudes at different frequencies for three g-level vibration tests.

Table 4.2.14 Fatigue life prediction between quasi-static and harmonic analyses (CDI = 0.5).

g-level	Quasi-static (I)	Harmonic (II)	Ratio(II/I)
3-g	1.10×10^8	1.43×10^8	1.30
5-g	2.57×10^7	3.27×10^7	1.27
10-g	2.61×10^6	2.45×10^6	0.94

results compared with experimental result. Therefore, such simplified models are the optimal and suitable models for the modal analysis of the FCOB assembly with saving computing time and resources and without loss of accuracy.

Both quasi-static analysis and harmonic analysis methods with submodeling technique have been developed for dynamic vibration FEA simulation. Vibration fatigue life prediction based on simulation results by using stress-based high cycle fatigue model and Miner's law has been demonstrated. Different CDI effect on fatigue life prediction was investigated and the fatigue life prediction results with CDI of 0.5 are recommended by comparing with experimental results.

4.2.4 Drop impact test and analysis

Portable electronics devices such as mobile phone, personal digital assistant, and laptop have to be designed to withstand repeated drops because it is common for them to be subjected to accidental drop impacts resulting in damage and hence failures. Therefore, the reliability characterization of electronic assembly subjected to drop impact is a major concern. When an electronic product drops on the ground, impact force is transmitted to the PCB, solder joints, and the packages. The packages are susceptible to solder joint failures induced by a combination of PCB bending and inertial force during impact. Drop tests are often substituted in qualification testing of microelectronic devices by shock tests. In shock tests, a short-term load with the given magnitude (i.e., a constant or half-sine load with the given maximum value of g, gravity) and duration is applied to the support structure of the device. The maximum acceleration of PCB board can reach to more than 1000-g when subjected to drop impact loading. The induced stress depends on local acceleration, deformation, and dynamic strength of the structure. In the drop test, the measured maximum acceleration is often used as a criterion of the strength of structure in microelectronic products. It is well known, however, that it is the maximum stress, not the maximum acceleration, which is responsible for the dynamic strength of a structure. The maximum acceleration and the maximum dynamic stress are affected differently by different factors and therefore the use of the maximum acceleration as a strength criterion for dynamic strength is not adequate. So understanding of the maximum dynamic stress is required to characterize the dynamic response of electronic product to shock excitation. The drop impact reliability of portable electronic equipment depends on several external and product factors, the forces, and accelerations during impact: drop height, housing material, weight, shape, orientation at impact, and surface onto which it drops. Typically, the accidental drops of a portable product are randomly oriented. Research showed that the horizontal impact case of PCB is the most dangerous case for reliability of solder joints. The first-order mode vibration is more dominant in dynamic fracture of solder joints. Usually, packages of larger size, stiffer construction, and lower solder join stand-off height are more vulnerable to drop impact and stress increases with increase of drop height, PCB length, package stiffness, and package size and decreases with increase of solder bump height, solder bump size, and solder bump number. There are mainly three types of drop tests in the electronic industry: (a) free fall product level; (b) free fall board-level; and (c) controlled pulse drop at board-level. Board-level drop test is convenient to characterize the solder joint performance, as it is more controllable than product level drop test. Some researchers had conducted board-level drop tests to understand the response of solder joints to impact loading using different boundary conditions such as 4 or 6 screw support [87], clamped-clamped boundary condition [88,89], and free fall drop [90]. Some researchers [91,92] had conducted product-level drop tests due to real drop events.

Lead-free solder usually exhibits less drop impact life than lead-based solder from many test results [88,93]. The study of impact reliability for solder joints conducted by Date et al. [93] showed Sn-Ag-Cu solder was more prone to fracture at interface than Sn/Pb solder under impact test because of higher bulk strength. Solder joint failures during drop testing is a complex failure interaction process between low cycle impact fatigue crack growth and the brittle fracture of the intermetallic interfaces. During a drop test event, dynamic hardening causes the yield stress in the solder to rise several times above the normal monotonic tensile test yield stress. The increase in dynamic strength in the solder joint can cause dynamic strain cycling in the solder material and lead to progressive low cycle impact drop fatigue failures. On the other hand, when the drop loading is excessive, impact failure strength of the intermetallic compound (IMC) interface will result in a brittle fracture of the solder joint. Charpy test was commonly used for evaluating dynamic strength of solder joints and assessing IMC interface strength and failure mechanism [94]. Recently, more and more FEA simulations were carried out for drop impact reliability study to understand the stress−strain response of solder joint under dynamic drop loading. The dynamic material properties of solder are different from static ones because the dynamic material properties are dependent on strain rate. For Sn-Ag-Cu solder, the yield stress at high strain rate is about three times higher than that at low strain rate [94]. However, the elastic modulus of the solder joints varies less with strain rate. The solder constitutive model used in FEA simulation will affect stress−strain behavior of solder significantly.

In this section, drop impact test is conducted for PBGA package with Sn-Ag-Cu solder joints. Failure mode and drop impact life are also investigated. FEA modeling and simulation of drop impact are conducted by considering different constitutive models of solder containing elastic, bilinear plastic, and rate-dependent plastic models to investigate board-level drop reliability of Sn-Ag-Cu lead-free solder joint.

4.2.4.1 Drop impact test approach

4.2.4.1.1 Test vehicle and set up

The Lansmont Model 65/81 drop impact tester is used to provide the drop impact load for board-level specimen drop test. This machine can provide half sine shock pulse. The acceleration level of the half sine pulse is related to the drop height and the mount of felt pads on drop base. The duration of the pulse is adjustable somewhat by the addition or subtraction of the felt pads. The basic drop machine structure consists of a steel base, two solid chrome-plated guide rods, drop table, and hoist positioning system as shown in Fig. 4.2.67. The table is an aluminum weldment and contains brakes and low friction bearings, which rid on the guide rods. An integral brake pressure regulator in the main control panel supplies nitrogen pressure to the brake.

The samples used in the drop test are 35 mm × 35 mm PBGA package (312 balls), 28 mm × 28 mm plastic quad flat package (PQFP) (208 leads), and 7 mm × 7 mm VQFN package. All packages are daisy chained for the purpose of interconnection failure monitoring. As shown in Fig. 4.2.68a, these packages are mounted on a 150 mm × 200 mm PCB board using the Sn-3.8Ag-0.7Cu lead-free solder. As shown in Fig. 4.2.68b, the PCB specimen is fixed on the top of drop table with a clamped-clamped boundary condition and with horizontal drop orientation and face-down packages. Two different pad surface finishes of ENIG and OSP are used on PCB pad finish to investigate the effect of pad finish on the drop impact reliability of solder joint interconnects. The microminiature piezoelectric accelerometer (Endevco model 22) is used in drop test. Its light weight, 0.14 g, effectively eliminates mass loading effects on samples during drop impact test. An accelerometer is mounted on the drop table near the fixture to measure the input acceleration. Another accelerometer is mounted at the center of the PCB to characterize the output acceleration response of the PCB.

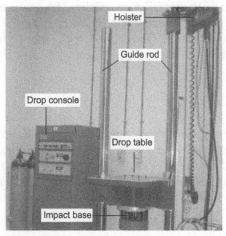

Figure 4.2.67 Lansmont drop impact tester.

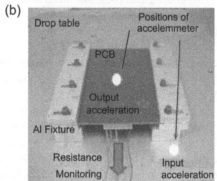

Figure 4.2.68 Board-level drop test setup and PCB specimen: (a) PCB specimen with different packages and (b) PCB mounted on the top of the drop table with the fixture.

Daisy chain is designed for each package, and the dynamic resistance measurement system as shown in Fig. 4.2.69 is used to measure the resistance of daisy-chained solder joints to determine the drop to failure in real time. According to this method, a constant resistor, R_0, is placed in series with the daisy-chain solder joints of R_x and connected to a DC power supply. The oscilloscope is used to measure the dynamic voltage dropped on daisy-chain loop. The dynamic resistance, R_x, can be expressed

$$R_x = \frac{R_0 V_x}{U - V_x} \tag{4.2.36}$$

where U is the voltage of the DC power supply and 1.8 V was used in this test, V_x is the dynamic voltage of daisy chain measured by oscilloscope, and R_0 is the constant resistance of 10Ω used in this test. When $V_x \to U$, $R_x \to \infty$ (implies an open circuit), it indicates the critical solder joint has failed with crack opening. The open circuit (infinite resistance) is adopted as a failure criterion in this study.

4.2.4.1.2 Drop reliability test and analysis

The experiment is performed with an input peak acceleration of about 580 g and 2 ms duration at the drop table as sown Fig. 4.2.70. The output acceleration of PCB center as

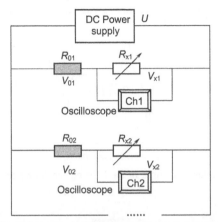

Figure 4.2.69 Schematic of dynamic resistance measurement system.

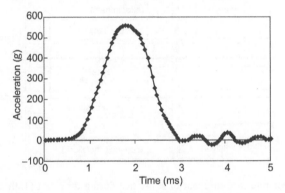

Figure 4.2.70 Acceleration measured on drop table during drop test.

shown in Fig. 4.2.71 is more than that of drop table due to deflection effect of PCB. As the PCB is experiencing the maximum bending at the center during drop test, only real-time resistances of packages labeled "A," "B," and "C" (Fig. 4.2.68a) are monitored by a dynamic measurement method as illustrated in Fig. 4.2.69.

The drop dynamic responses of resistance for daisy-chain loop are illustrated in Fig. 4.2.72. With the PBGA solder balls mounted onto an OSP pad finish, the first failure was observed at the 15th drop. From the graph, the resistance increases from an initial value of 1.4–1.8 V after impact. The dynamic resistance does not reached the 1.8 V level immediately but fluctuates for a short period of time before reaching 1.8 V permanently. This observation is first postulated that on impact, a solder joint crack is initiated followed by a rapid crack propagation till total failure (dynamic resistance reaches infinity, $V \rightarrow E = 1.8$ V). Manual probing after the drop confirmed that the static resistance reached infinity and hence permanent failure was registered. Cross sectioning showed that no solder joint crack were found. Copper trace failure was observed. From the dynamic curve response, the copper trace fatigued after 14 impact

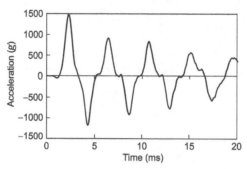

Figure 4.2.71 Output acceleration curve of the PCB center during drop test.

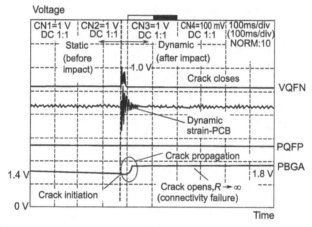

Figure 4.2.72 Dynamic measurement of resistance during drop test (15th drop). Sampling rate = 10,000 samples/s.

drops. At the 15th drop, the copper trace tears in a ductile manner and the failure is not immediate. PCB board flexing causes the copper trace to rupture and results in a permanent failure. As for the VQFN package, some intermittent failures were observed in the dynamic resistance curve. Crack in the VQFN interconnects has started to initiate and propagate through the joint interface at the 15th drop. However, the crack is only partial; hence, the resistance peaks to about 1.0 V instead of 1.8 V. When the PCB bends downwards, the outermost solder joints are subjected to tensile stress, thus leading to an open mode in the crack. When the board bends up, the joints are under compression, which helps the crack to close up. Toward the end of a drop, the partial crack would most probably close up when the flexing of the board is over. At this instance, manual probing after the drop test would not register any failure. As such, no failure is considered for the VQFN package after the 15th drop. When probed manually, the static resistance after drop is 7.1 Ω as compared with a static resistance of 5.1 Ω before drop. No change in the resistance response is observed for PQFP after 15 drops.

A total drop of 50 times was conducted for each test board, and 12 test boards were tested. Being the smallest in size and weight and with the center pad holding the entire body firmly onto the PCB, the VQFN package proves to be most robust under drop test and still survives after 50 drops. Fig. 4.2.73 shows a comparison of the drop life for different packages based on the averaged test results. The test results show that OSP board finish gives better drop reliability than ENIG finish for the PBGA package. With the compliant gull-wing leads, the PQFP package outperforms the PBGA in drop impact reliability where the solder balls are more rigid. Fig. 4.2.74 shows the failure mode for both PBAG and PQFP packages. The PQFP package failure is not due to solder joint crack but the lead fingers breaking off (Fig. 4.2.74c) from mold compound due to a large impact force during drop impact. For the PBGA package with different PCB surface finishes, there are several failure modes such as Cu trace broken and resin crack (Fig. 4.2.74a for OSP finish case) and interface brittle failure (Fig. 4.2.74b for ENIG finish case) besides solder joint cracking failure. Most of failure locations happen at the PCB side from experimental results. Usually, a binary IMC of Cu/Sn is formed for Sn–Ag–Cu solder on OSP-finished PCB Cu pads, while a ternary IMC of Cu-Ni-Sn and a binary IMC of Ni/Sn are formed on ENIG finished PCB Cu pads [95,96]. Previous studies showed that Cu/Sn IMC has stronger adhesion strength than Ni/Sn IMC [97]. For a given thickness, the fracture toughness of Cu/Sn appears to be greater than that of Ni/Sn [98]. Cu-Ni-Sn and Ni/Sn IMCs have a higher hardness and are much more brittle compared with Cu/Sn IMC [52]. Therefore, for the PBGA package with ENIG finish, the common failure mode is IMC layer brittle cracking. For the PBGA package with OSP finish, due to stronger IMC adhesion strength, the failure site is being migrated to the solder joint, copper traces, and PCB resin layers [87,88].

Interface failure and IMC brittle failure are important failure modes in the drop test, which require further investigations on the failure mechanism and material characterization, such as IMC toughness, IMC morphology, and failure mechanism of solder/IMC interface under highly dynamic load. However, there is a big challenge to model IMC in the simulation accurately in terms of its morphology and material properties. Study on interface strength and toughness of solder joints under dynamic loading is

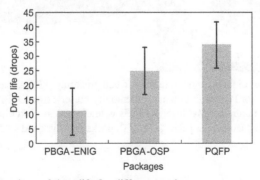

Figure 4.2.73 Comparison of drop life for different packages.

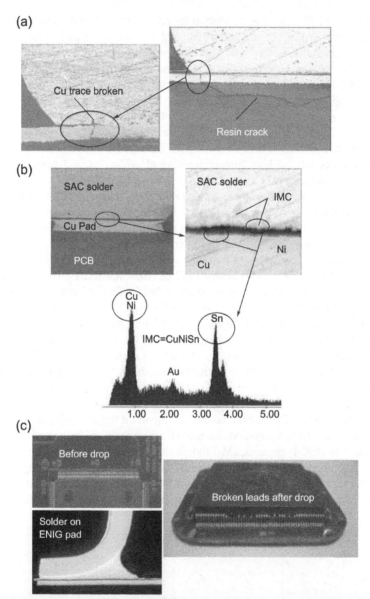

Figure 4.2.74 Images of failure modes: (a) plastic ball grid array (PBGA) package with Cu trace and PCB rinse crack for PCB with organic solderability preservative (OSP) finish, (b) PBGA package with brittle failure of IMC layer for PCB with electroless nickel immersion gold (ENIG) finish, (c) plastic quad flat package (PQFP) packages with good solder joint intact and broken leads.

necessary to investigate the effect of board surface finish and IMC on the drop reliability of electronic package. Charpy test is usually used for investigating interface and bulk material toughness, which is discussed in the following section.

4.2.4.1.3 Charpy impact test and analysis

The Charpy impact test helps determine the impact energy absorbed and failure mode. Two different Charpy specimens of bulk Sn-Ag-Cu specimen and soldered specimen as shown in Fig. 4.2.75 are used in the test. For soldered specimen, two copper blocks are soldered together with Sn-Ag-Cu solder after reflow process. Two cases are studied for soldered specimen: copper block with and without Ni/Au plating. Three samples are tested for each type of Charpy specimen. Fig. 4.2.76 shows the fracture modes for different specimens. The bulk specimen exhibits ductile failure, and soldered specimen shows all brittle interface failure. Table 4.2.15 lists the impact toughness for different specimens. The bulk solder specimen has the largest toughness value. The impact toughness of soldered specimen without Ni/Au plating is twice higher than that with Ni/Au plating. It implies that electronic assembly with Sn-Ag-Cu solder and Ni/Au finish is liable to impact failure compared with the case without Ni/Au finish, which is consistent with drop test results for the PBGA package mentioned above. Cu/Sn IMC is formed between copper block and Sn-Ag-Cu solder, while Cu-Ni-Sn IMC is formed when Ni/Au plating is used. The Charpy impact test results prove that when the drop failure mode shows IMC brittle failure, package with ENIG board surface finish has lower drop lifetime than that with OSP board surface finish.

4.2.4.2 Finite element analysis for drop impact test

4.2.4.2.1 Material property of solder

Solder joint failure (bulk solder or IMC brittle cracking) is one dominant failure mode for electronic assembly under drop test. Material properties of solder are much sensitive to strain rate. Therefore, it is important to know material response of solder under different strain rates. The mechanical properties of Sn-3.8Ag-0.7Cu lead-free solder are characterized by performing two types of testing: tensile test and impact test. Dog bone-shaped bulk solder specimens as shown in Fig. 4.2.77 are machined from a solder bar. The specimen for impact test is threaded at both ends for screwing onto the test system. Before testing, the specimens are annealed at 60°C for 24 h to eliminate the

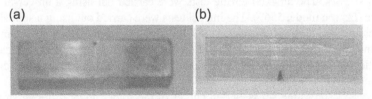

Figure 4.2.75 Charpy impact test specimens: (a) soldered specimen, and (b) bulk Sn-Ag-Cu specimen.

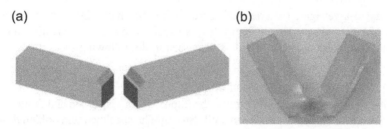

Figure 4.2.76 Failure mode under Charpy impact test: (a) brittle failure for soldered specimen, and (b) ductile failure for bulk Sn-Ag-Cu specimen.

Table 4.2.15 Impact toughness for different specimens.

Specimens	Bulk Sn-Ag-Cu	Soldered sample without Ni/Au	Soldered sample with Ni/Au
Impact toughness (J)	74.0	0.261	0.118

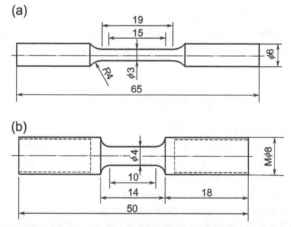

Figure 4.2.77 Geometry of solder specimen (in mm): (a) tensile test specimen and (b) impact test specimen.

residual stresses. The uniaxial tensile tests were carried out using a universal testing machine (Instron model 5569). The impact tests were carried out using a split Hopkinson pressure bar (SHPB) tester to study the dynamic response of solder material. The principle of the SHPB test is based on the theory of stress wave propagation in an elastic bar and the interaction between a stress pulse and a specimen of different impedance.

Fig. 4.2.78 shows the stress–strain curves of the solder under different strain rate conditions. The test data under low strain rate are from normal tensile tests [37] while

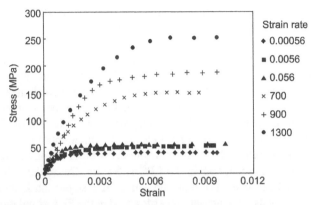

Figure 4.2.78 Rate-dependent stress—strain curves of Sn-Ag-Cu solder.

the test data under high strain rate are from the SHPB impact tests. The results show that the stress—strain behavior of the solder is highly dependent on the strain rate. For each test condition, three to five specimens are tested and the average mechanical properties are calculated. Table 4.2.16 summarizes the mechanical properties of the Sn-3.8 Ag-0.7Cu solder under different strain rates. The tangent modulus is estimated based on a bilinear plastic model. The strain rate affects the yield stress and tangent modulus more significantly than the elastic modulus. Solder material becomes harder under high strain rate and consequently exhibits higher yield stresses. Under drop impact load, the strain rate of solder material varies significantly from several tens to several hundred per second. Therefore, the strain rate—dependent material properties are essential for solder material in the drop impact modeling to achieve the reasonable and accurate stress results. The stress—strain curves under different strain rate tests as shown in Fig. 4.2.78 can be implemented in modeling, and the actual strain rate—dependent material data can be achieved through data interpolation in FEA modeling.

4.2.4.2.2 Finite element model

The FEA simulation is conducted for PBGA package to assess stress response of solder joints under drop impact test. Some assumptions are made in the FEA modeling.

Table 4.2.16 Material properties of solder under different strain rates.

Strain rate (1/s)	5.6×10^{-4}	5.6×10^{-3}	5.6×10^{-2}	7×10^2	9×10^2	1.3×10^2
Young's modulus (GPa)	44.4	50.3	52.8	63.2	67.7	69.4
Yield stress (MPa)	35.1	40.8	51.1	129	175	216
Tangent modulus (MPa)	194.3	207.3	230.7	1279	1796	2165

For instance, the dynamic response of each package mounted on the PCB is not affected by the other packages. In addition, the effect of individual packages on the global response of the PCB is ignored. As such, a quarter FEA model for the PBGA assembly is created as shown in Fig. 4.2.79 with PBGA package facing down at the center of PCB board. A symmetric boundary condition is applied on the symmetric planes. The input acceleration from experimental measurements as shown in Fig. 4.2.70 is applied on the clamped PCB edge as an input-g loading. The critical corner solder joint is modeled using fine meshed elements and other solder joints are modeled using effective cuboid elements, which could help to reduce the number of elements significantly for the entire model without losing accuracy. The materials used in the FEA model contain FR4 PCB, Sn-Ag-Cu solder, copper pads on both board and package sides, BT substrate, silicon die, solder mask, and mold compound. Table 4.2.17 lists the material properties used in the FEA simulation. A total of 15 μs are simulated with 150 substeps for the general result file and 750 substeps for the history result file. Solder material is modeled with different properties for comparison including elastic, bilinear plastic, and rate-dependent models.

4.2.4.2.3 Simulation results and discussion
4.2.4.2.3.1 Results based on elastic model of solder In the FEA modeling and simulation for the drop test of the PBGA assembly with lead-free solder joint, the important and critical material is solder. Therefore, it is essential to model solder material in the high strain rate loading condition. In this study, the elastic model and rate-dependent inelastic model are implemented for the Sn-Ag-Cu lead-free

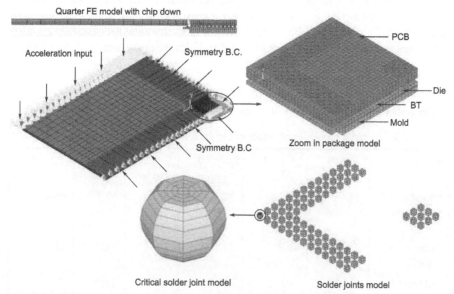

Figure 4.2.79 Quarter finite element model of plastic ball grid array (PBGA) assembly for drop impact simulation.

Table 4.2.17 Material properties used in the finite element analysis (FEA) simulation for drop test.

Materials	Young's modulus (GPa)	Poisson's ratio	Density (kg/m³)
Sn-Ag-Cu solder	52.8	0.35	7500
FR4 PCB	20(x,z),9.8(y)	0.11(x,z),0.28(y)	1900
BT substrate	26(x,z),11(y)	0.11(x,z),0.39(y)	2000
Copper pad	120	0.34	8900
Silicon chip	131	0.278	2330
Mold compound	16	0.3	1970
Solder mask	2.4	0.3	1300

solder to investigate the effect of the constitutive model on the dynamic response of the solder joint. Firstly, the elastic model developed under the low strain rate tensile test is used for the solder joint. It was known that the damping effect is important for dynamic behavior of the PCB under drop impact load. Consequently, a calibration study is carried out in the FEA simulation to obtain an optimal damping ratio by comparing the simulation results with experimental data. An optimal damping ratio of 2% is determined by comparing the output acceleration of the PCB center as shown in Fig. 4.2.80 from the FEA simulation result with the acceleration measured in the drop test as shown in Fig. 4.2.71.

Fig. 4.2.81 shows a comparison of the peeling stress for the critical nodes of solder joints at the interfaces of both board and package sides. It is obvious that solder stress at the PCB board side is larger than that at the package side, which indicates that the solder/PCB interface is prone to failure for large PBGA assembly due to dynamic bending effect of the PCB board. Most of failure locations happen at the PCB side from experimental results, which are consistent with simulation results having higher stress at the PCB board side. When package size becomes smaller, e.g., less than 15 mm × 15 mm, solder joint failure at chip side becomes very common [87,99].

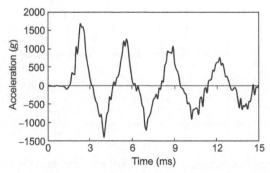

Figure 4.2.80 Simulation result of acceleration at the center of PCB board.

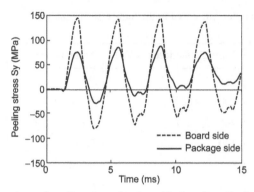

Figure 4.2.81 Comparison of peeling stress for the critical nodes at both board and package sides.

The failure location is related not only to high stress concentration but also to IMC composition and morphology at both joint sides. However, solder stress is still a reasonable indicator to study the solder joint drop impact reliability. Fig. 4.2.82 shows the peeling stress in the vertical direction and the first principal stress of the critical node at the PCB board side. It can be seen that the peak stress value between the peeling stress and principal stress is quite close, which implies that the peeling stress is the dominant factor inducing crack initiation and propagation at the interface of solder/PCB. The elastic model of the solder can be used in the FEA simulation for drop test to investigate the relative trend of drop life. However, using the elastic model to simulate the dynamic behavior of solder material under drop tests could lead to errors. Usually, elastic model of solder leads to high stress and low life prediction accordingly. As such, the predicted drop life based on the elastic model underestimates drop reliability.

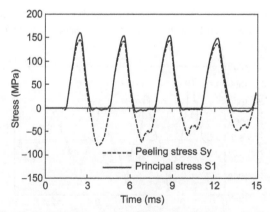

Figure 4.2.82 Comparison of peeling stress and first principal stress for the critical node of solder joint.

4.2.4.2.3.2 Results based on plastic model of solder

Plastic deformation of the solder joint can be expected in a dramatic dynamic loading such as drop impact. It is imperative to simulate for the solder behavior using strain rate–dependent plastic model when subjected to drop impact loading. The rate-dependent plastic model was developed from the characterization tests of the solder under different strain rates as shown in Fig. 4.2.78 mentioned above. The rate-dependent material properties of the solder used in the FEA simulation include the Young's modulus, yield stress, and tangent modulus. The bilinear elastic-plastic model of the solder is also simulated for comparison. The bilinear elastic-plastic model used in the simulation is mostly obtained from the tensile tests of the solder under low strain rate such as from 10^{-4} to 10^{-2} 1/s, which is always used in the FEA simulation of TC load. The material properties of the bilinear elastic-plastic model used in this work are obtained from tensile tests under a strain rate of 5.6×10^{-2}, as shown in Table 4.2.16.

Fig. 4.2.83 shows the effect of the constitutive model of the solder on the peeling stress history of the critical node in solder joints under drop impact load. It can be seen that the peak value of stress reduces significantly when considering solder with dynamic plastic behavior compared with the elastic model. The rate-dependent plastic model leads to slightly higher peeling stress of the solder compared with the bilinear plastic model. Fig. 4.2.84 shows a comparison of the Von Mises stress of solder when using different constitutive models of the solder. The error of the peak Von Mises stress between the elastic and plastic models is more than 100%. The stress level of the solder joint would be overestimated when only the elastic material behavior is modeled for the solder under dynamic loading. The bilinear elastic-plastic model results in lower Von Mises stress compared with the rate-dependent plastic model because the effect of strain rate on solder strain hardening is not considered when using the bilinear elastic-plastic model. The solder exhibits more ductile properties and lower yield stress in the bilinear elastic-plastic model, which results in lower Von Mises stress. On the other hand, the bilinear elastic-plastic model will result in larger plastic deformation compared with the strain rate–dependent plastic model, which is verified by the

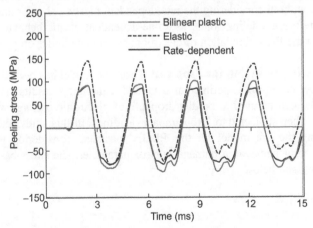

Figure 4.2.83 Effect of constitutive models of solder on peeling stress of solder under drop test.

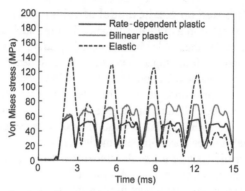

Figure 4.2.84 Effect of constitutive models of solder on Von Mises stress of solder under drop test.

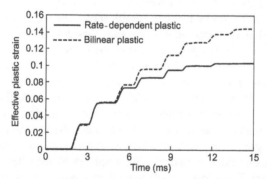

Figure 4.2.85 Comparison of effective plastic strain of solder for different plastic models.

effective plastic strain results as shown in Fig. 4.2.85. The plastic deformation of the solder joint will be overestimated when the bilinear elastic-plastic behavior is modeled for the solder without considering the rate-dependent plastic behavior under dynamic loading. Therefore, modeling solder as a rate-dependent plastic behavior is essential and needed in the FEA simulation of electronic assemblies under drop test.

4.2.4.2.3.3 Life prediction for drop test Drop life model has been developed for lead-free solder based on experimental data and simulation results [100]. The developed model can be used to predict drop life of electronic package with lead-free solder joints when subjected to drop impact loading. In this study, the drop life is predicted using the life model for the PBGA package considering the simulated maximum peeling stress as a damage control parameter. The stress-based drop life model is expressed below [100]:

$$N_d = 9.9 \times 10^9 \times \sigma^{-3.9} \tag{4.2.37}$$

where N_d is drop life and σ is the maximum peeling stress of the solder joint, which can be obtained from dynamic drop simulation. The maximum peeling stress variation is calculated from simulation results as shown in Fig. 4.2.83. When the stress values are substituted into above equation, the drop life can be predicted for the PBGA package. Fig. 4.2.86 shows the drop life prediction with 50% error bar for the PBGA package when using different solder constitutive models in drop test dynamic simulation. The elastic model underestimates the drop life of package significantly because large peeling stress is obtained from the dynamic simulation when plastic behavior of solder is not modeled. The bilinear plastic model overestimates the drop life of package compared with the rate-dependent model. Compared with experimental results of PBGA drop life under drop test as shown in Fig. 4.2.73, numerical life prediction has a good agreement with experimental data when modeling solder with rate-dependent material model.

In terms of IMC effect on drop life, some assumptions are usually used to model the IMC layer, including modeling IMC as a flat layer and elastic material property [53,101]. Modeling an IMC layer in the solder joint interface results in a high stress level in drop impact simulation because the stiffness of the solder joint increases. However, modeling IMC layer in drop test constrains the plastic deformation of the solder, which results in lower plastic strain of the solder [101]. Therefore, just considering the plastic deformation or stress of the solder joint is not sufficient for investigating reliability of electronic assembly subjected to a drop impact load. Study on interface strength and toughness of solder joints under dynamic loading is necessary to investigate brittle failure or transition failure.

4.2.4.3 Design-for-reliability methodology for drop test

A flowchart of a design-for-reliability (DFR) methodology for drop impact reliability analysis is shown in Fig. 4.2.87. Material characterization for lead-free solder subjected to dynamic high strain rate conditions can be carried out by SHPB test and Charpy impact test. Strain rate—dependent elastic-plastic model is then developed

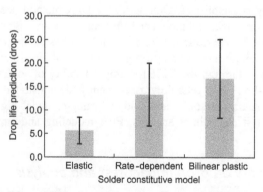

Figure 4.2.86 Drop life prediction for plastic ball grid array (PBGA) package with using different solder constitutive models.

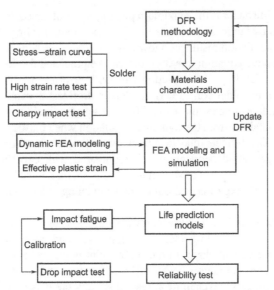

Figure 4.2.87 Flowchart of a design-for-reliability methodology for drop impact reliability.

for solder material. The high strain rate solder material properties are implemented in the dynamic FEA simulation for board-level solder joint drop impact test to study dynamic stress–strain behavior of the solder joint. Dynamic peeling stress or plastic strain in the solder can be used as an impact damage driving force for analyzing high strain and low cycle impact fatigue failures of the solder. Solder fatigue models developed from strain-based and stress-based tests with slow strain rates or frequency tests cannot be used for drop impact fatigue analysis where high strain rate plastic deformation is important. An impact fatigue model based on FEA simulation results with rate-dependent plastic behavior for solder material and drop impact testing data is necessary for the drop life prediction and drop performance optimization of electronic assembly under drop impact load. Drop test data are used to validate the numerical prediction results. When experimental data and simulation results are consistent, the systematic DFR methodology is established. Otherwise, the DFR method needs to be updated and modified. Accurate material model and FEA modeling are essential for evaluation and optimization of drop performance of solder joints. Physical drop test is still needed to study different failure modes and drop performance of electronic assembly. Such topics can be linked through the DFR for drop reliability of solder joints. The DFR method for board-level drop reliability assessment is demonstrated above, including material characterization of solder and soldered samples under high strain rate test, board-level drop test and failure analysis, FEA modeling and simulation, and life prediction.

4.2.4.4 Summary on drop impact test and analysis

The methodology on DFR of lead-free electronic assembly under drop test has been developed by conducting material characterization test, FEA and life prediction, board-level drop test, and Charpy impact test for bulk solder specimen and soldered specimen. The link among each part is also discussed. Some conclusions can be summarized below:

Board-level drop impact tests were conducted for electronic assembly with different packages. The PBGA package is more prone to drop impact failure than the leaded PQFP package. The PQFP package failure shows the lead fingers breaking off from mold compound. The PBGA package with OSP board finish exhibited higher drop reliability than that with Ni/Au board finish due to different IMC formation and IMC toughness. Drop life of the PBGA package is used to validate the simulation. Charpy impact test results show that the soldered specimen with Ni/Au plating has lower impact toughness than the specimen without Ni/Au plating, which is consistent with drop test results of that PBGA package with OSP board finish exhibited higher dropt reliability than PBGA package with ENIG board finish.

Drop impact FEA simulations for PBGA assembly were carried out using an input-G modeling method with using different solder constitutive models including the elastic model and strain rate-dependent plastic model obtained from material characterization tests. The stresses of the solder joints at the board side are higher than those at the package side for large PBGA package, which is consistent with drop failure location in the drop test where most of failure happened at the PCB side. The elastic model overestimates the stress level of the solder compared with the strain rate-dependent plastic model under drop impact load. Life prediction based on simulation results was carried out for the PBGA package. The predicted drop life time is consistent with drop test results when modeling solder with rate-dependent material behavior. It is necessary and accurate to implement the rate-dependent plastic model for the solder joint in the drop impact modeling.

4.3 Case study of design-for-reliability

4.3.1 Design-for-reliability using finite element method

Manufacturers of electronic products face with demands for design with high reliability and performance at lower costs. Reliability tests of electronic assemblies are very important concerns to ensure long-term reliability of electronic products. In service, electronic products failures can occur within a year but often much longer. Therefore, it is not practical to test at service load level and it is uneconomical with respect to test time and cost. ALT methods are employed for the purpose of accelerating reliability test results.

Reliability tests aim at revealing and understanding the physics of failure. Another objective of reliability tests is to accumulate representative failure statistics. Accelerated tests use elevated stress levels and/or higher stress cycle frequency to precipitate failures over a much shorter time. Stress loading includes constant stress, step stress, progressive stress, and random stress. Typical accelerating stresses are temperature, mechanical load, TC or TS, cyclic bend, drop impact, and vibration. Such tests can facilitate physics of failure for reliability tests that cost effectively, shorten the product process, and improve long-term product reliability. Accelerated tests can be performed at any level, such as part level, component level, board-level, equipment level, and system level. The degree of stress acceleration is usually determined by an AF. This factor is defined as the ratio of lifetime under normal service conditions to the lifetime under accelerated conditions.

Computational modeling and simulation can reduce the product development time to market in a competitive electronic product sector. Commercial FEA software, such as ANSYS, ABQUS, and NASTRAN, have been used extensively to simulate reliability test, design and service operation loads on electronic components, board assemblies, and product systems. By using FEA simulations, industry can minimize the requirement for extensive and time-consuming physical testing. This can further reduce product development costs, increase reliability, and reduce the product time to market. Further miniaturizations of IC component size and higher I/O counts are expected trends in electronic packaging applications. Thus, conventional application of finite element modeling technique will become more difficult as the geometry features become smaller and require higher number of elements so that FEA simulation will require higher speed computing, larger memory size, and hard disk storage space. These critical requirements can limit the use of full 3D model applications for finite element reliability analysis of the electronic assemblies. To reduce the element size in the FEA simulation for reliability modeling, some reduced models based on reasonable assumption are used by researchers, including slice model, one-eighth model, and 2D model. Some trade-off in accuracy is expected in these simple models. Such simplified models can be used for quick assess and evaluate packaging design and material selection. In the previous section, an example is provided for a DFR methodology for drop impact reliability analysis (Fig. 4.2.87). Successful DFR methodology includes accurate material characterization and material property data, life model and damage prediction parameter selection, FEA modeling and simulation details, modeling results analysis for reliability assessment and design improvement, and reliability test data used to validate the numerical prediction results. When experimental data and simulation results are consistent, the systematic DFR methodology is established. Otherwise, the DFR method needs to be updated and modified.

4.3.2 Optimization of package design

Following Moore's law, the scaling of the IC goes to a very small physical dimension, e.g., less than 10 nm, which makes IC design and manufacturing to face several significant challenges such as lithography technology, circuit design, yield, and cost. Recently, more than Moore technology applied in advanced packaging has been attracting more and more attention and becoming a cost-effective way to overcome challenges of Moore's law. System-in-package (SiP) technology is one promising way to cope with requirements for microelectronic systems. There are two typical types of SiP technologies. First, multiple chips are integrated in one package with side-by-side placement, which is referred to 2.5D integration, such as interposer technology. The other is chips integrated in one package through stacking each other, which is referred to 3D integration, such as chip stacking and interconnection by through-silicon via (TSV) and package-on-package.

SiP technology has many advantages in terms of electrical and thermal performance, small package profile, and high integration density. However, package design and assembly process and long-term reliability are critical to ensure such technology cost effective. Package layout and assembly process become complicated compared with traditional packaging technology. Design cycle time and cost are critical for SiP packaging technology. In addition, chip package interaction becomes a critical concern in advanced packaging because advanced IC technology requires low-k or ultra−low-k dielectrics in BEOL to improve electrical performance and advanced packaging technology with 3D stacking for thinned IC chips with TSVs, and Cu wire bonding process cause high stress on package. Therefore, package needs to be carefully designed and assembled. Physical design of experiment for package design and reliability assessment is time and cost consuming and sometimes it is not practical. Numerical modeling and simulation is a powerful tool to numerically assess the reliability of the designed package and to provide optimal solutions for an improper design. Materials also can be quick assessed through FEA modeling and simulation, and optimal material selection can be determined based on data of material properties, loading condition, package design, and stress level of package. Recommendation from simulation results is adopted for reduction of material candidates, optimizing package structure design, assembly, and packaging process, and long-term reliability evaluation.

In the following section, case study is presented to show how to use FEA modeling and simulation to improve package design and assess reliability for 3D stacked package with TSV chips.

4.3.3 Case study

4.3.3.1 Problem statement

In this case study, the reliability of a pyramidal shape 3-die stacked package with TSVs was demonstrated experimentally and numerically. The initial designed microbumps locate peripherally along the edge of the TSV die, which induces a concentrated bending force on the lower die when the upper die is stacked. The experimental results show that the bottom die cracks when the middle die is stacked and the middle die cracks when the top die is stacked even with a small stacking force. FEA results show that such bump layout induces large stress and deflection in the lower die under the die stacking process. Three-point bend tests are conducted to determine the die strength first. Consistent results have been obtained among FEA simulation, die strength bend test, and die stacking experiments. An optimal bump layout design is proposed based on DFR methodology, which adds some dummy bumps on the central area of the die to support the bending force induced by the die stacking process. The optimal design significantly reduces the die stress level and deflection. Finally, a successful die stacking process is achieved even using a larger stacking force.

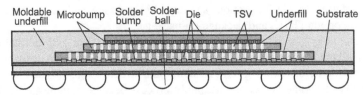

Figure 4.3.1 Schematic of a 3-die stacked package.

Fig. 4.3.1 shows a schematic of a 3-die stacked package with TSVs. The diameter and depth of Cu TSV are 20 μm and 100 μm for the middle die, 50 μm and 200 μm for the bottom die, respectively. All TSVs are located at the edge of the die peripheral. The package has three stacked dies and the structure has a pyramidal shape. The top die is 5 mm × 5 mm × 0.1 mm in size with 176 microbumps; the middle die is 7 mm × 7 mm × 0.1 mm with 212 microbumps; and the bottom die is 10 mm × 10 mm × 0.2 mm with 360 solder bumps. Fig. 4.3.2 shows the die dimension and bump layout in the initial design for three chips. Bumps are distributed at one, two, and three peripheral rows for the top, middle, and bottom die, respectively. In die stacking process, the bottom die is mounted on the substrate through solder bumps, and then the middle die is stacked onto the bottom die through microbumps. Finally, the top die is stacked onto the middle die. Reflow process was conducted for joint permanent formation. Underfill is used in between all dies after 3-die stacking process. The moldable underfill is used to fill the gap between the bottom die and substrate and then full package encapsulation.

A die stacking process was conducted using flip-chip bonder FB35. When mounting the middle die onto the bottom die, the bottom die cracks with cracking path close to the inner solder bumps as the mounting force increases to about 5 kgf (1 kgf = 9.8 N). When stacking the top die on the middle die, the middle die also cracks as the stacking force increases to 3.5 kgf. Fig. 4.3.3 shows the failure location in the bottom die. Such die stacking results show that the maximum die stress induced in the die stacking process exceeds the die strength. A concentrated bending force will be applied on the lower die when a smaller upper die is stacked due to peripherally distributed bumps. This concentrated force induces large stress and deflection in the lower die. Improper solder bump layout design results in the large die stress induced during

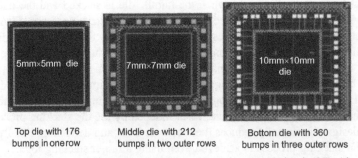

Figure 4.3.2 Geometry and bump layout design of three integrated circuit (IC) chips.

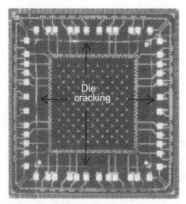

Figure 4.3.3 Die cracking in the die stacking process.

the die stacking process. DFR using FEA modeling and simulation was conducted to assess the package reliability and optimize the bump layout design for overcoming the reliability problem of the package under the die stacking process.

4.3.3.2 Die strength characterization

To assess the package design and die strength and reliability, die strength measurement test needs to be carried out. In this work, three-point bend test [102] was conducted for a thin die with TSVs. The dimension of the sample is 10 mm × 10 mm × 0.2 mm, which has same size as the actual bottom die used in a 3-die stacked package. Fig. 4.3.4 shows a schematic of a three-point bend test. The maximum die stress can be determined by

$$\sigma_{3p} = \frac{3lF}{2bh^2} \qquad (4.3.1)$$

where l is the support span, F the applied load, b the die width, and h the die thickness.

Five samples were tested for each type of specimen to determine the die strength through averaging the testing data. Fig. 4.3.5 shows the typical loading curve of 3-point bend test of the die. When the load head contacts with die surface, the bending

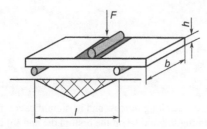

Figure 4.3.4 Schematic of three-point bend test.

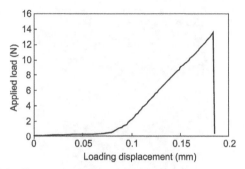

Figure 4.3.5 Typical loading curve of three-point bend test.

force increases linearly with the loading displacement. The loading force drops drastically after the die breaks. For each test, the breaking force is measured and the die strength is determined using Eq. (4.3.1). The averaged die strength value is 279 MPa. This die strength value can be used to evaluate the reliability of die under the die stacking and other packaging processes, which will be discussed in the next sections. In a three-point bend test, the maximum tensile stress occurs at the centerline of the die surface so that die cracks along the centerline.

4.3.3.3 Finite element analysis and optimization

4.3.3.3.1 Finite element model

In this work, FE simulation was carried out to evaluate and optimize the package design. Fig. 4.3.6 shows a schematic of the die stacking process and its simplification for FEA simulation. It shows that a mounting pressure added on the upper die is transferred to the lower die in a type of concentrated force. Fig. 4.3.7 shows FEA models for the die stacking process simulation. When the middle die is stacked, a mounting pressure applied on the middle die is transferred to the bottom die as a concentrated force at

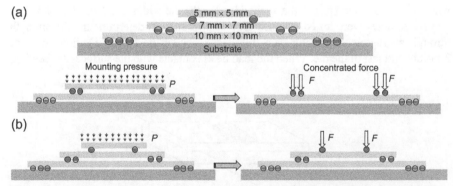

Figure 4.3.6 Schematic of die stacking process and its simplification for finite element analysis (FEA) simulation: (a) mounting the middle on the bottom die and (b) mounting the top die on the middle die.

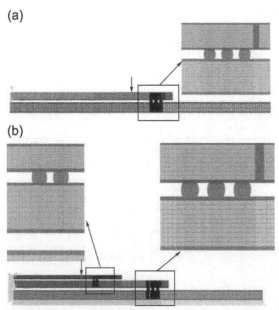

Figure 4.3.7 Finite element analysis (FEA) models: (a) stacking the middle die on the bottom die and (b) stacking the top die on the middle die.

a location corresponding to the microbumps of the middle die. Therefore, concentrated force can be applied onto each corresponding bump location instead of constructing a model with the middle die. Similar technique can be used for stacking the top die onto the middle die. For convenience, a 2D axisymmetric FEA model is used. A symmetric boundary condition is applied on a symmetric plane, and the bottom surface of the substrate is constrained in the vertical direction. In the FEA simulation for the die stacking process, three load conditions, i.e. 5 g, 10 g, and 20 g per bump, are simulated. The concentrated force applied on the die is calculated by considering various stacking load conditions and the number of microbumps in the stacking die. For solder and copper materials, a bilinear elastoplastic constitutive model is implemented. Other materials are considered as having elastic behavior. The orthotropic material properties are implemented for the substrate material.

4.3.3.3.2 Analysis for stacking the middle die

Fig. 4.3.8 shows the simulation results of die deflection and stress when the middle die is stacked on the bottom die using stacking force of 10 g per bump. The bottom die has large deflection due to the bending force. The maximum bending stress occurs at the top surface of the die, and its location is corresponding to the inner bumps of the bottom die. The deflection of the beam is proportional to the bending moment, which is also proportional to the bending force. An approximately linear relationship between the maximum die deflection and stacking force is shown in Fig. 4.3.9a. Large

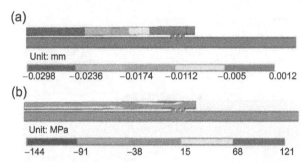

Figure 4.3.8 Simulation results for stacking force of 10 g per bump: (a) deflection results and (b) bending stress.

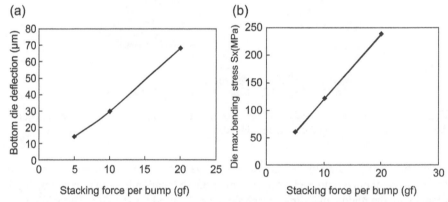

Figure 4.3.9 Effect of stacking force on results of bottom die: (a) the deflection of the bottom die and (b) the maximum stress of the bottom die (1gf = 0.0098 N).

deflection is dangerous for a thin die because the die will crack under large deflection and stress.

An approximately linear relationship between the maximum die stress and stacking force is shown in Fig. 4.3.9 b. Based on the beam theory, the normal stress of the beam with rectangular cross section can be determined by the following equation:

$$\sigma_{max} = \frac{My_{max}}{I} = \frac{6M}{b \cdot h^2} \quad (4.3.2)$$

where M is the bending moment, y the perpendicular distance to the neutral axis, I the area moment of inertia, b the width of the cross section, and h the height of the cross section. The maximum bending stress is proportional to the bending moment but inversely proportional to the square of the beam thickness. Thus, the maximum stress is more sensitive to the thickness of the beam. There is no bump at the bottom die to support the bending force directly due to improper bump layout design. Therefore, the bottom die is subjected to a large bending moment and die stress when the middle die is stacked.

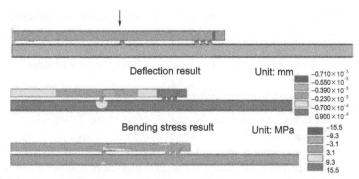

Figure 4.3.10 New design and simulation results of the bottom die under the stacking force of 20 g per bump.

Based on the simulation result, the bump layout of the bottom die is optimized by adding one more row of solder bumps (see Fig. 4.3.10) to support the bending force induced by stacking the middle die. Die deflection and die bending stress reduce significantly by introducing the optimal bump layout design for the bottom die. For example, for a stacking force of 20 g per bump, optimal bump layout design yields 0.71 μm die deflection, which is 98.9% lower in magnitude comparing to that of the initial bump layout design (68.1 μm). Maximum stress reduces from 237.8 to 15.5 MPa.

4.3.3.3.3 Analysis for stacking the top die

Fig. 4.3.11 shows the simulation results of die deflection and bending stress when the top die is stacked on the middle die for the stacking force of 20 g per bump. Compared with the case of stacking the middle die, both die deflection and stress are higher for the case of stacking the top die. The thickness of the middle die is 100 μm, which is half

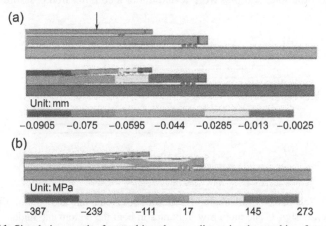

Figure 4.3.11 Simulation results for stacking the top die under the stacking force of 20 g per bump: (a) die deflection and (b) die bending stress.

that of the bottom die. The die deflection and stress are sensitive to die thickness. When the top die is stacked, the middle die has larger deflection and stress than the bottom die. Bump layout should be carefully designed for both bottom die and middle die to overcome the large deflection and stress resulting from stacking process.

4.3.3.3.4 Optimization of bump layout design

For the initial bump layout design, the bumps are close to the TSVs, which are distributed in the die peripherally. There is no bump at the central area of the die to support the bending force introduced by the stacking of upper die, and thus large die deflection and stress occur during the die stacking process. In the optimal design as shown in Fig. 4.3.12, dummy bumps are added to support the concentrated bending force induced in the die stacking process.

Fig. 4.3.13 shows a comparison of maximum bending stress for different bump layout designs when the top die is stacked. The optimal designs help to reduce bending stress significantly compared with the initial design, which can improve die reliability significantly. New design 2 is adopted, which is shown in Fig. 4.3.14 by a schematic of cross-sectional view for the optimized bump layout design in the 3-die stack package.

The 3-die stacking process is achieved successfully for the TSV dies with the optimized bump layout design even under high stacking force because the die deflection and stress is reduced significantly. Fig. 4.3.15 shows a pyramidal shape 3-die stacked package after the die stacking process. After that, solder reflow, underfilling and molding processes can be carried out.

4.3.3.4 Experimental validation

To obtain the accurate die cracking failure data, the compression test on the stacked dies was carried out. Flip-chip bonder was used to stack the dies in the sample preparation. The 3-die stack samples were tested under a compression pressure to simulate

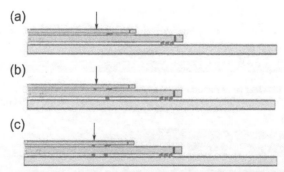

Figure 4.3.12 Schematic of microbump layout for both the middle and bottom dies: (a) initial design, (b) new design 1: one more row of bumps in both the bottom and middle dies, and (c) new design 2: one more row of bumps in the middle die and two more rows of bumps in the bottom die.

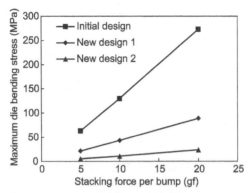

Figure 4.3.13 Comparison of the maximum stress for different designs when stacking the top die.

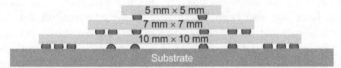

Figure 4.3.14 Schematic of the optimized bump layout for 3-die stacked package.

Figure 4.3.15 3-die stacked package after die stacking process.

the die stacking condition. Fig. 4.3.16 shows a typical loading curve of displacement versus compression load for 3-die stack sample. The applied compression load increases with increasing displacement. The sharp drop of compression load occurs when the load increases to 35.64 N, which indicates a die cracking failure load. We found that the middle die cracks in 3-die compression test. Such failure data can be used to verify FEA simulation results.

FEA simulation, stack die compression test and die strength test have been carried out to investigate the reliability of 3-die stacked package with the initially designed

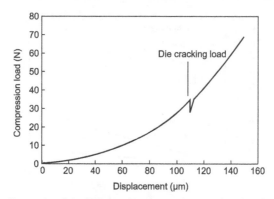

Figure 4.3.16 Loading curve of the 3-die stack compression test showing the die cracking force.

bump layout. The die strength determined from die strength test is 279 MPa. Simulation results show that the die stress increases linearly with increasing stacking force. Die cracking force has also been determined by the compression test. Combining the simulation results and die cracking force results, the die stress corresponding to die cracking failure can be determined, which can be used to compare with die strength to verify the simulation results. In the compression test of 3-die stack, the middle die cracks at a compressive force of 35.64 N, which is equivalent to 20.64 g per bump by considering a total of 176 bumps in the top die. Die stress can be determined by extrapolating simulation results and this failed stacking force. A comparison of results is presented in Table 4.3.1 for 3-die stacking condition. It can be seen from the table that die failure stress determined by FEA simulation and die compression test is similar to the die strength value by three-point bending test (281.74 MPa vs. 279 MPa). Thus, the FEA modeling methodology and simulation results have been validated experimentally.

4.3.3.5 Summary

In this case study, FEA simulation, die stacking experiment, stacked die compression test, and die strength 3-point bending test for a pyramidal shape 3-die stacked package have been carried out. FEA modeling and simulation is proved as one effective and

Table 4.3.1 Comparison of results for 3-die stacking condition.

3-Die stacking	Failure load(N)	Stacking load (gf/bump)	Die stress (MPa)	Die strength (MPa)
Simulation result	—	20	273	—
Compression test	35.64	20.64	281.74	281.74
3-P Bending test	—	—	—	279

powerful tool in the DFR methodology. The initial bump layout design induces a concentrated bending force on the lower die when the upper die is stacked, and thus the large die stress and deflection occur under the die stacking process. The die deflection and stress linearly increase with increasing stacking force, which is verified by simulation results. Thus, this design limits stacking force and leads to a potential die cracking failure. FEA simulation was conducted to investigate the reliability of die stacking process and to optimize the bump layout. Under the help of FEA modeling results, the bump layout design has been optimized by adding some dummy bumps on the dies to support the bending force induced by stacking the upper die. The die deflection and stress are reduced significantly under the die stacking process in optimal bump layout design. A 3-die stacking process has been achieved successfully by optimizing bump layout design.

References

[1] N.E. Dowling, Mechanics Behavior of Materials: Engineering Methods for Deformation, Fracture, and Fatigue, 4th Edition, Prentice-Hall International Editions, Boston, 2013.
[2] R. Darveaux, K. Banerji, Constitutive relations for tin-based solder joints, IEEE Trans. Compon. Hybrids and Manuf. Technol. 15 (1992) 1013–1024.
[3] Z.N. Cheng, G.Z. Wang, L. Chen, et al., Viscoplastic anand model for solder alloys and its application, Solder. Surf. Mt. Technol. 12 (2) (2000) 31–36.
[4] G.Z. Wang, Z.N. Cheng, K. Becker, et al., Applying anand model to represent the viscoplastic deformation behavior of solder alloys, J. Electron. Packag. 123 (2001) 247–253.
[5] R. Darveaux, Effect of simulation methodology on solder joint crack growth correlation, in: Proceedings of 50th Electronic Components and Technology Conference, 2000, pp. 1048–1058.
[6] L. Anand, Constitutive equations for hot-working of metals, Int. J. Plast. 1 (1985) (1985) 213–231.
[7] T.C. Chiu, C.L. Kung, H.W. Huang, Y.S. Lai, Effects of curing and chemical aging on warpage characterization and simulation, IEEE Trans. Device Mater. Reliab. 11 (2) (2011) 339–348.
[8] D.J. O'Brien, P.T. Mather, S.R. White, Viscoelastic properties of an epoxy resin during cure, J. Compos. Mater. 35 (10) (2001) 883–904.
[9] M.H.R. Ghoreishy, Determination of the parameters of the Prony series in hyperviscoelastic material models using the finite element method, Mater. Des. 35 (2012) 791–797.
[10] G.S. Zhang, H.Y. Jing, L.Y. Xu, J. Wei, Y.D. Han, Creep behavior of eutectic 80Au/20Sn solder alloy, J. Alloy. Comp. 476 (2009) 138.
[11] S. Wen, L.A. Keer, S. Vaynman, L.R. Lawson, A constitutive model for a high lead solder, IEEE Trans. Compon. Packag. Technol. 25 (2002) 23.
[12] F.X. Che, W.H. Zhu, S.W. Edith, X. Poh, R. Zhang, X.W. Zhang, T.C. Chai, et al., Creep properties of Sn-1.0Ag-0.5Cu lead-free solder with the addition of Ni, J. Electron. Mater. 40 (3) (2011) 344–354.
[13] H.L.J. Pang, C.W. Seetoh, Z.P. Wang, CBGA solder joint reliability evaluation based on elastic-plastic-creep analysis, J. Electron. Packag. 122 (2000) 255–261.

[14] Y.W. Wang, Y.W. Lin, C.T. Tu, C.R. Kao, Effects of minor Fe, Co, and Ni additions on the reaction between SnAgCu solder and Cu, J. Alloy. Comp. 478 (2009) 121–127.

[15] Y. Shi, J. Tian, H. Hao, Z. Xia, Y. Lei, F. Guo, Effects of small amount addition of rare earth Er on microstructure and property of SnAgCu solder, J. Alloy. Comp. 453 (2008) 180–184.

[16] B.I. Noh, J.H. Choi, J.W. Yoon, S.B. Jung, Effects of cerium content on wettability, microstructure and mechanical properties of Sn-Ag-Ce solder alloys, J. Alloy. Comp. 499 (2010) 154–159.

[17] M.F.M. Sabri, D.A.A. Shnawah, I.A. Badruddin, S.B.M. Said, F.X. Che, T. Arigad, Microstructural stability of Sn-1Ag-0.5Cu-xAl ($x=1$, 1.5, and 2 wt.%) solder alloys and the effects of high-temperature aging on their mechanical properties, Mater. Char. 78 (2013) 129–143.

[18] D.A.A. Shnawah, S.B.M. Said, M.F.M. Sabri, I.A. Badruddin, F.X. Che, Novel Fe-containing Sn–1Ag–0.5Cu lead-free solder alloy with further enhanced elastic compliance and plastic energy dissipation ability for mobile products, Microelectron. Reliab. 52 (2012) 2701–2708.

[19] P. Liu, P. Yao, J. Liu, Evolutions of the interface and shear strength between SnAgCu–xNi solder and Cu substrate during isothermal aging at 150 °C, J. Alloy. Comp. 486 (2009) 474–479.

[20] W.W. Lee, L.T. Nguyen, G.S. Selvaduray, Solder joint fatigue models: review and application to chip scale packages, Microelectron. Reliab. 40 (2000) 231–244.

[21] F.X. Che, J.H.L. Pang, Vibration reliability test and finite element analysis for flip chip solder joint, Microelectron. Reliab. 49 (2009) 754–760.

[22] J.H. Lau, Solder joint reliability of flip chip and plastic ball grid array assemblies under thermal mechanical, and vibrational conditions, IEEE Trans. Compon. Packag. Manuf. Technol. B 19 (4) (1996) 728–735.

[23] X.Q. Shi, H.L.J. Pang, W. Zhou, et al., A modified energy-based Low cycle fatigue model for eutectic solder alloy, Scr. Mater. 41 (3) (1999) 289–296.

[24] H.L.J. Pang, D.Y.R. Chong, T.H. Low, Thermal cycling analysis of flip-chip solder joint reliability, IEEE Trans. Compon. Packag. Technol. 24 (4) (2001) 705–712.

[25] A. Syed, Accumulated creep strain and energy density based thermal fatigue life prediction models for SnAgCu solder joints, in: Proceedings of 54th Electronic Components and Technology Conference, Las Vegas, 2004, pp. 737–746.

[26] A. Schubert, R. Dudek, E. Auerswald, et al., Fatigue life models for SnAgCu and SnPb solder joints evaluated by experiments and simulation, in: Proceedings of 53rd Electronic Components and Technology Conference, New Orleans, 2003, pp. 603–610.

[27] J.H.L. Pang, F.X. Che, Isothermal cyclic bend fatigue test method for lead-free solder joints, J. Electron. Packag. 129 (4) (2007) 496–503.

[28] X.Q. Shi, H.L.J. Pang, W. Zhou, et al., Low cycle fatigue analysis of temperature and frequency effects in eutectic solder alloy, Int. J. Fatigue 22 (2000) 217–228.

[29] H.D. Solomon, E.D. Tolksdorf, Energy approach to the fatigue of 60/40 solder: part I-influence of temperature and cycle frequency, J. Electron. Packag. 117 (2) (1995) 130–135.

[30] H.L.J. Pang, K.H. Tan, X.Q. Shi, et al., Microstructure and intermetallic growth effects on shear and fatigue of solder joints subjected to thermal cycling aging, Mater. Sci. Eng. A307 (2001) 42–50.

[31] D.H. Kim, P. Elenius, S. Barrettt, Solder joint reliability and characteristics of deformation and crack growth of Sn-Ag-Cu versus eutectic Sn-Pb on a WLP in a thermal cycling test, IEEE Trans. Electron. Packag. Manuf. 25 (2) (2002) 84–90.

[32] R.R. Tummala, Fundamentals of Microsystems Packaging, Mc Graw Hill, 2001.
[33] W. Nelson, Accelerated Testing: Statistical Models, Test Pla-ns, and Data Analyses, John Willey and Sons Publication, New York, 1990.
[34] K.C. Norris, A.H. Landzberg, Reliability of controlled collapse interconnections, IBM J. Res. Dev. 13 (1969) 266–271.
[35] JEDEC Standard, JESD22-A104D: Temperature Cycling, 2005.
[36] JEDEC Standard, JESD22-A106B: Thermal Shock, 2004.
[37] H.L.J. Pang, B.S. Xiong, Mechanical properties for Sn-3.8Ag-0.7Cu lead-free solder alloy, IEEE Trans. Compon. Packag. Technol. 28 (2005) 830–840.
[38] H.L.J. Pang, B.S. Xiong, T.H. Low, Creep and fatigue characterization of lead free Sn-3.8Ag-0.7Cu solder, in: Proceedings of 54th Electronic Components Technology Conference, 2004, pp. 1333–1337.
[39] M. Amagai, Characterization of chip scale packaging materials, Microelectron. Reliab. 39 (1999) 1365–1377.
[40] H.L.J. Pang, B.S. Xiong, T.H. Low, Low cycle fatigue models for lead-free solders, Thin Solid Films 462–463 (2004) 408–412.
[41] M. Amagai, M. Watanabe, M. Omiya, K. Kishimoto, T. Shibuya, Mechanical characterization of Sn-Ag-based lead-free solders, Microelectron. Reliab. 42 (2002) 951–966.
[42] H. Mavoori, J. Chin, S. Vaynman, B. Moran, L. Keer, M. Fine, Creep, stress relaxation, and plastic deformation in Sn-Ag and Sn-Zn eutectic solders, J. Electron. Mater. 26 (7) (1997) 783–790.
[43] B.Z. Hong, L.G. Burrell, Modeling thermally induced viscoplastic deformation and low cycle fatigue of CBGA solder joints in a surface mount package, IEEE Trans. Compon. Packag. Manuf. Technol. A 20 (3) (1997) 280–285.
[44] A.G. Guédon, E. Woirgard, C. Zardini, Reliability of lead-free BGA assembly: correlation between accelerated ageing tests and FE simulations, IEEE Trans. Device Mater. Reliab. 8 (3) (2008) 449–454.
[45] J. Pyland, R.V. Pucha, S.K. Sitaraman, Thermomechanical reliability of underfilled BGA packages, IEEE Trans. Electron. Packag. Manuf. 25 (2002) 100–106.
[46] X.J. Fan, M. Pei, P.K. Bhatti, Effect of finite element modeling techniques on solder joint fatigue life prediction of flip-chip BGA packages, in: Proceedings of 56th Electronic Components Technology Conference, 2006, pp. 972–980.
[47] F.X. Che, J.E. Luan, D. Yap, K.Y. Goh, X. Baraton, Thermal cycling fatigue model development for FBGA assembly with Sn-Ag-based lead-free solder, in: Proceedings of 33rd International Electronics Manufacturing Technology Conference, 2008, pp. 1–6.
[48] T.T. Nguyen, D. Yu, S.B. Park, Characterizing the mechanical properties of actual SAC105, SAC305, and SAC405 solder joints by digital image correlation, J. Electron. Mater. 40 (6) (2001) 1409–1415.
[49] H.T. Chen, M. Mueller, X. Liu, K.-J. Wolter, M. Paulasto-Krockel, Localized recrystallization and cracking of lead-free solder interconnections under thermal cycling, J. Mater. Res. 26 (16) (2011) 2103–2116.
[50] J. Li, T.T. Mattlia, J.K. Kivilahti, Multiscale simulation of microstructural changes in solder interconnections during thermal cycling, J. Electron. Mater. 39 (1) (2010) 77–84.
[51] J.H.L. Pang, L.H. Xu, X.Q. Shi, W. Zhou, S.L. Ngoh, Intermetallic growth studies on Sn-Ag-Cu lead-free solder joints, J. Electron. Mater. 33 (2004) 1219–1226.
[52] Y.C. Chan, P.L. Tu, C.W. Tang, K.C. Hung, J.K.L. Lai, Reliability studies of μBGA solder joints-effect of Ni-Sn intermetallic compound, IEEE Trans. Adv. Packag. 24 (2001) 25–32.

[53] F.X. Che, J.H.L. Pang, Characterization of IMC layer and its effect on thermomechanical fatigue life of Sn-3.8Ag-0.7Cu solder joints, J. Alloy. Comp. 541 (2012) 6–13.
[54] D.R. Frear, S.N. Burchett, H.S. Morgan, J.H. Lau, The Mechanics of Solder Alloy Interconnects, Van Nostrand Reinhold, New York, 1994, 60.
[55] Y. Yao, L.M. Keer, M.E. Fine, Modeling the failure of intermetallic/solder interfaces, Intermetallics 18 (2010) 1603–1611.
[56] S. Shetty, V. Lehtinen, A. Dasgupta, et al., Fatigue of chip scale package interconnects due to cyclic bending, J. Electron. Packag. 123 (2001) 302–308.
[57] S. Shetty, T. Reinikainen, Three- and four-point bend testing for electronic packages, J. Electron. Packag. 125 (2003) 556–561.
[58] L.L. Mercado, B. Philips, S. Sahasrabudhe, et al., Use-condition-based cyclic bend test development for handheld components, in: Proceedings of 54th Electronic Components and Technology Conference, Las Vegas, 2004, pp. 1279–1287.
[59] P. Geng, P. Chen, Y. Ling, Effect of strain rate on solder joint failure under mechanical load, in: Proceedings of 52nd Electronic Components and Technology Conference, 2002, pp. 974–978.
[60] JEDEC Standard, JESD22-B111: Board Level Drop Test Method of Components for Handheld Electronic Products, 2003.
[61] JEDEC Standard, JESD22B113: Board Level Cyclic Bend Test Method for Interconnect Reliability Characterization of Components for Handheld Electronic Products, 2004.
[62] IPC/JEDEC Standard, IPC/JEDEC-9702: Monotonic Bend Characterization of Board-Level Interconnects, 2004.
[63] F.X. Che, J.H.L. Pang, Fatigue reliability analysis of Sn-Ag-Cu solder joints subject to thermal cycling, IEEE Trans. Device Mater. Reliab. 13 (1) (2013) 36–49.
[64] A. Skipor, L. Leicht, Mechanical bending fatigue reliability and its application to area array packaging, in: Proceedings of 51st Electronic Components and Technology Conference, 2011, pp. 606–612.
[65] D.S. Steinberg, Vibration Analysis for Electronic Equipment, second ed., John Wiley & Sons, New York, 1998.
[66] D.B. Barker, Y.S. Chen, A. Dasgupta, Estimating the vibration fatigue life of quad leaded surface mount components, J. Electron. Packag. 115 (1993) 195–200.
[67] H.L.J. Pang, F.X. Che, T.H. Low, Vibration fatigue analysis for FCOB solder joints, in: Proc. 54th Electronic Components Technology Conference, 2004, pp. 1055–1061.
[68] Q.J. Yang, H.L.J. Pang, Z.P. Wang, G.H. Lim, F.F. Yap, R.M. Lin, Vibration reliability characterization of PBGA assemblies, Microelectron. Reliab. 40 (2000) 1097–1107.
[69] C. Basaran, R. Chandaroy, Nonlinear dynamic analysis of surface mount interconnects: part I-theory, J. Electron. Packag. 121 (1999) 8–11.
[70] C. Basaran, R. Chandaroy, Nonlinear dynamic analysis of surface mount interconnects: part II-application, J. Electron. Packag. 121 (1999) 12–17.
[71] Y. Zhao, C. Basaran, A. Cartwright, T. Dishongh, Thermomechanical behavior of micron scale solder joints under dynamic loads, J. Mech. Mater 32 (2000) 161–173.
[72] R.S. Li, A methodology for fatigue prediction of electronic components under random vibration load, J. Electron. Packag. 123 (2001) 394–400.
[73] T.E. Wong, B.A. Reed, H.M. Gohen, D.W. Chu, Development of BGA solder joint vibration fatigue life prediction model, in: Proc. 49th Electronic Components Technology Conference, 1999, pp. 149–154.
[74] Y.B. Kim, H. Noguchi, M. Amagai, Vibration fatigue reliability of BGA-IC package with Pb-free solder and Pb-Sn solder, Microelectron. Reliab. 46 (2006) 459–466.

[75] F. Stepniak, Failure criteria of flip chip joints during accelerated testing, Microelectron. Reliab. 42 (2002) 1921−1930.
[76] Y.S. Chen, C.S. Wang, Y.J. Yang, Combining vibration test with finite element analysis for the fatigue life estimation of PBGA components, Microelectron. Reliab. 48 (2008) 638−644.
[77] S.F. Wong, P. Malatkar, C. Rick, V. Kulkarni, I. Chin, Vibration testing and analysis of ball-grid array package solder joints, in: Proc. 57th Electronic Components and Technology Conference, 2007, pp. 373−380.
[78] A. Perkins, S.K. Sitaraman, A study into the sequencing of thermal cycling and vibration tests, in: Proc. 58th Electronic Components and Technology Conference, 2008, pp. 584−592.
[79] IPC Test Standards, IPC-9701: performance test methods and qualification requirements for surface mount solder attachments, IPC-Assoc. Connecting Electron. Ind. (2002) 13.
[80] L. Meirovitch, Elements of Vibration Analysis, second ed., McGraw-Hill, New York, 1986.
[81] Q.J. Yang, Z.P. Wang, G.H. Lim, H.L.J. Pang, F.F. Yap, R.M. Lin, Reliability of PBGA under out-of-plane vibration excitation, IEEE Trans. Compon. Packag. Technol. 25 (2002) 293−300.
[82] M.A. Miner, Cumulative fatigue damage, J. Appl. Mech. 12 (1945) A159−A164.
[83] A. Perkins, S.K. Sitaraman, Analysis and prediction of vibration-induced solder joint failure for a ceramic column grid array package, J. Electron. Packag. 130 (2008), 011012−1−11.
[84] C. Basaran, R. Chandaroy, Thermomechanical analysis of solder joints under thermal and vibrational loading, J. Electron. Packag. 124 (2002) 60−66.
[85] F.X. Che, H.L.J. Pang, W.H. Zhu, W. Sun, A.Y.S. Sun, C.K. Wang, H.B. Tan, Development and assessment of global-local modeling technique used in advanced microelectronic packaging, in: Proc. International Conference on Thermal, Mechanical and Multi-Physics Simulation Experiments in Microelectronics and Micro-systems, EuroSimE, 2007, pp. 375−381, 2007.
[86] Q.Z. Yao, J.M. Qu, S.X. Wu, Solder fatigue life in two chip scale packages, in: Proc. IEEE-IMAPS International Symposium on Microelectronics, 1999, pp. 563−570.
[87] D.Y.R. Chong, F.X. Che, J.H.L. Pang, L.H. Xu, B.S. Xiong, H.J. Toh, B.K. Lim, Evaluation on influencing factors of board-level drop reliability for chip scale packages (Fine-pitch ball-grid array), IEEE Trans. Adv. Packag. 31 (1) (2008) 66−75.
[88] D.Y.R. Chong, F.X. Che, J.H.L. Pang, K. Ng, J.Y.N. Tan, P.T.H. Low, Drop impact reliability testing for lead-free and leaded soldered IC packages, Microelectron. Reliab. 46 (2006) 1160−1171.
[89] Y.Q. Wang, K.H. Low, H.L.J. Pang, K.H. Hoon, F.X. Che, Y.S. Yong, Modeling and simulation for a drop-impact analysis of multi-layered printed circuit boards, Microelectron. Reliab. 46 (2006) 558−573.
[90] K. Mishiro, S. Ishikawa, M. Abe, et al., Effect of the drop impact on BGA/CSP package reliability, Microelectron. Reliab. (42) (2002) 77−82.
[91] L.P. Zhu, Drop impact reliability analysis of CSP packages at board and product levels through modeling approaches, IEEE Trans. Compon. Packag. Technol. 28 (3) (2005) 449−456.
[92] C.T. Lim, C.W. Ang, L.B. Tan, et al., Drop impact survey of portable electronic products, in: Proceedings of 53rd Electronic Components and Technology Conference, New Orleans, 2003, pp. 113−120.

[93] M. Date, T. Shoji, M. Fujiyoshi, K. Sato, K.N. Tu, Impact reliability of solder joints, in: Proc. 54th Electron. Compon. Technol. Conf., Las Vegas, 2004, pp. 668–674.

[94] F.X. Che, J.H.L. Pang, Study on board-level drop impact reliability of Sn-Ag-Cu solder joint with considering strain rate dependent properties of solder, IEEE Trans. Device Mater. Reliab. 15 (2) (2015) 181–190.

[95] L.H. Xu, J.H.L. Pang, F.X. Che, Impact of thermal cycling on Sn-Ag-Cu solder joints and board-level drop reliability, J. Electron. Mater. 37 (6) (2008) 880–886.

[96] G.Y. Li, B.L. Chen, Formation and growth kinetics of interfacial intermetallic in Pb-free solder joint, IEEE Trans. Compon. Packag. Technol. 26 (3) (2003) 651–658.

[97] E. Bradley, K. Banerji, Effect of PCB finish on the reliability and wettability of ball-grid array packages, IEEE Trans. Compon. Packag. Manuf. Technol. 19 (2) (1996) 320–330.

[98] S.M. Hayes, N. Chawl, D.R. Frear, Interfacial fracture toughness of Pb-free solders, Microelectron. Reliab. 49 (3) (2009) 269–287.

[99] T.T. Mattila, P. Marjamäki, J.K. Kivilahti, Reliability of CSP components under mechanical shock loading, IEEE Trans. Compon. Packag. Technol. 29 (4) (2006) 787–795.

[100] F.X. Che, J.E. Luan, K.Y. Goh, X. Baraton, Drop impact life model development for FBGA assembly with lead-free solder joint, in: Proc. 11th Electron. Packag. Technol. Conf., Singapore, 2009, pp. 656–662.

[101] J.H.L. Pang, F.X. Che, Drop impact analysis of Sn-Ag-Cu solder joints using dynamic high-strain rate plastic strain as the impact damage driving force, in: Proc. 56th Electron. Compon. Technol. Conf., San Diego, California, 2006, pp. 49–54.

[102] J.H. Zhao, J. Tellkamp, V. Gupta, D.R. Edwards, Experimental evaluations of the strength of silicon die by 3-point-bend versus ball-on-ring tests, IEEE Trans. Electron. Packag. Manuf. 32 (4) (2009) 248–255.

Reliability and failure analysis of encapsulated packages

5.1 Typical integrated circuit packaging failure modes

In the last fifty years, IC packages experienced the major packaging development in terms of typical wire bonded package, flip-chip substrate package, fan-in, and fan-out wafer-level packages by using Cu pillar and 2.5D/3D through-silicon vias package. Chip failures are described by many cracking at different locations of the die and package, such as the die corner, the lateral side of the die, die backside, solder bump, under bump metallization (UBM), wire bond pad low-k, etc. The thermal mechanical stress—induced package failures are mainly characterized by the cracking with underfill, substrate, and bump near substrate. Chip package interaction (CPI) failures come from thermal mechanical stress during assembly and field applications under various environmental conditions. CPI associated failures are shown in Figs. 5.1.1—5.1.3, respectively [1].

The materials, design, and assembly process may also have an impact on the CPI reliability issues. Most of the time, materials play a critical role, including the CTE mismatch between materials, stress concentration caused by material defects, low-k die, underfill, or lead-free bumps. From design perspective, people may ignore the effects of polyimide opening size, UBM thickness, metal structure, ELK, substrate routing, and bump density, etc. Process steps, such as wafer dicing parameter setup, wire bond process window setup, and flip-chip attachment profile, could also contribute to the CPI issues.

The IC package failures are closely related to the package stress, which results from the multiphysics interaction. Typical thermal, thermal—mechanical, hygro-, and hygro—thermal—mechanical stress and the corresponding fracture mechanics will be elaborated in the following sections [2—9].

5.2 Heat transfer and moisture diffusion in plastics integrated circuit packages

The typical "popcorn" cracking phenomenon of plastic integrated circuit (IC) packages during reflow soldering was attributed to the effect of moisture absorbed by the plastic molding compound resin before the solder reflow process [10—14]. Fig. 5.2.1 illustrates the mechanism of popcorn cracking, which was proposed by Fukuzawa et al. [10]. If the plastic IC package is exposed to a humid environment before the solder reflow process, moisture would diffuse into the plastic resin encapsulating the pad—die assembly. The longer the exposure is, the greater the moisture level in the package is. Moisture may accumulate along the tiny pores at the pad—resin

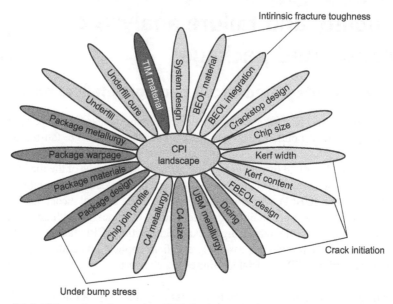

Figure 5.1.1 Chip package interaction (CPI) landscape for integrated circuit (IC) package.

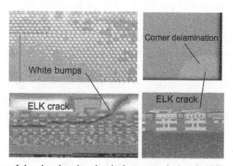

Figure 5.1.2 Die corner delamination in plastic integrated circuit (IC) package.

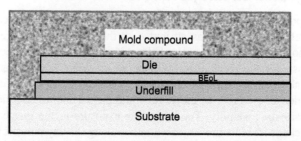

Figure 5.1.3 Edge delamination due to CTE mismatch during thermal cycling (substrate package).

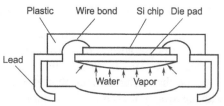

Figure 5.2.1 Mechanism of popcorn cracking of the small-outline J-leaded (SOJ) package.

interface. During reflow soldering, the temperature of the package is raised to about 215°C. At such a high temperature, the water along the pad—resin interface would vaporize and cause delamination in conjunction with the induced thermo—mechanical stress and eventual popcorn cracking of the plastic resin.

Two types of moisture preconditioning were used, namely, moisture absorption and moisture desorption. Kitano et al. [11] conducted experiments in which moisture-preconditioned plastic small-outline J-leaded (SOJ) packages were placed in a vapor phase reflow (VPR) oven for several minutes and then examined for evidence of package cracking. In the moisture absorption preconditioning process, completely dried IC packages were subjected to an 85°C/85%RH environment for varying periods. In the moisture desorption preconditioning process, the IC packages were first exposed to an 85°C/85%RH environment until saturated and then baked in dry air at 80°C. They found that cracking occurred in packages that had been either preconditioned by moisture absorption for longer than 34 h or by moisture desorption for shorter than 26 h. A primary conclusion from these experiments was that package cracking was not controlled by the absolute moisture level in the package but by the moisture concentration at the critical pad—resin interface. They then performed a one-dimensional analysis of the moisture distribution within the layer of plastic resin adjacent to the bottom of the die pad during the preconditioning moisture absorption and desorption processes. They found that the critical moisture concentration level at the pad—resin interface, which corresponded to the critical 34 h of absorption, was 63% and that compared with the critical 26 h of desorption was 60%. The 1D solution of Kitano et al. [11] is applicable for the central region of the die pad. However, it is not clear how moisture concentration and rates of diffusion vary along the entire width and length of the die pad. As the thermomechanical stress is the greatest at the edge of the die pad, the moisture there could prove to be more critical than that at the center. This is particularly true for small-outline packages with very small thickness of resin around the edge.

This section describes two-dimensional (2D) finite element (FE) simulations of the heat transfer and moisture diffusion in the IC packages during moisture absorption, desorption, and VPR. The results show good agreement with experimental data. The limitations of 1D analyses are also examined. The transient buildup of water vapor pressure in the delaminated pad—resin interface during VPR soldering is also obtained, which is necessary for package cracking analysis.

5.2.1 Governing equations for heat and moisture diffusion

When moisture enters the resin, it exists in many different forms [12]. Two diffusion theories, single-phase and two-phase absorption, have been applied to describe the moisture absorption process [12–18]. In this section, we use the single-phase absorption/desorption theory that assumes that moisture diffusion occurs freely without any bonding with the resin. For 2D diffusion of moisture with no local sources or sinks, the governing differential equation obeys Fick's second law of diffusion, given by

$$\frac{\partial}{\partial x_i} \text{DM}_{ij} \left(\frac{\partial C}{\partial x_j} \right) = \frac{\partial C}{\partial t} \tag{5.2.1}$$

and for an isotropic region of the resin

$$\nabla \cdot \text{DM} \nabla C = \frac{\partial C}{\partial t} \tag{5.2.2}$$

where C is the moisture concentration, DM is the moisture diffusivity, x is the Cartesian coordinate, and t is the time. For moisture diffusion processes accompanied by simultaneous heat conduction in the same region, the analogous equations governing thermal diffusion are as follows:

$$\frac{\partial}{\partial x_i} \left(k_{ij} \frac{\partial T}{\partial x_j} \right) = \rho c_p \frac{\partial T}{\partial t} \tag{5.2.3}$$

$$\nabla \cdot k \nabla T = \rho c_p \frac{\partial T}{\partial t} \tag{5.2.4}$$

where T is temperature, k is thermal conductivity, ρ is density, and c_p is specific heat. DM and k are assumed to be functions of temperature.

The water absorption process within the resin includes both solubility and diffusivity, respectively. Diffusion is described by the above Fick's laws of diffusion. Solubility of water molecules in resin obeys Henry's law in the linear regime. The diffusivity DM is determined by the following Arrhenius equation:

$$\text{DM} = \text{DM}_0 \exp\left(-\frac{Q_{\text{act}}}{R_u T}\right) \tag{5.2.5}$$

where T is absolute temperature, R_u is the universal gas constant, Q_{act} is the activation energy (4.84×10^4 J/mol), and $\text{DM}_0 = 47.2$ mm^2/s [11]. The solubility s can also be described by the Arrhenius equation

$$s = s_0 \exp\left(-\frac{Q_s}{R_u T}\right) \tag{5.2.6}$$

where $Q_s = -3.87 \times 10^4$ J/mol and $s_0 = 4.96 \times 10^{-7}$ mg·mm^{-3}·MPa^{-1} [11].

5.2.2 Constitutive relations for water vapor

The conservation of mass of water vapor is governed by the following equation:

$$\int_V \frac{dC}{dt} dV + \int_s n \cdot J ds = 0 \tag{5.2.7}$$

where V is the volume with the surface s, n is the outward normal to s, and J is the concentration flux of the diffusing phase leaving s, given by

$$J = -n \cdot \mathrm{DM} \nabla C = -s \cdot \mathrm{DM} \nabla \cdot \phi \tag{5.2.8}$$

where $\phi = C/s$ is the normalized concentration. In mass diffusion studies involving regions of dissimilar materials, computations are usually carried out in terms of the normalized concentration ϕ, as this is continuous across interfaces of dissimilar materials [19].

The total mass of water in the delaminated pad–resin interface from time t_1 to t_2 is determined as follows:

$$m = m_0 + \int_A \int_{t_1}^{t_2} \mathrm{DM} \left(\frac{\partial C}{\partial x} + \frac{\partial C}{\partial y} \right) dt dA \tag{5.2.9}$$

where m_0 is the initial mass of water in the delaminated interface and A is area of the chip pad. Treating the delaminated layer of resin below the pad as a thin plate deforming under the uniform vapor pressure, Kitano et al. [11] showed that the maximum clearance of the gap between the pad and the resin is given by

$$\delta = 0.249 \frac{a^{2.91}}{E^{0.93} H^{1.97}} p + \delta_0 \tag{5.2.10}$$

where a is length of chip pad in the short direction, E is the modulus of elasticity of the plastic, H is the thickness of the plastic, p is the pressure, and δ_0 is the initial clearance. The units of δ, δ_0, a, and H are mm and those of p and E are MPa. The volume of the delamination (gap) V is given by Ref. [11].

$$V = \eta \cdot \delta \cdot A \tag{5.2.11}$$

where η is a constant. The specific volume of the vapor is given by

$$v = \frac{V}{m} \tag{5.2.12}$$

The specific volume v is a function of vapor pressure p and temperature T. The relationship is described by an equation of state [11] as follows:

$$v = \frac{461.2T}{p} - \frac{0.668}{\left(\dfrac{T}{100}\right)^2} - \frac{437 + 6.44 \times 10^6 p - 2 \times 10^{-34} p^5}{\left(\dfrac{T}{100}\right)^{8.4}} - \frac{1.77 \times 10^{-5} p^5}{\left(\dfrac{T}{100}\right)^{30.5}}$$

$$- \frac{5.4 \times 10^{-68} p^{25}}{\left(\dfrac{T}{100}\right)^{147}}$$

(5.2.13)

5.2.3 Initial and boundary conditions

The die and die pad are impermeable to moisture. Hence, the initial and boundary conditions for concentration are as follow:

Absorption:

$$C(t=0) = 0 \qquad (5.2.14)$$

$$C_{\Gamma_1} = \text{RH} \cdot p_s \cdot s \qquad (5.2.15)$$

$$\left.\frac{\partial C}{\partial n}\right|_{\Gamma_2} = \left.\frac{\partial C}{\partial x}\right|_{x=0} = 0 \qquad (5.2.16)$$

Desorption:

$$C(t=0) = C_0(x,y) \qquad (5.2.17)$$

$$C_{\Gamma_1} = 0 \qquad (5.2.18)$$

$$\left.\frac{\partial C}{\partial T}\right|_{\Gamma_2} = \left.\frac{\partial C}{\partial x}\right|_{x=0} = 0 \qquad (5.2.19)$$

where RH is the relative humidity, p_s is the saturated pressure, and $C_0(x,y)$ is the initial concentration distribution, and S has been defined in Eq.(5.2.6). Γ_1 is the outer top, right, and bottom boundary of the package body shown in Fig. 5.2.2. Γ_2 is the outer boundary of the die/pad assembly, excluding the portion along the plane of symmetry $x = 0$.

The initial and boundary conditions for temperature during preconditioning are

$$T(t=0) = T_0(x,y) \qquad (5.2.20)$$

$$T = T_s \quad \text{at } \Gamma_1 \qquad (5.2.21)$$

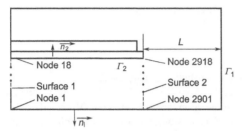

Figure 5.2.2 Finite element (FE) model for absorption and desorption.

$$\left.\frac{\partial T}{\partial x}\right|_{x=0} = 0 \tag{5.2.22}$$

where $T_0(x, y)$ is the initial temperature distribution and Γ_s is the VPR temperature.

5.2.4 Boundary condition during vapor phase reflow

The thermal conductivity of water vapor in the delaminated region is much lower than that of the resin. Hence, the heat flux along Γ_2 and Γ_3 (Fig. 5.2.3) may be neglected. In addition, the outer boundary is exposed to surroundings at 215°C. Thus, $T = 215$°C at Γ_1. The moisture concentration along the outer boundary Γ_1 is assumed to be zero and Γ_3 is impermeable to moisture.

5.2.5 Solution procedure

The procedure employed to solve the simultaneous heat transfer and moisture diffusion problem is detailed as follows, assuming that $T(x, y, t)$ and $C(x, y, t)$ are already known:

(1) Compute the temperature distribution in the package $T(x, y, t + \Delta t)$ at $t + \Delta t$.
(2) Obtain $D(T)$ and $s(T)$.
(3) Assume *a* value for $p(t + \Delta t)$, the water vapor pressure in the delamination (gap).

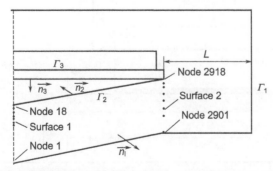

Figure 5.2.3 Finite element (FE) model of integrated circuit (IC) packages during vapor phase reflow (VPR) with delamination.

(4) Calculate the concentration along Γ_2 from

$$\Gamma_2 = p \cdot s \qquad (5.2.23)$$

(5) Compute the moisture distribution $C(x, y, t + \Delta t)$.
(6) Calculate the pressure p from Eqs. (5.2.9)–(5.2.13) and compare this value with that assumed in step (3). If the difference is less than 0.001, then the assumed $p(t+\Delta t)$ is correct. Otherwise, iterate for $p(t+\Delta t)$ between steps (3)–(6). Following steps (1)–(6), the transient distributions of temperature and moisture during VPR are obtained.

5.2.6 Finite element analysis

The 2D heat transfer and moisture diffusion analysis were carried out on the SOJ package shown in Fig. 5.2.4, using the general purpose FE code ABAQUS. Owing to symmetry, only half of the package needs be analyzed. The FE mesh employed consisting of 1332 nodes and 1260 eight-node quadrilateral elements is shown in Fig. 5.2.5. The properties of materials constituting the IC packages are listed in Table 5.2.1.

5.2.7 Results and discussion

5.2.7.1 Moisture distributions during preconditioning

The moisture absorption and desorption preconditioning processes performed by Kitano et al. [11] for their IC package were simulated. The variation of level of moisture saturation of the resin layer adjacent to the central pad region after varying hours of preconditioning are shown in Figs. 5.2.6 and 5.2.7. The results compare well with those of Kitano et al. suggesting that at the pad central, the 1D model is adequate. This can also be seen in Figs. 5.2.8 and 5.2.9, which show typical moisture distributions obtained after 34 h of absorption and 26 h of desorption, respectively. However, Figs. 5.2.8 and 5.2.9 also show some 2D effects around the edge of the pad.

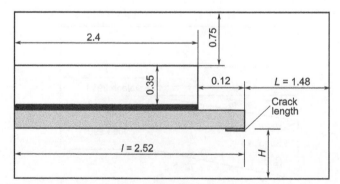

Figure 5.2.4 Geometry of small-outline J-leaded (SOJ) integrated circuit (IC) package. Unit: mm.

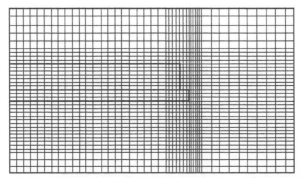

Figure 5.2.5 Finite element (FE) mesh of small-outline J-leaded (SOJ) package.

Table 5.2.1 Properties of packaging materials.

Component	Material	Thermal conductivity $(W \cdot m^{-1} \cdot K^{-1})$	Specific heat $(J \cdot kg^{-1} \cdot K^{-1})$	Density(kg/m^3)
Chip	Silicon	150	761	2340
Lead frame	Cu	385	380	8930

Component	Material	Thermal conductivity $(W \cdot m^{-1} \cdot K^{-1})$	Specific heat $(J \cdot kg^{-1} \cdot K^{-1})$	Density(kg/m^3)
Lead frame	Alloy	13.4	502	8250
Plastic	Epoxy	0.8	1052	1800

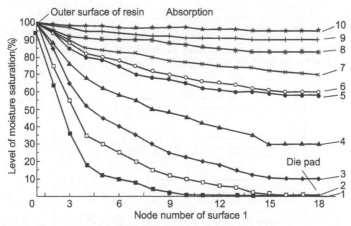

Figure 5.2.6 Moisture absorption from 0 to 100 h: (1) 2 h (not crack); (2) 4 h (not crack); (3) 8 h (not crack); (4) 16 h (not crack); (5) 30 h (not crack); (6) 34 h(crack); (7) 40 h (crack); (8) 60 h (crack); (9) 80 h (crack); and (10) 100 h (crack).

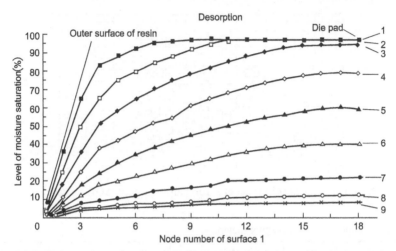

Figure 5.2.7 Moisture desorption from 0 to 100 h: (1) 2 h, crack; (2) 4 h, crack; (3) 8 h, crack; (4) 16 h, crack; (5) 26 h, crack; (6) 40 h, not crack; (7) 60 h, not crack; (8) 80 h, not crack; and (9) 100 h, not crack.

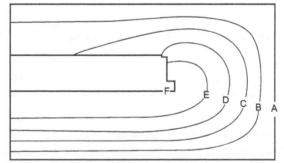

Figure 5.2.8 Moisture concentration distribution after 34 h of absorption. Concentration(kg/m^3): (A) 11; (B)10.2; (C) 9.39; (D) 8.58; (E) 7.67; and (F) 0.

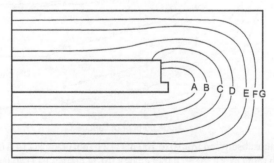

Figure 5.2.9 Moisture concentration distribution after 26 h of desorption. Concentration (kg/m^3):(A) 6.17×10^{-3};(B) 5.65×10^{-3};(C) 4.62×10^{-3}; (D) 3.59×10^{-3}; (E) 2.05×10^{-3}; (F) 2.02×10^{-3}; and (G) 0.

The effect of the length L (see Fig. 5.2.4) on the moisture concentration around the pad edge after 34 h of absorption was studied. The results are shown in Figs. 5.2.10—5.2.13. Figs. 5.2.10—5.2.12 show the transient buildup of moisture concentration at two representative pad—resin interface locations, node 18 at the center of the pad, and node 2918 at the edge of the pad (Fig. 5.2.3).

As can be seen, when L is large at 1.48 mm, there is very little difference between the rate of moisture buildup at the two locations. However, as L decreases (Figs. 5.2.11 and 5.2.12), the rate of moisture buildup at the pad edge is much faster than that at the center of the pad. In Fig. 5.2.13, the level of moisture saturation is plotted against the distance from the bottom surface of the resin, represented by nodes 2901 to 2918 (Fig.5.2.3). As can be seen, when L is large at 1.48 mm, there is significant difference between the moisture saturation level at the pad corner and that at the resin surface. However, this difference decreases as L decreases. The above results show that when L is about the same or larger than the thickness of the resin layer below the pad, 1D analysis will suffice. However, if L is significantly smaller than the thickness of the resin layer below the pad, 2D analysis has to be performed to obtain good accuracy.

The level of moisture saturation for the package was computed and plotted against the period of preconditioning in Figs. 5.2.14 and 5.2.15. The experimental results and the results from the 1D analysis of Kitano et al. [11] are also plotted in Figs. 5.2.14 and 5.2.15. It can be seen that for the case of desorption, the results obtained using 2D analysis gave better agreement with the experimental results than those obtained using 1D analysis. For the case of absorption, the agreement between 1D and 2D analysis is about the same.

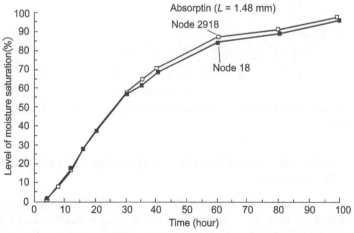

Figure 5.2.10 Transient rise of moisture saturation level at nodes 18 and 2918 during absorption.

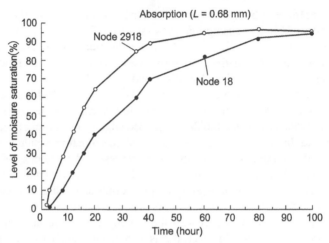

Figure 5.2.11 Transient rise of moisture saturation level at nodes 18 and 2918 during absorption ($l = 0.68$ mm).

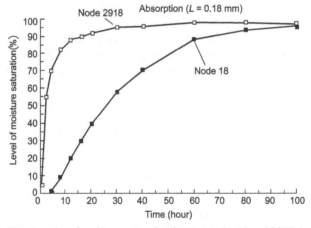

Figure 5.2.12 Transient rise of moisture saturation level at nodes 18 and 2918 during absorption ($L = 0.18$ mm).

5.2.7.2 Temperature and moisture distribution during vapor phase reflow

The experiments performed by Kitano et al. [11] in which IC packages were first preconditioned with moisture for varying periods before being subjected to a VPR process were simulated using FE analysis. It was found that the diffusion of heat in the IC package during VPR was very much faster than that of moisture. Thermal equilibrium in the package was reached rapidly within 10 s. Fig. 5.2.16 shows the temperature distribution after 3 s of VPR. Fig. 5.2.17 shows the moisture distribution after 20 s of

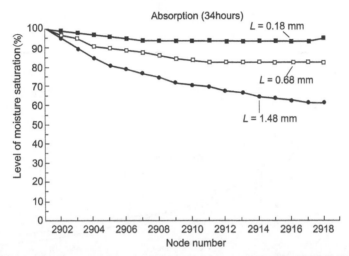

Figure 5.2.13 Distribution of moisture saturation level across resin adjacent to the die pad after 34 h of absorption.

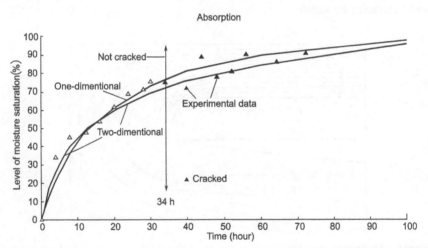

Figure 5.2.14 Comparison between the results of 1D and 2D finite element (FE) analysis and experiment during absorption.

VPR for a package that has been preconditioned with 34 h of moisture absorption. Fig. 5.2.18 shows the moisture distribution after 30 s of VPR for a package that has been preconditioned with 26 h of moisture desorption. Figs. 5.2.19 and 5.2.20 give the transient moisture distribution across the resin thickness. It can be seen that the moisture diffusion is much slower than the heat diffusion.

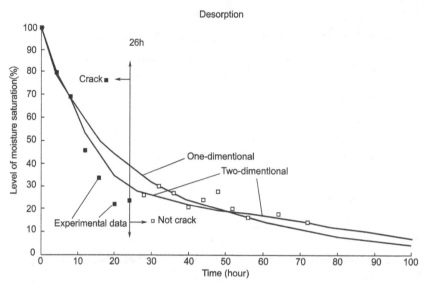

Figure 5.2.15 Comparison between the results of 1D and 2D finite element (FE) analysis and experiment during desorption.

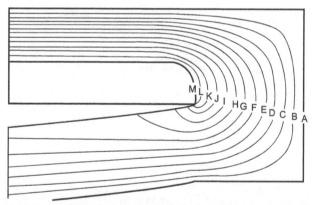

Figure 5.2.16 Temperature distribution in integrated circuit (IC) package 3 s after vapor phase reflow (VPR) soldering, with contour temperature levels in Kelvin: (A) 488; (B) 485: (C) 483; (D) 481; (E) 479.8; (F) 476; (G) 474; (H) 472.6; (I) 469; (J) 467; (K) 466; (L) 463.6; and (M) 459.

5.2.7.3 Vapor pressure buildup in the delaminated gap during vapor phase reflow

Fig. 5.2.21 shows the transient rise of temperature of the water vapor in the gap during VPR for a package that has been preconditioned with 34 h of absorption and another package that has been preconditioned with 26 h of desorption.

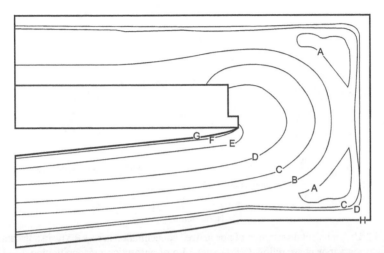

Figure 5.2.17 Moisture concentration distribution in integrated circuit (IC) package after 34 h of absorption preconditioning and 20 s of vapor phase reflow (VPR) soldering, with concentration levels in kg/m^3: (A) 11.9; (B) 11; (C) 10.1; (D) 9.22; (E) 8.29; (F) 7.378; (G) 6.456; and (H) 0.0.

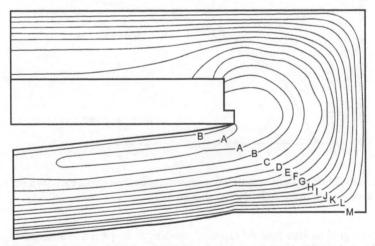

Figure 5.2.18 Moisture concentration distribution in integrated circuit (IC) package after 26 h of desorption preconditioning and 30 s of vapor phase reflow (VPR) soldering. Concentration (kg/m^3): (A) 6.94; (B) 6.36; (C) 5.78; (D) 5.21; (E) 4.68; (F) 4.05; (G) 3.49; (H) 2.99; (I) 2.31; (J) 1.73; (K) 1.15; (L) 0.78; and (M) 0.00.

Figs. 5.2.22 and 5.2.23 show the transient buildup of water vapor pressure in the delaminated gap during VPR for packages that have been subjected to varying hours of preconditioning by absorption and desorption, respectively. As can be seen, the buildup of water vapor pressure in the gap was rapid during the first 10 s of VPR

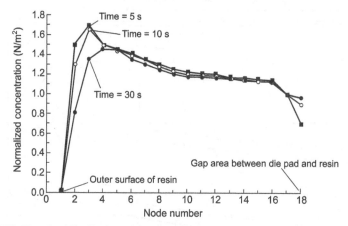

Figure 5.2.19 Transient distribution of normalized concentration across the resin adjacent to the die pad during vapor phase reflow (VPR) after 34 h of absorption preconditioning.

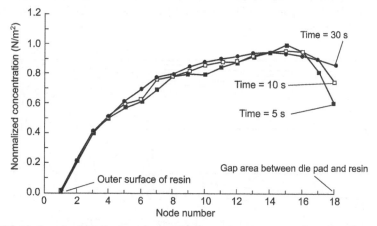

Figure 5.2.20 Transient distribution of normalized concentration across the resin adjacent to the die pad during vapor phase reflow (VPR) after 26 h of desorption preconditioning.

but became gradual after that. As expected, the longer the period of absorption preconditioning and the shorter the period of desorption preconditioning, the higher the final vapor pressure. A higher final vapor pressure will tend to cause package cracking. This is consistent with experimental observations that the greater the amount of moisture in the package, the greater the likelihood of package cracking during VPR.

Fig. 5.2.24 shows the transient rise of water vapor pressure in the gap for a package that has been preconditioned with 34 h of absorption and another package that has been preconditioned with 26 h of desorption. These two states of preconditioning are being extensively studied in this chapter as they represent the two critical states of preconditioning to lead to package cracking during VPR [11]. It is interesting to

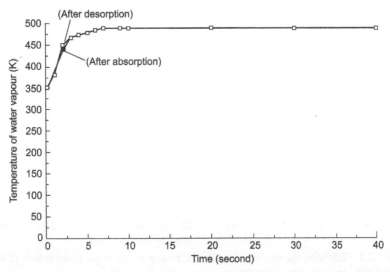

Figure 5.2.21 Transient rise of temperature of water vapor in the gap during vapor phase reflow (VPR) soldering.

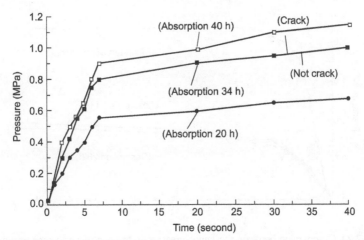

Figure 5.2.22 Transient rise of vapor pressure in the gap during vapor phase reflow (VPR) soldering after absorption preconditioning.

note that both these critical states of preconditioning gave virtually the same final vapor pressure in the gap of about 1 MPa. This pressure would appear to be the critical vapor pressure for the package under study which would lead to cracking.

The saturated vapor pressure corresponding to the temperature of the vapor in the gap is also plotted in Fig. 5.2.24. As can be seen, the water vapor in the gap during VPR is far from saturated, and judging from the very slow rate of increase of vapor pressure after the initial 10 s, it is unlikely that saturation pressure will be reached

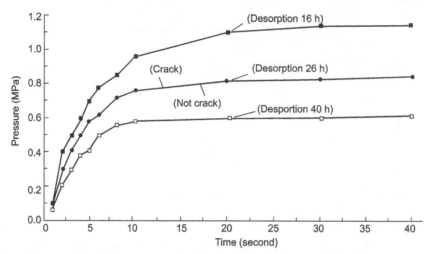

Figure 5.2.23 Transient rise of vapor pressure in the gap during vapor phase reflow (VPR) soldering after desorption preconditioning.

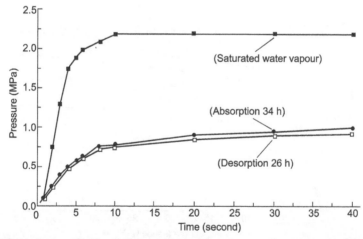

Figure 5.2.24 Comparison between the vapor pressure in the gap during vapor phase reflow (VPR) soldering and the saturation vapor pressure.

during the critical period during which solder is reflowed and resolidified. Thus, any analysis of package fracture that assumes that the water vapor in the gap is saturated is likely to be erroneous. The actual vapor pressure is a result of the dynamic influence of temperature, the rate of moisture diffusion into the gap, and the rate of increase in the volume of the gap due to the vapor pressure.

5.3 Thermal- and moisture-induced stress analysis

When plastic IC packages are subjected to high temperatures during solder reflow, large thermal stresses are induced along the interfaces within the package owing to differences in the thermal expansion coefficient of the different materials constituting the package. If the plastic package has been previously exposed to moisture, the problem is compounded by the additional hygrostress that is induced due to the swelling of the mold compound arising from moisture absorption. During solder reflow, the hygrostress adds to the thermal stress and increases the likelihood of delamination.

5.3.1 Development of thermal stress

When a plastic IC package is placed in a solder reflow oven, the environment temperature is higher than the surface temperature of the package, and heat will diffuse into the package. The temperature distribution in the plastic IC package is governed and described by previous equations Eq. (5.2.3).

For the boundary condition, it is assumed that the surface temperature of the package is the same as that of the environment temperature. This is particularly applicable for VPR processes where the surface heat transfer coefficient is very large. In this section, the molding compound was cured at 175°C while the package is in a zero-stress state. The thermal strain is defined as

$$\varepsilon_t = \int_{T_{cure}}^{T} \alpha \, dT \tag{5.3.1}$$

where α is the coefficient of thermal expansion (CTE) and T is temperature.
Thermal stress is given by

$$\sigma_t = E\varepsilon_t \tag{5.3.2}$$

where E is the modulus, and the mold compound was cured at 175°C while the package is in zero-stress state.

5.3.2 Development of hygrostress

When a plastic-encapsulated IC package is exposed to a humid environment, moisture diffuses through pores in the plastic causing the plastic encapsulate to swell. But moisture does not penetrate the silicon chip or metallic lead frame and hence does not affect their dimensions. This differential expansion between the plastic and the other materials results in the development of hygrostress.

Hygrostrain

$$\varepsilon_h = SC \times C \tag{5.3.3}$$

where SC is the swelling coefficient, or the coefficient of hygro expansion, and C the moisture concentration (wt.%) as shown below.

Hygrostress

$$\sigma_h = E\varepsilon_h \tag{5.3.4}$$

For moisture diffusion, governing Eqs. (5.2.1) and (5.2.2) clearly define moisture concentration C.

5.3.3 Case study for hygrothermal stress analysis

In this section, the plastic stacked die package was constructed in TQFP with 100 I/Os, denoted by 100TQFP, and temperature cycles after JEDEC level 3 (60°C at 60% relative humidity for 40 h) were performed. The temperature cycle results showed that 100TQFP with stacked die construction could pass 1000 temperature cycles with the electrical function test. However, it was found that the full delamination occurred at 100 temperature cycles at the lead frame paddle/mold compound interface where conventional failure modes were observed and could induce the popcorn cracking [20–23]. The actual evaluations were performed in terms of the comparisons between single-die and stacked packages by the fishbone diagrams implemented by process, material, design, and reliability assessments. The main results revealed that die attach paste voids initiated the small delamination at the die attach edge and propagated down to the lead frame paddle/mold compound interface due to the stress concentration and the weak adhesion strength (without dimple structure at the backside of the lead frame). The finite element analysis (FEA) was applied for the relative stress comparison between single and stacked die packages and finally verified by scanning acoustic microscopy (SAM). The assessment results demonstrated that the selection of packaging material and robust process was very crucial for maintaining the excellent package integrity and reliability.

5.3.3.1 Observed failure modes of TQFP package

Reliability evaluation and SAM were performed on TQFP100 with a stacked die after 100 temperature cycles. Based on the JEDEC/IPC standard, 1000 cycles are usually taken by industry to be a criterion for reliability assessment. The results of SAM showed that 16/82 failed at the mold compound/lead frame paddle interface with full delamination and popcorn cracking that was validated by SAM results, as shown in Figs. 5.3.1 and 5.3.2.

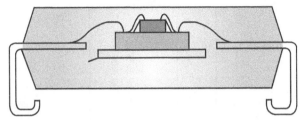

Figure 5.3.1 The schematic failure mode of 100TQFP with the stacked die at the lead frame paddle–mold compound interface.

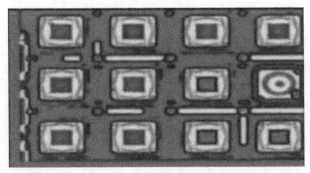

Figure 5.3.2 Photos of results on the acoustic microscopy performed by the typical through-mode scanning.

5.3.3.2 Design and material properties

Table 5.3.1 shows the material items containing the mold compound, lead frame, and die attach paste that was used for 100TQFP test vehicle. Based on Table 5.3.1, it was found that the die attach paste used for the stacked die was totally different from the single die because EP2 had higher die shear strength than EP1. The same lead frame and mold compound were used for single and stacked die packages.

For the structure of single and stacked die construction, there were a lot of differences, as shown in Table 5.3.2. In this study, the stacked die used had the size of 3.17 mm × 3.8 mm (top die) and 5 mm × 5 mm (bottom die) with a thickness of 10 mil, and the single die had the size of 3.17 mm × 3.8 mm (die thickness was 13 mil). It could be assumed that the stiffness and stress distribution for single and stacked die packages had large gaps due to the different die sizes and thicknesses. During the reliability assessment, the moisture soaking process and temperature cycles could result in different stress concentration for single and stacked die package. It is necessary to build a finite element model for single and stacked die packages to address the stress distribution and possible failure locations.

Table 5.3.1 Materials used for 100TQFP test vehicle.

Test vehicles	Mold compound (body): 14 mm × 14 mm × 1.4 mm	Lead frame (pad): 6 mm × 6 mm × 5 mil[*]	Die attach paste
100TQFP with single die	C1	Full silver plating without dimple structure	EP1
Controlled vehicle 100TQFP with stacked die	C1	Full silver plating without dimple structure	EP2 at the bottom die attach level, EP3 (nonconductive paste) at the top side

[*] 1 mil = 0.0254 mm.

Table 5.3.2 The impact of die size with 13 mil die thickness on package design.

	Bottom die	Top die	Thickness
100TQFP with single die (control lot)	3.17 mm × 3.8 mm	None	13 mil
100TQFP with stacked die	5 mm × 5 mm	3.17 mm × 3.8 mm	10 mil

5.3.3.3 Finite element analysis

The objective of this evaluation is to address the difference of the stress level between single and the stacked die packages. The results will be used to justify the possible failures. The thermal stress was simulated based on the temperature variations from 175 to 25°C using commercial FEA code ANSYS. It was noted that same mold compound (C1) and die attach paste (EP2) were used in both single and stacked die packages. For stacked die package, another die attach paste EP3 was used at the top die. The top die attach delamination was not found in this reliability assessment due to the minimum die-to-die CTE mismatch. The material properties are listed in Table 5.3.3.

Figs. 5.3.3 and 5.3.4 showed FEA model mesh for 100TQFP with single and stacked die, respectively. Densified mesh was applied at the delamination interfaces.

The simulation results (maximum principal stress) are shown in Figs. 5.3.5–5.3.8. The actual field observation by SAM was concurrently shown in Table 5.3.4. Note that the values shown in Table 5.3.4 were selected from the edge because there always were defects or microvoids or stress concentration areas in which the delamination could occur. Based on this experimental observation, the delamination always was found at the edge and propagated to the center areas.

As seen in Table 5.3.4, two critical facts revealed that the delamination occurred at both single and stacked die packages with different locations that were, respectively, corresponding to the higher stress level. In addition to the high-stress factor, the

Table 5.3.3 Material properties, 100TQFP stacked die: EP3 die attach paste was used at the top die and EP2 paste was used at the bottom die, 100TQFP single die: only EP2 paste was applied; the Poisson ratios of die and lead frame are 0.3 and 0.35, respectively.

Properties	C1 compound	EP2 die attach paste (bottom)	EP3 die attach paste (top)
CTE1 (ppm/°C)	11	85	45
CTE2 (ppm/°C)	47	140	136
Modulus (GPa)	21.5	2.4	0.72
Poisson ratio	0.25	0.25	0.25

Figure 5.3.3 Finite element analysis (FEA) model for 100TQFP with a single die, focusing on single-die and lead frame areas.

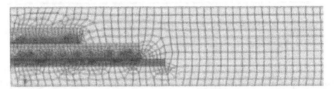

Figure 5.3.4 Finite element analysis (FEA) model for 100TQFP with stacked die, focusing on stacked die areas.

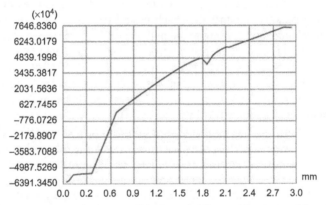

Figure 5.3.5 Maximum principal stress (Pa) along paddle−mold compound interface for TQFP100 with stacked die.

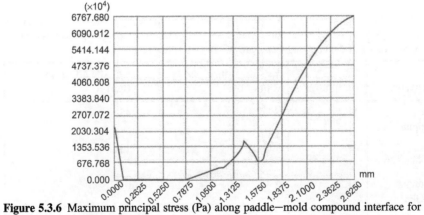

Figure 5.3.6 Maximum principal stress (Pa) along paddle−mold compound interface for TQFP100 with a single die.

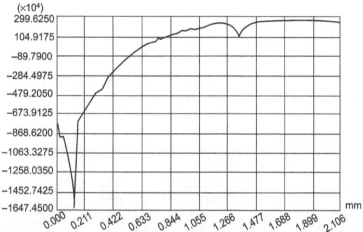

Figure 5.3.7 Maximum principal stress (Pa) along paddle die attach paste interface for TQFP100 with stacked die.

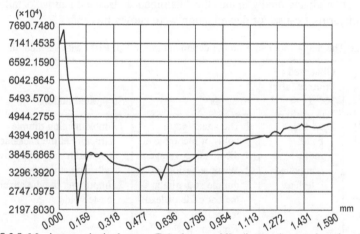

Figure 5.3.8 Maximum principal stress (Pa) along paddle die attach paste interface for TQFP100 with a single die.

adhesion strength was weak because there was no dimple structure at the backside of the lead frame, which could contribute to the delamination. It was noted that the above stress simulation was only representing the static status of the thermal stress from 175 to 25°C, which could not really describe the stress state of the reflow and temperature cycle process. However, if we considered the real stress state of the reflow process and temperature cycles, the stress or strain accumulations differences were actually not 14% for lead frame paddle/mold compound interface or 32% for bottom die attach/ lead frame interface if the stress and strain correlation were not proportional. They

Table 5.3.4 Comparison between the stress simulation and experimental observation by scanning acoustic microscopy(SAM) for 100TQFP test vehicle, compound C1, and die attach paste EP2.

Locations	100TQFP stacked die	100TQFP single die	Difference (%)
Lead frame/mold compound interface	76 MPa (edge corner)	67 MPa (edge corner)	14% (76−67/67)
Field observation	Delamination on LF/MC	No delamination	
Die attach/lead frame interface	30 MPa (edge corner)	44 MPa (edge corner)	32% (44−30/44)
Field observation	No delamination	Delamination on/ DA/LF	

could be very high and finally induce the delamination. Based on above detailed analysis, the major root causes of delamination were shown in Table 5.3.5.

Table 5.3.5 The major root cause on the delamination of TQFP100 with stacked die.

Package type	Observed failures (after 100 TCs)	Reasons	Suggested actions
100TQFP stacked die	Lead frame paddle/ mold compound interface	1. Die attach paste voids or incomplete die attach coverage 2. Weak adhesion (no dimple structure at the backside of lead frame) 3. High-stress concentration at the backside of lead frame	1. Implementing dimple structure at the backside of the lead frame to improve the adhesion strength 2. Improving the die attach paste voids

5.4 Fracture mechanics analysis in integrated circuits package

When plastic IC packages are exposed to high temperatures during solder reflow, large thermal stresses are developed along the interfaces within the package owing to CTE mismatch of the package. Voids or defects are often present at interfaces in plastic IC packages [23]. Very high-stress concentrations then develop at the edges of these interfacial defects or cracks. Stress intensity factor (SIF), stain energy release rate and J-integrals are basics for applying fracture mechanics theory. If the SIF exceeds a critical value, delamination or crack extension will occur along the interface [4,24,25]. If the plastic package has been previously exposed to moisture, the problem is compounded by the additional hygrostress that is induced due to the swelling of the mold compound arising from moisture absorption. As pointed out by Lin and Tay [4], during solder reflow, the hygrostress adds on to the thermal stress giving a higher stress intensity level and increasing the likelihood of delamination. Apart from increasing the SIF at the crack tip, moisture also degrades the adhesion between the pad and the mold compound [4,26,27], which further increases the likelihood of delamination propagation during solder reflow. Tanaka et al. [20] proposed a new adhesion test methodology, which can separate the residual stress from the true adhesion strength. This is useful especially for measuring the adhesion strength of moisturized specimens as there is no need to determine the hygrostress induced at the crack tip. Later Tanaka and Nishimura [20] used the method to study the effect of temperature on fracture toughness of interfaces between Alloy 42 lead frames and two mold compounds. They also considered the effect of moisture on interfacial fracture toughness. They placed specimens that were initially dry in an environment of 70°C/62%RH for varying periods of time. They then let the specimens cooled down to room temperature and measured the interfacial fracture toughness as a function of absorption time. They found that the fracture toughness decreases with absorption time, that is, with increasing moisture content.

At present, there is still very little experimental data available on the effect of temperature and humidity on interfacial fracture toughness. In this section, the method of Nishimura et al. [20] is employed to measure the fracture toughness of interfaces of copper lead frames and a typical mold compound. The influence of temperature as well as humidity is investigated.

During the 1980's, there were some publications on stress-induced delamination and cracking, especially thermal stress—induced interfacial failure in microelectronics. Barber and Comninou [21] applied fracture mechanics principles to analyze an interface penny-shaped crack subjected to a steady heat flux. Fisher [22] studied the effect of cyclic thermal shock on cracking in the interface region between ceramic and metal. Kokini [28] analyzed the interfacial crack between a ceramic-to-metal bond subjected to a transient thermal loading. He found that the transient strain energy release rate would be significantly different for edge and center cracks. In this calculation model, he considered the effects of thickness ratio of the ceramic-to-metal bond and thermal properties.

The interfacial fracture mechanism is controlled by many factors. In recent years, the interfacial fracture theory of thin films was developed by Hutchinson and Suo [29]. Elastic-plastic interfacial fracture was studied by Shih [30]. Hutchinson and Suo [29], Wang and Suo [31], and Liu et al. [32] have shown that interfacial fracture toughness is a function of mode mixity.

The propagation of the small cracks or defects at the edge and at the center of the pad—encapsulant interface in plastic IC packages during solder reflow is analyzed using 2D FEA. The modified-integral or strain energy release rate due to thermal stress effects was computed and then the SIF K obtained based on the J-K relationship. Finally, K was compared with the critical interfacial SIF and the possibility of delamination discussed.

5.4.1 Measurement of interfacial fracture toughness

The principle of a novel three-point bending test of two-layer end-notched flexural (ENF) specimens for measuring interfacial adhesion strength was described by Nishimura et al. [20]. Critical loads were measured for identical bimaterial specimens (Fig. 5.4.1) by loading from the molding compound side and from the lead frame material side when the precrack started to propagate during loading. The apparent SIFs due to the critical loads P_{b1} and that due to P_{b2} were calculated. It was shown by Nishimura et al. that the average of these two SIFs effectively eliminates the effect of residual stresses and gives the true interfacial fracture toughness.

Following the method of Nishimura et al. described above, three-point bending tests were carried out on bimaterial specimens of length 40 mm, width 6 mm, and overall thickness 4.15 mm made by transfer molding after placing 0.15 mm-thick strips of a copper lead frame material into a mold. Precracks of length 10 mm were created at one end of the specimens by painting the required portion of the copper strips with silicone grease. Specimens of the above dimensions were used as the mold was available. A three-point testing machine was used, which had a test chamber where the temperature could be maintained at any set temperature up to 300°C. The span of the two supports for the three-point test was 32 mm. The properties of the mold compound and the lead frame material employed are given in Tables 5.4.1 and 5.4.2.

To obtain the effect of moisture and temperature on interfacial fracture toughness, three different groups of specimens were preconditioned at three different moisture levels and tested at three different temperatures, namely, 180°C, 220°C, and 260°C. One group was fully dried. The second group was preconditioned at 85°C/60%RH for 168 h and the third group at 85°C/85%RH for 168 h. Specimens from each group were taken one at a time from the preconditioning chambers and immediately set up for the three-point test in the test chamber. About 10 min was allowed for the specimens to fully attain the temperature of the test chamber.

During the approximately 15-min period from when the specimens were taken out of the preconditioning chambers to the time when the loading was applied in the three-point bending test, it is possible that some moisture could have diffused out of or into

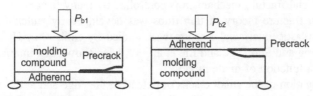

Figure 5.4.1 Three-point bending test.

Table 5.4.1 Properties of encapsulant and lead frame material: *at 25°C/at 240°C and **below T_g/above T_g ($T_g = 160°C$).

Constituent material	Encapsulant	Lead frame
Young's modulus (GPa)	12.3/1.23*	115
Coefficient of thermal expansion (ppm/°C)	16.7/70.2**	17
Poisson's ratio	0.25	0.35
Hygroswelling coefficient (ppm/wt%)	2600	–

Table 5.4.2 Moisture absorption properties for molding compound.

s_0 (mg·mm^{-3}·MPa^{-1})	1.9×10^{-8}
n	1
Q_s (J/mol)	-4.83×10^4
D_0 (mm²/s)	0.21
Q_{act} (J/mol)	3.52×10^4

the specimens, depending on the humidity conditions surrounding the specimen during this period. However, it has been shown that moisture diffusion is a very slow process, even at relatively elevated temperatures [4]. Hence, it is unlikely that the moisture diffusion out of or into the specimen will significantly affect the moisture concentration at the crack tip of the test specimens during this relatively short period of 15 min. Moreover, as almost the entire edge of the crack tip is hidden, any such effect will only be limited to the ends of the edge of the crack tip.

For each combination of temperature and relative humidity, three measurements were carried out and the average value of the critical load obtained. The measured values are given in Table 5.4.3. The critical loads measured were then used to calculate the interfacial fracture toughness using both FEA and beam theory [20]. The results obtained are shown in Fig. 5.4.2. As can be seen the interfacial fracture toughness decreases with increase in temperature and also with increase in moisture content.

5.4.2 Interfacial fracture analysis

This section describes the effect of the location of interfacial voids or defects (cracks) on delamination at the pad–encapsulant interface that is the precursor of much popcorn failure of plastic IC packages. Because of imperfect manufacturing processes, small defects could exist at interfaces in IC packages. Such defects could also arise due to damage mechanisms during the initial period of service of packages or during the initial temperature cycling packages under evaluation and are subjected to before moisture preconditioning and solder reflow processes. It is assumed that the initial

Table 5.4.3 Critical loads measured in three-point bending test.

T(°C)	Loading side (copper or encapsulant)	85°C/85% RH (for 168 h) Critical load (N)		85°C/60% RH (for 168 h) Critical load (N)		Dry Critical load (N)	
		Each test	Average value	Each test	Average value	Each test	Average value
100	Copper	−50	−50.00	−56	−54.6	−112	−85.33
	Copper	−38		−50		−82	
	Copper	−62		−58		−62	
	Encapsulant	−72	−61.33	−73	−75	−110.5	−103.31
	Encapsulant	−62		−75		−97.74	
	Encapsulant	−50		−77		−101.68	
220	Copper	−27.5	−27.50	−30.5	−29	−45.4	−44.91
	Copper	−33		−27.5		−43.59	
	Copper	−22		−28.9		−45.73	
	Encapsulant	−34	−35.00	−46.5	−21.9	−46.88	−56.19
	Encapsulant	−36		−43		−65.55	
	Encapsulant	−35		−39		−56.15	
260	Copper	−23.2	−21.73	−21.1	−21.9	−24.69	−26.76
	Copper	−20		−21.5		−26.82	
	Copper	−22		−23.2		−28.77	
	Encapsulant	−28.5	−24.83	−25	−25.4	−28.74	−26.71
	Encapsulant	−21		−23		−26.04	
	Encapsulant	−25		−26.2		−25.34	

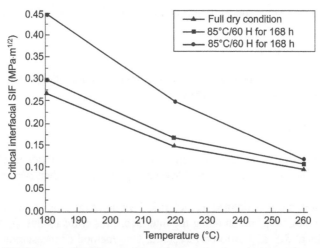

Figure 5.4.2 Variation of critical interfacial stress intensity factor with temperature and humidity.

sizes of such defects are of the order of 0.2 mm [4]. Such defects may be located at any point in the interfaces. It would be important to be able to predict which location would be more serious in terms of the growth of such defects. This is the subject of this section where FE linear elastic fracture mechanics will be employed to analyze the SIFs of defects at various locations on the interface. The two obvious extremes in terms of the location of the interfacial defects would be at the boundaries (edges) or at the center of the interfaces. Hence, in this study, the effect of the location of the interfacial defects (cracks) will be studied by considering 0.2 mm-sized cracks located at the center and at the edge of the interface. Only thermal stress effects are considered in this analysis. Surface temperature in the range of 175–290°C was assumed on the boundary. The effect of the buildup of water vapor pressure in the 0.2 mm interfacial voids was also investigated. The package used in this study was an 80-pin quad flat package with outer dimensions of 20 mm × 14 mm × 2.7 mm and a die pad size of 10.8 mm × 8.4 mm [20].

5.4.2.1 Criterion for delamination growth

During solder reflow, the interfacial crack will propagate when the strain energy release rate G_t at the crack tip exceeds the interface toughness G_c, i.e., when

$$G_t(T, C, \varphi) > G_c(T, C, \varphi) \tag{5.4.1}$$

or alternatively, when the SIF K_t at the cracktip exceeds the critical interfacial SIF K_c, i.e., when

$$K_t(T, C, \varphi) > K_c(T, C, \varphi) \tag{5.4.2}$$

where C is the moisture concentration and φ the mode mixity defined as

$$\varphi = \arctan(K_{II} / K_{I}) \tag{5.4.3}$$

where K_I and K_{II} are the mode I (opening mode) and mode II (shearing mode) components of the complex SIF. Thus,

$$K_t = K_{I,t}/iK_{II,t} \tag{5.4.4}$$

where φ is a measure of the relative proportion of the mode II component to the mode I component. When $\varphi = 0$ degree, we have pure mode I, and when $\varphi = \pm 90$ degrees, we have pure mode II.

Interface toughness is defined as the critical value of the strain energy release rate required for crack propagation along the interface. This critical value G_c or the corresponding critical SIF K_c has to be measured. One method of measurement has been described in the previous section where it was found that K_c decreases with temperature and moisture concentration. However, it has been found that K_c also depends strongly on the mode mixity φ. Thus, in making comparison between K_t and K_c in Eq. (5.4.2), one must ensure that the mode mixity is the same value for both parameters. A typical variation of G_c with φ is illustrated in Fig. 5.4.3. It can be seen that G_c has a minimum value at $\varphi = 0$ degree (pure mode I) and a maximum value at $\varphi = 90$ degrees (pure mode II).

The strain energy release rate G_t and the interfacial SIF K_t at the tip of an interfacial crack due to the effects of thermal stress has been obtained by Wilson and Yu [33] as follows:

$$G_t = E' K_t^2 = J_T \int_{\Gamma+\Gamma-} T_j \frac{\partial D_{p_j}}{\partial x_1} ds = \frac{E\alpha}{1+2\nu} \int_{A_0} \left[\frac{1}{2} \frac{\partial}{\partial x_1}(T\varepsilon_{ij}) + \varepsilon_{ij}\frac{\partial T}{\partial x_1} \right] dA \tag{5.4.5}$$

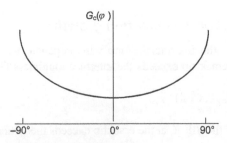

Figure 5.4.3 Typical variation of $G_c(\varphi)$ with φ.

where $T_j = \sigma_{jk}n_k$ is the traction along the crack surfaces, J_T the J-line integral computed along a contour around the crack tip, D_{p_j} is the displacement, x_1 is the coordinate along the interface, e is the strain, and

$$E' = \begin{cases} \dfrac{1+\nu_1}{4G_1} + \dfrac{1+\nu_2}{4G_2} & \text{(plane strain)} \\[8pt] \dfrac{1}{4G_1(1+\nu_1)} + \dfrac{1}{4G_2(1+\nu_2)} & \text{(plane stress)} \end{cases} \qquad (5.4.6)$$

where G_i and ν_i are the shear modulus and Poisson's ratio of the material forming the interface, respectively. As the analysis in this chapter is a 2D FEA of half the plastic IC package sectioned along the central plane parallel to the long side, plane strain conditions may be assumed.

The above values of G_t and K_t obtained from Eqs. (5.4.4)–(5.4.5) are the magnitudes of the energy release rate and SIF, respectively, at the tip of the interfacial crack due to thermal stresses. However, as the critical SIF K_c is a function of the mode mixity φ, this must also be determined. This is done using an extrapolation method due to Yuuki and Cho [34], where

$$\tan \varphi = K_{II}/K_I = \lim_{r \to 0} \frac{(\tau_{xy}/\sigma_y)\tan X}{1 + (\tau_{xy}/\sigma_y)\tan X} \qquad (5.4.7)$$

where

$$X = \lambda \ln\left(\frac{r}{L}\right)$$

$$\lambda = \frac{1}{2\pi} \ln\left[\left(\frac{K_1}{G_1} + \frac{1}{G_2}\right) \Big/ \left(\frac{K_2}{G_2} + \frac{1}{G_1}\right)\right] \qquad (5.4.8)$$

$$K_i = 3 - 4\nu_i \quad i = 1, 2 \qquad (5.4.9)$$

where L is the crack length, r is the distance from the crack tip, and σ_y and τ_{xy} are normal and shear stresses along the interface ahead of the crack tip.

5.4.2.2 Finite element analysis

Finite element analyzes were carried out to determine the crack tip SIF for 0.2 mm interfacial cracks located at the edge and center of the pad–encapsultant interface. This is illustrated in Fig. 5.4.4. Crack tip elements were used to capture the singularity

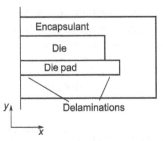

Figure 5.4.4 Diagram of package cross section illustrating location of delaminations.

at the crack tip. Ten concentric layers of elements surrounding the crack tip had been employed to obtain the value of the J-integral. The finite element model consisted of 1283 eight-noded elements and 3935 nodes. The mesh is shown in Fig. 5.4.5. Transient heat transfer, thermal stress, and the energy release rate due to thermal stresses were computed using the ABAQUS finite element code. The material properties employed are shown in Table 5.4.3 [20]. Although the Young's modulus decreases drastically at typical solder reflow temperatures, it was found by Yang et al. [34] that the failure of the mold compound is still largely brittle in nature. Hence, linear elastic fracture mechanics may be employed.

5.4.3 Results and discussion

Fig. 5.4.6 shows the temperature distribution along the pad–encapsultant interface for a package of initial temperature 175°C whose surface temperature is suddenly raised to 290°C. As may be seen the temperature is essentially uniform along the interface at any time after thermal loading is applied. This shows that the thermal load history experienced by a crack located anywhere along the pad–encapsultant interface is about the same, and that in fact, only a steady state thermal stress analysis needs to be done determine whether an edge crack or a central crack will propagate first. However, it should be noted that if this is not the case, then a transient analysis would be essential as different crack locations will then experience different rates of thermal loading and hence different values of SIF. This is important in studying the effect of crack location on delamination of the pad–encapsultant interface.

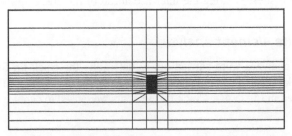

Figure 5.4.5 Typical finite element mesh for crack located at edge of pad–encapsulant interface.

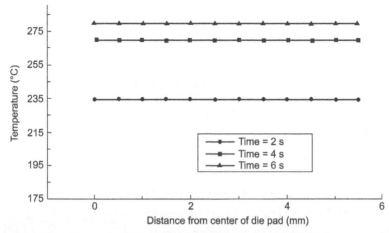

Figure 5.4.6 Temperature distribution along the pad–encapsulant interface.

Fig. 5.4.7 shows that the temperature at the interface virtually reached equilibrium after about 10 s. It also shows that the length L off the mold compound at the edge of the pad has some initial effect on the temperature, but the difference disappears after 10 s.

Fig. 5.4.8 shows that the value of the strain energy release rate G_t at the crack tip of a 0.2-mm edge crack increases with temperature increment. It also shows that equilibrium is reached in about 10 s. Fig. 5.4.9 shows that for the same thermal load, the strain energy release rate for an edge crack is almost double that for a central crack of the same size.

Fig. 5.4.10 shows the variation of the crack tip SIF K_t with temperature for 0.2-mm cracks at two different locations along the pad–encapsulant interface. The critical SIF K_c as a function of temperature measured by Tanaka and Nishimura [20] using the

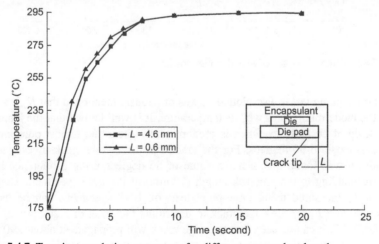

Figure 5.4.7 Transient crack tip temperature for different encapsulant length.

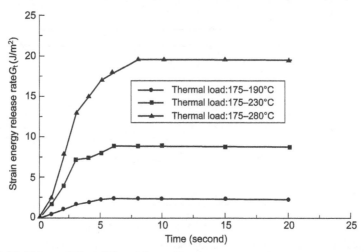

Figure 5.4.8 Values of G_t at different thermal loads.

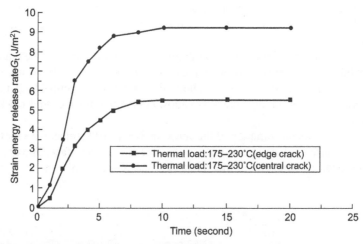

Figure 5.4.9 Values of G_t for edge and center cracks.

three-point bending test is also plotted. It was previously mentioned that K_c is a function of the mode mixity φ as well as temperature. It is well known that the loading at the crack tip of an ENF specimen is predominantly shear and almost pure mode II where φ is close to 90 degrees. For the crack tip at the edge crack investigated in this study, it was found that φ has a value of 86 degrees using a reference length [29] corresponding to the precrack length (10 mm) of the ENF specimen. Thus, the values of K_c measured using three-point tests on ENF specimens can be used in conjunction with K_t for the edge crack to determine the onset of crack propagation. From Fig. 5.4.7, it can be seen that the edge crack will propagate at about 290°C.

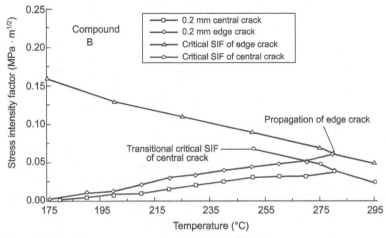

Figure 5.4.10 Variation of stress intensity factor with temperature.

For the central crack of 0.2 mm, it was found that φ has a value of 22 degrees for a reference length of 10 mm. To predict the onset of crack propagation for the central crack, we need to have the variation of interface fracture toughness or K_c with temperature for $\varphi = 22$ degrees. From the experimental data of Wang and Suo [31] and Liu et al. [32], the interface toughness for $\varphi = 22$ degrees is practically equal to that for pure mode I crack where $\varphi = 0$ degree. Unfortunately, data on the interface toughness for the pad—encapsultant bimaterial interface for the case where $\varphi = 0$ degrees were not available from Ref. [20]. However, it may be noted that whether the edge crack or the central crack will propagate first depends on the ratio K_{IIc}/K_{Ic} at about 290°C. The transitional case is shown in Fig. 5.4.10 where the ratio of K_{IIc} at 290°C is found to be 1.64. It is clear that if the ratio K_{IIc}/K_{Ic} is smaller than 1.64 the edge crack will propagate first as temperature is raised from room temperature.

The combined effect of thermal stress and that of water vapor pressure buildup in the delamination was also investigated. The water vapor pressure developed in the delamination during solder reflow was calculated using the method developed by Kitano et al. [11]. This pressure was found to be 0.31 MPa for a temperature change from 175 to 290°C. The results are given in Fig. 5.4.8, where it is seen that the water vapor pressure buildup at the delamination does not significantly increase the J-integral. This is probably due to the fact that the delamination considered here is very small.

5.4.4 Summary

Interfacial fracture toughness has been measured using a three-point bending test with ENF specimens. It was found to decrease with increase in temperature and moisture content. FEA was employed to calculate the transient temperature distribution, modified integral, and SIF at small defects located at the edge and at the center of the pad—

encapsultant interface of a plastic IC package. The methodology for predicting the likelihood of interfacial delamination propagating from defects located at various parts of the interface has been established. However, as data on the variation of interfacial fracture toughness with mode mixity at high temperature are not available, the effect of defect location could not be conclusively determined. It was found that as the temperature of the package was increased, the SIF of the edge crack was higher than that of the center crack. However, whether the edge crack will propagate first as temperature is increased depends on the ratio of mode II interface toughness to that of the mode I interface toughness. For the package under investigation, it was established that when this ratio is less than 2.69 (or alternatively when K_{IIc}/K_{Ic} is smaller than 1.64), the edge crack will propagate first, otherwise the center crack will. For small defects, it was found that the water vapor pressure developed at the interface did not have a significant effect of the value of the strain energy release rate and hence the crack tip SIF.

5.5 Reliability enhancement in PBGA package

The electronics industry is moving to lead- and halogen-free electronic products, driven by market pressure and environmental regulations. As a result, they are moving to "green" IC packaging technologies, and manufacturers are facing critical challenges with respect to cost and reliability, including the use of nongreen IC packages in lead-free solder reflow processes [7,35–37]. The melting temperature of the Sn-Ag-Cu-based lead-free alloy is about 30°C higher than the traditional melting temperature (183°C). As a result, the peak solder reflow temperature for IC assembly will increase as well. The maximum solder reflow temperature (up to maximum 260°C) can also negatively impact the useable floor life performance designated by the moisture sensitivity levels (MSL) of the IC packages [38,39]. The common failure mechanisms of traditional large size plastic ball grid array (PBGA) packages include pop-corning, as well as delamination and cracks between solder mask/copper interface in the multiple layer substrates [12,27]. In fact, one of the challenges with the move to lead-free and halogen-free IC packages is the degradation of MSL performance for large size PBGAs [40].

To achieve higher or comparable MSL performance for green PBGA packages, the die attach, mold compound, and substrate must be concurrently improved to reduce hygrothermally induced delamination and popcorn cracking [40–47]. The reliability enhancement of substrates is one of the critical improveements for PBGA packages. From a material perspective, the solder mask material must be sufficiently ductile but strong to provide high fracture and crack resistance; it also needs to provide adhesion between mold compound to solder mask and the solder mask to copper interface under various temperature and humidity conditions. In addition, the substrate needs to withstand the influence of high temperature (260°C) for the duration of reflow and have sufficient warpage resistance [48,49] so that the assembly yield is at least comparable with traditional PBGA packages.

A surface finish treatment must be introduced onto the copper surface during the green substrate manufacturing process to ensure adequate adhesion. The traditional jet scrubbing will not fully meet the adhesion requirements at the solder mask/copper interface. The advanced chemical microetch (MEC) with fine surface treatment (down to 1 μm) can make much fine anchor holes on the copper surface that can provide a better locking mechanism between solder mask/copper interface for better adhesion. In addition to the substrate enhancement, the integrity of green PBGA packages must be evaluated for each package assembly process, such as die attach, wire bonding, molding, ball attachment, and singulation. The material, process, and packaging integrity characterization are major tasks for achieving higher MSL performance and acceptable assembly yield with minimum cost penalty.

Here, the performance of nongreen PBGA packages was evaluated at 260°C reflow temperature. It was found that delamination and cracks could occur at the solder mask/copper interface. To mitigate these failure mechanisms, a solder mask material and a MEC etching process was introduced. Comparative assessments were performed between the old and new solder mask and surface treatment process. The new solder mask material and mechanic etching process could greatly improve the adhesion between the solder mask to copper interface. Finally, the package integrity of green PBGA was evaluated with the new solder mask and surface etching process. It was observed that the substrate failures were reduced; however, some failures still occurred at the solder mask/mold compound interface. This indicates that improvements on the adhesion strength between solder mask/mold compound interfaces are still needed to have more robust green PBGA packages.

5.5.1 Characterization and improvement of moisture sensitivity of performance of PBGA

The MSL performance of nonhermetic packages is typically characterized using the joint IPC-JEDEC standard J-Std-20C Moisture/Reflow Sensitivity Classification for Nonhermetic Solid State Surface Mount Devices [38]. The reflow condition for lead-free solders in the standard is 250°C for package thickness <2.5 mm and package volume <350 mm^3 and 245°C for package thickness ⩽2.5 mm and package volume ⩾350 mm^3. The test conditions used in this study are more stringent than the standard, to determine limits of reliability and provide margins.

5.5.1.1 Impact of reflow soldering temperature (260°C) on MSL of nongreen PBGA

Here a 27 mm PBGA with two-layer substrate and 35 mm PBGA with four-layer substrate combined with nongreen mold compound and die attach paste (silver filled epoxy) was used to assess the impact of reflow temperature (maximum 260°C) on MSL performance. Fig. 5.5.1 shows the schematic of the BGA under consideration. The assessment flow chart is shown in Fig. 5.5.2 [7]. For each MSL level assessment, a new set of packages are used. The detailed MSL assessment results were shown in

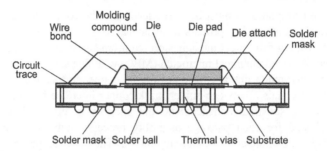

Figure 5.5.1 Schematic of the ball-grid array (BGA) package under consideration.

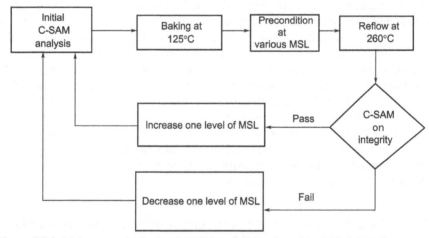

Figure 5.5.2 Moisture sensitivity level (MSL) assessment flow for plastic ball grid array (PBGA) packages.

Table 5.5.1. In this table, a floor life at 30°C/60% RH conditions is assumed with a life of 1 year at MSL 2, 1 month at MSL 2a, 1 week at MSL 3, and 3 days at MSL 4. It was found that 27 and 35 mm PBGA both can pass MSL 2a at the normal reflow temperature (maximum 225°C). However, 27 mm PBGA failed at MSL 2a and passed MSL 3 at the lead-free reflow temperature (260°C). The 35 mm PBGA performed worse at the lead-free reflow temperature. They failed at MSL 3 and passed MSL 4. As such, 27 and 35 mm PBGA degraded one and two MSLs, respectively. The failure mechanisms were cracks and delamination at the solder mask/copper trace interface as shown in Fig. 5.5.3.

5.5.1.2 Enhancing adhesion strength at Cu to solder mask interface

The failures of nongreen PBGA packages at the lead-free reflow temperature revealed that the adhesion between solder mask and copper interface could be weak.

Table 5.5.1 Impact of lead-free temperature (260°C) on moisture sensitivity level (MSL) performance (27 and 35 mm plastic ball grid array [PBGA], sample size: 40units per package per MSL test).

	Peak reflow temperature	
MSL performance	225°C	260°C (lead-free temperature)
2	Fail (only for 35 mm PBGA)	N/A
2a	Pass (both 35 and 27 mm PBGA)	Fail (27 mm PBGA)
3	N/A	Fail (35 mm PBGA)/pass (27 mm PBGA)
4	N/A	Pass (35 mm PBGA)

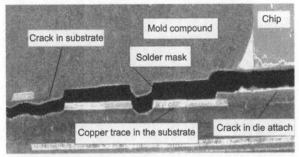

Figure 5.5.3 Failure mechanism of interfacial delamination between solder mask/copper trace in 35 mm.

The requirements of the solder mask material include low moisture absorption, high crack resistance, high fracture strength, and elongation with low elastic modulus. As such, the advanced material and processing technology must be developed and enhanced in the substrate manufacturing to provide a robust PBGA package.

Fig. 5.5.4 shows the SEM image of the copper surface after the copper surface was treated by jet scrubbing, which is the typical surface treatment process during the substrate manufacturing process. However, from a peel test perspective, jet-scrubbing technologies cannot fully meet the adhesion requirement for lead-free packaging solution. As a consequence, a new etching technology was developed to surface treat the copper that provides better adhesion with finer structure, which is indicated in Fig. 5.5.5.

It was found that the "MEC" etching process could provide a fine surface finish of copper and provide a stronger locking mechanism between copper and the solder mask. Atomic force microscopy (AFM) shows that the finer surface treatment with fine anchor holes could provide better adhesion. Fig. 5.5.6 shows the MEC etching

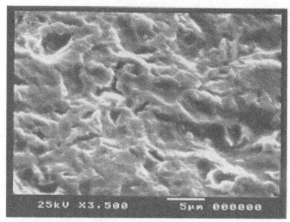

Figure 5.5.4 Scanning electron microscopy (SEM) image of surface treatment by old jet scrubbing technologies for copper surface finishing (rough surface crystal structure).

Figure 5.5.5 Scanning electron microscopy (SEM) image of the surface treatment by the chemical microetch (MEC) etching technologies for copper surface finishing (fine surface crystal structure).

with 1 and 2 μm surface treatment. It was seen that the 1 μm surface treatment had a finer surface crystal structure compared with the 2 μm surface treatment.

The adhesion between the copper to solder mask was evaluated using peel tests, after the samples were subjected to pressure cooker test (PCT at 121°C, 100%RH, 2 atm (1 atm = 10^5 Pa) for 100 h) for both the traditional material and new material with jet scrub and MEC etching process (1 μm etch) as seen in Table 5.5.2. An adhesion index corresponding from the best (index: 1) to poor (index: 7) level was introduced to evaluate the performance of the adhesion strength. Index 1–7 indicates the level of peeling-off at the solder mask to copper interface. Index 1 indicates the best

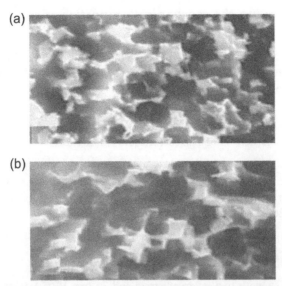

Figure 5.5.6 Atomic force microscopy (AFM) plot of the surface treatment by the chemical microetch (MEC) etching 1 μm (a) and 2 μm (b) on copper surface finishing.

Table 5.5.2 Impact of solder mask material and surface finishing process on the adhesion strength on copper/solder mask interface (Peel test after PCT 100 h).

	Old solder mask		New solder mask	
	MEC etching	Jet scrub	MEC etching	Jet scrub
Post bake 150°C for 5 h	4	7	2	4
Post bake 170°C for 5 h	4	5	2	4
Coating thickness (5 μm)	1	1	1	1
Coating thickness (10 μm)	1	1	1	1
Coating thickness (15 μm)	2	4	4	2
Coating thickness (20 μm)	2	4	2	3

adhesion that shows "no peeling or blistering". Index 2 showed slight peeling. Index 7 corresponds to delamination at the copper to solder mask interface.

In Table 5.5.2, it is shown that the new solder mask material together with MEC etching process of 1 μm could improve the adhesion strength at the copper to solder mask interface. Nevertheless, a thicker coating and post baking at high temperature with longer time degraded the adhesion strength.

Index 1–4: good adhesion with subsequent low peel strength.

5.5.1.3 Improving the mechanical properties of solder mask materials

The new surface treatment process works together with the new solder mask material to provide greater adhesion strength. The development of solder mask materials focuses on the following areas:

- better ductile properties to overcome solder mask cracking;
- reduction in the CTE mismatch to reduce stress;
- reduction in the moisture absorption coefficient to reduce stress.

Comparisons between the old and new solder mask materials are shown in Table 5.5.3. The new solder mask material is more ductile and has a low CTE value and lower water absorption ratio. These improvements help overcome the solder mask cracks at 260°C reflow temperature.

5.5.2 Reliability assessment for PBGA packages

The 27 mm PBGA with two-layer substrate and 35 mm PBGA with four-layer substrate was selected as the test vehicles. These PBGA packages were built by normal IC manufacturing processes. C-SAM was performed on all the packages at receipt. C-SAM was performed again after baking, moisture soaking at 60°C/60% RH for 120 h (MSL 2a) and 60°C/60% RH for 40 h (MSL 3), and reflow at 260°C, respectively. The new substrate with new solder mask material together with etching (1 µ surface treatment) was used for the sample builds. In addition, two new green compounds suppliers (A and B) provided their optimized green mold compound for the reliability assessment. The same green die attach paste was used for all the samples.

The assessment results are shown in Table 5.5.4. It was found that delamination at the mold compound to solder mask interface was observed for both the test vehicles. The delamination was observed by C-SAM using the through-scan function after reflow at 260°C and was verified by SEM. The 27 mm PBGA demonstrated more

Table 5.5.3 Comparison on the mechanical properties of old and new solder mask materials.

	Old solder mask	**New solder mask**
Young's modulus	3500 MPa	2400 MPa
Elongation at facture	30 MPa	43 MPa
CTE1 (ppm/°C)	50	60
CTE1 (ppm/°C)	140	130
T_g (°C)	104	101
Water absorption ratio (%)	1.5	1.3

Table 5.5.4 Impact of lead-free temperature on the reliability of green plastic ball grid array (PBGA) packages.

Test vehicles	Fail/pass	Failure site
27 mm PBGA	Compound supplier A:9/12failures	Solder mask/mold compound interface
	Compound supplier B:2/12failures	Solder mask/mold compound interface
35 mm PBGA	Compound supplier A:6/12failures	Solder mask/mold compound interface
	Compound supplier B:0/12pass	NA

failures than the 35 mm PBGA because the package has higher moisture susceptibility. The thicker substrate (four-layer substrate) and large size molding package have more resistance to moisture. In addition, the adhesion between mold compound and substrate plays a critical role in reducing the delamination. Table 5.5.4 shows that the 35 mm packages with molding compound from supplier B did not show any delamination. This result indicated that compound supplier B provided the compound with higher moisture resistance and better adhesion with the solder mask material.

Mold compounds A and B have similar properties, e.g., Young's modulus and CTE values (CTE1/CTE2). However, the adhesion between the compound and solder mask is different under moisture condition as indicated by the hygroscopic properties of mold compound.

5.5.3 Summary

The new solder mask material together with the special MEC etching copper surface treatment can provide stronger adhesion between the solder mask to copper interface and reduce the substrate failures that occur at lead-free reflow temperature for nongreen PBGA packages. The detailed surface analysis and PCT resistance comparisons on new and old solder mask material and process were performed and addressed by SEM, AFM, and peel tests.

The package integrity was also assessed for two green compounds, one green die attach paste and a new substrate after reflow at 260°C. It is found that the substrate failures are no longer observed. However, delamination still occurs at the solder mask/mold compound interface. The assessment results suggest that high moisture resistant molding compounds and better adhesion between solder mask to the mold compound is necessary to maintain good packaging integrity. There is continued effort on further improving the multi interfacial adhesion and MSL performance of the green mold compounds.

References

[1] R. Rao, Introduction-What An Advanced IC Looks Like, NCAP Open day, 2016.

[2] A.A.O. Tay, T.Y. Lin, Moisture diffusion and heat transfer in plastic IC packages, IEEE Trans. Compon. Packag. Manuf. Technol. A 19 (2) (1996) 186–193.

[3] T.Y. Lin, A.A.O. Tay, A J-integral criterion for delamination of bi-material interfaces incorporating hygrothermal stress, Am. Soc. Mech. Eng. EEP 19 (2) (1997) 1421–1428.

[4] T.Y. Lin, A.A. O Tay, Dynamics of moisture diffusion, hygrothermal stresses and delamination in plastic IC packages, ASME Adv. Electron. Packag. 19 (2) (1997) 1429–1436.

[5] T.Y. Lin, A.A.O. Tay, Influence of temperature, humidity, and defect location on delamination in plastic IC packages, IEEE Trans. Compon. Packag. Technol. 22 (4) (1999) 512–518.

[6] T.Y. Lin, et al., Failure analysis of full delamination on the stacked die leaded packages, J. Electron. Packag. 125 (2003) 392–399.

[7] T.Y. Lin, et al., Bottlenecks and strategies of green mold compound, IEEE Trans. Compon. Packag. Technol. 26 (2) (2003) 492–494.

[8] S. Sengupta, T.Y. Lin, et al., Assessment for thermal-mechanical damage of electronic parts due to solder dipping, IEEE Trans. Electron. Packag. Manuf. 30 (2) (2006) 128–137.

[9] T.Y. Lin, et al., The influence of substrate enhance on moisture sensitivity level (MSL) performance for green PBGA packages, IEEE Trans. Compon. Packag. Technol. 29 (3) (2006) 522–527.

[10] I. Fukuzawa, S. Ishiguro, S. Nanbu, Moisture resistance degradation of plastic LSI's by reflow soldering, Proc. IRPS (1985) 192–197.

[11] M. Kitano, A. Nishimura, S. Kawai, Analysis of package cracking during reflow soldering process, Proc. IRPS (1988) 90–95.

[12] M.G. McMaster, D.S. Soane, Water sorption in epoxy thin films, IEEE Trans. Comp. Hybrids Manufact. Technol. 12 (3) (1989) 373–385.

[13] I. Anjoh, K. Nishi, M. Kitano, T. Yoshida, Analysis of the package cracking problem with vapor phase reflow soldering and corrective action, Brazing Solder. (1988) 48–51.

[14] H. Miura, Temperature distribution in IC plastic packages in the reflow soldering process, IEEE Trans. Comp. Hybrids Manufact. Technol. 4 (1988) 499–505.

[15] R. Lin, E. Blackshear, P. Serisky, Moisture induced package cracking in plastic encapsulated surface mount components during solder reflow process, Proc. IRPS (1988) 83–89.

[16] M. Tanaka, K. Nishi, A Novel Test Method of Resistance to Soldering Heat of Plastic Encapsulated Surface Mount LSI's, Tech. Rep., Hitachi Ltd., 1990.

[17] S. Ito, Development of epoxy encapsulants for surface mounted devices, Nitto Tech. Rep. (1987) 78–82.

[18] P.C. Liu, D.W. Wang, Moisture absorption behavior of printed circuit laminate materials, Adv. Electron. Packing. 4 (1) (1993) 435–442.

[19] H. Karlsson, Sorensen, Inc., ABAQUS User's Manual, Version 5.4, 1994.

[20] N. Tanaka, A. Nishimura, Measurement of IC molding compound adhesion strength and prediction of interface delamination within package, Adv. Electron. Packag. ASME 10 (2) (1995) 765–773.

[21] J.R. Barber, M. Comninou, The penny-shaped interface crack with heat flow, Part 2: imperfect contact, ASME J. Appl. Mech. 50 (1983) 770–776.

[22] G. Fisher, Advanced ceramics continue progress to products, Am. Ceram. Soc. Bull. 63 (2) (1984) 249–252.
[23] D.L. Hartsock, D.F. McLean, What the designer with ceramic needs, Am. Ceram. Soc. Bull. 63 (2) (1984) 266–270.
[24] A.A.O. Tay, G.L. Tan, T.B. Lim, Predicting delamination in plastic IC packages and determining suitable mold compound properties, IEEE Trans. Compon. Packag. Manuf. Technol. B 17 (1984) 201–208.
[25] K.X. Hu, C.P. Yeh, X.S. Wu, K. Wyatt, An interfacial delamination analysis for multichip module thin film interconnects, Trans. ASME J. Electron. Packag. 118 (1996) 206–213.
[26] R. Lin, E. Blackshear, P. Serisky, Moisture induced package cracking in plastic encapsulated surface mount components during solder reflow process, in: Proc. Int. Reliab. Phys. Symp, 1988, pp. 83–89.
[27] G.S. Ganesan, H.M. Berg, Model and analysis for solder reflow cracking phenomenon in SMT plastic packages, Proc. IEEE ECTC (1993) 653–660.
[28] K. Kokini, Interfacial cracks in ceramic-to-metal bonds under transient thermal loads, J. Am. Ceram. Soc. 70 (12) (1987) 855–859.
[29] J.W. Hutchinson, Z. Suo, Mixed mode cracking in layered materials, Adv. Appl. Mech. 29 (1992) 63–191.
[30] C.F. Shih, Cracks on biomaterial interfaces: elasticity and plasticity aspects, Mater. Sci. Eng. A143 (1991) 77–90.
[31] J.S. Wang, Z. Suo, Experimental determination of interfacial toughness curves using Brazil-nut sandwiches, Acta Metall. 38 (1990) 1279–1290.
[32] S. Liu, Y. Mei, T.Y. Wu, Bimaterial interfacial crack growth as a function of mode-mixity, IEEE Trans. Compon. Packag. Manuf. Technol. A 18 (1995) 618–626.
[33] W.K. Wilson, I.W. Yu, The use of the J-integral in thermal stress crack problems, Int. J. Fract. 15 (4) (1979) 377–387.
[34] J.C. Yang, C.W. Leong, J.S. Goh, C.K. Yew, Effects of molding compound properties on lead-on-chip (LOC) package reliability during IR reflow, Electron. Compon. Technol. Conf. (1996) 48–55.
[35] C.M. Garner, V. Gupta, V. Bissessur, A. Kumar, R. Aspandiar, Challenges in converting to lead-free electronics, in: Proc. 3rd Electron Packag. Technol. Conf., Singapore, 2000, pp. 6–9.
[36] M. Dittes, Green packages-requirements, materials, results, in: Proc. 3rd Electron. Packag. Technol. Conf., Singapore, 2000, pp. 1–5.
[37] T.Y. Lin, C.M. Fang, Y.F. Yao, K.H. Chua, Development of the green plastic encapsulation for high density wirebonded leaded packages, Microelectron. Reliab. 43 (2003) 811–817.
[38] R. Ciocci, J. Wu, M. Pecht, Impact of environmental regulations on electronics manufacture, use and disposal, in: Proc. Eur. Microelectron. Packag. Interconnect. Symp., Cracow, Poland, 2002, pp. 73–80.
[39] J-STD-020C, Moisture/Reflow Sensitivity Classification for Nonhermetic Solid State Surface Mount Devices, Jul. 2004.
[40] E. Stellrecht, B. Han, M. Pecht, Measurement of hygroscopic swelling in mold compounds and its effect on PEM reliability, in: Proc. Int. Electron. Packag. Tech. Conf. Exhibition (IPACK'03), Maui, HI, 2003, pp. 6–11.
[41] T.Y. Lin, Reliability Assessment for PBGA and CSP Packages during Lead-free Reflow Soldering, Internal Report, Agere Systems, 2000.

[42] M. Pecht, H. Ardebili, A. Shukla, J. Hagge, D. Jennings, Moisture ingress into organic laminates, IEEE Trans. Compon. Packag. Technol. 22 (1) (1999) 104–110.
[43] A.A.O. Tay, T.Y. Lin, Moisture-induced interfacial delamination growth in plastic IC packages during solder reflow, in: Proc. 48th Electron. Comp. Technol. Conf., Seattle, WA, 1998, pp. 371–378.
[44] H. Ardebelli, E.H. Wong, M. Pecht, Hygroscopic swelling and sorption characteristics of epoxy molding compounds used in electronics packaging, IEEE Trans. Compon. Packag. Technol. 26 (1) (2003) 206–214.
[45] Characterization of hygroscopic swelling behavior of mold compounds and plastic packages, IEEE Trans. Compon. Packag. Technol. 27 (3) (2004) 499–506.
[46] E.H. Wong, K.C. Chan, R. Rajoo, T.B. Lim, The mechanics and impact of hygroscopic swelling of polymeric materials in: electronic packaging, in: Proc. 50th Electron. Comp. Technol. Conf., Las Vegas, NV, 2000, pp. 576–580.
[47] E.H. Wong, S.W. Koh, R. Rajoo, T.B. Lim, Underfill swelling and temperature-humidity performance of flip chip BGA package, in: Proc. 3rd Electron. Packag. Technol. Conf. (EPTC'00), Singapore, 2000, pp. 258–262.
[48] T.Y. Lin, B. Njoman, D. Crouthamel, K.H. Chua, S.Y. Teo, The impact of moisture in mold compound performs on the warpage of PBGA packages, Microelectron. Reliab. 44 (4) (2004) 603–609.
[49] M. Pecht, Moisture sensitivity characterization of build-up ball-grid array substrates, IEEE Trans. Adv. Packag. 22 (3) (1999) 515–523.

Further reading

[1] L.L. Marsh, R. Lasky, D.P. Seraphim, G.S. Springer, Moisture solubility and diffusion in epoxy and epoxy-glass composite, IBM J. Res. Dev. 28 (1984) 655–661.
[2] J.E. Ritter, K. Conley, Moisture-assisted crack propagation at polymer/glass interfaces, Int. J. Adhesion Adhes. 12 (1992) 245–250.

Thermal and mechanical tests for packages and materials

6.1 Package thermal tests

6.1.1 Introduction to thermal tests

In the package development stage, various thermal tests are involved to qualify the package. The thermal-related package tests include both the performance metrics and qualification tests. Performance metrics can be determined based on different test conditions such as natural convection and forced convection, whereas the qualification test during manufacturing involves tests such as power cycling test, burn-in test, and system-level test. Thermal test and characterization of packages can be conducted by either steady-state or transient measurement techniques. Test standards are available to define the measurement techniques, boundary conditions or test environments, and the test devices along with the measured thermal metrics. Since 1990, the Joint Electron Device Engineering Council (JEDEC) under the Electronic Industries Association has been devoting to create thermal measurement standards for semiconductor device packages. The JEDEC thermal testing standards from JEDEC JC-15 committee recommend specific environmental conditions, measurement techniques, fixture, heating power, and data reporting guidelines for electronic packaging test. The basic thermal testing standards are available in the JEDEC JC-15 websites [1].

Nonetheless, the steady-state test of the junction-to-case thermal resistance, which is of concern for package designers and suppliers, is not available in JEDEC standard. An approximate test method was documented in the method 1012.1 in MIL-STD-883H based on the cold plate test [2]. For the transient measurement of the junction-to-case thermal resistance, a new test standard has been added recently, known as the JESD51-14 standard based on the transient dual interface (TDI) test technique. Some other advanced tests such as the power and temperature cycling test documented in JEDS 22-A105C are also thermal-related. Besides, for the qualification and manufacturing tests such as burn-in test, a more detailed description is given in the military test standard MIL-STD-883H, method 1015.1 [3]. A summary of pertinent thermal test standards are shown in Fig. 6.1.1. The steady-state and transient tests of the packages, together with the burn-in test, are to be elaborated in the ensuing parts of this section.

6.1.2 Steady-state thermal test

In the semiconductor device, the most commonly used temperature-sensitive sensor is the voltage drop across a forward-biased diode. This diode is specifically designed into the thermal test die and usually fabricated as a parasitic device in most integrated circuit devices.

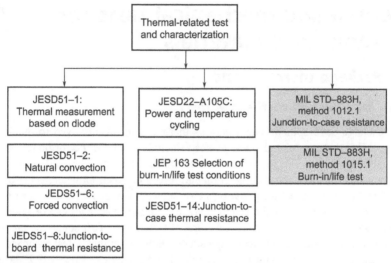

Figure 6.1.1 Thermal−related package test standards.

Thermal test die is specially designed to provide a uniform heating structure (i.e., a resistor or active device) for heating purpose and one or more small diodes for temperature sensing. In conformance with the JEDEC thermal measurement standard, the total heating structure must occupy 85% or more of the total die surface. The resistor may be configured as a single unit or as two or more isolated resistors that together meet the surface coverage requirement. The latter structure is usually preferred because it provides great flexibility in setting up the heating power supply for the desired power dissipation. If a single diode is built into the die, it is usually placed in the center of the die. To simulate a die heating with complex scenarios, it can be composed of multiple smaller thermal die with the different heating power to simulate the local heating pattern such as hot spot.

A commercially available thermal test die has varying size. The minimal die size of 1 mm is supplied from TEA associate. Fig. 6.1.2 illustrates a thermal test die of 5.08 mm × 5.08 mm made of four small dies, each with a size of 2.54 mm × 2.54 mm and built in with diodes and resistors. The on-die thermal resistors are connected with a DC power supply during the thermal test. To validate that the power dissipation is consumed on the chip instead of the interconnects, the four-wire test can be conducted to analyze the electrical resistance at the die level, package level, and board-level. An illustration of the four-wire test is exemplified in Fig. 6.1.3.

Standard thermal characterizations include the junction-to-ambient thermal resistance (R_{ja}) and junction-to-case thermal resistance (R_{jc}). Fig. 6.1.4 illustrates the thermal test of the junction-to-ambient thermal resistance under natural convection condition. The fixture is suggested to follow the standard given in the JEDEC standard 51-2. More importantly, an enclosure made of acrylics with wall thickness no less than 1/8″ or 3.175 mm should be used to minimize the disturbance especially in an air-conditioned room. The power levels at which devices are tested should be governed

Thermal and mechanical tests for packages and materials

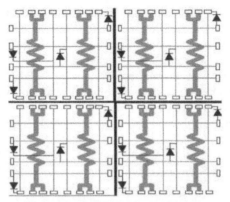

Figure 6.1.2 A thermal test die formed with four small dies, each die having a resistor and four diodes located at the center and peripheral. The marked diodes, D1–D6, are connected for package assembly and test.

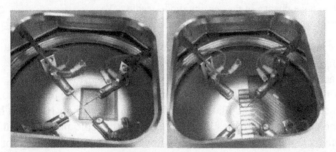

Figure 6.1.3 Four-wire test of the package and board electrical resistance for the through-silicon via (TSV) package based on a four-wire probe station.

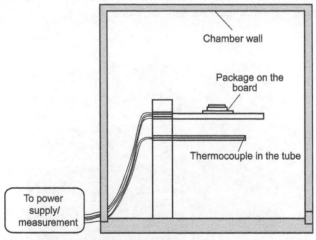

Figure 6.1.4 Thermal test of 2.5D package according to the Joint Electron Device Engineering Council (JEDEC) standard 51-2 for the junction-to-ambient thermal resistance.

by actual use conditions. The minimum recommended junction temperature rise for testing is 20°C. The typical junction temperature rise during testing is between 30 and 60°C, which is the normal range of use for most devices. To obtain the thermal equilibrium, record the die temperature in consecutive 5 min. If the temperature deviation is less than 0.2°C, the thermal equilibrium is considered reached. It is also feasible to put a pin fin heat sink on the top of the package to characterize the thermal resistance at the dissipated heat range of 4−6 W.

The junction-to-ambient thermal resistance at thermal equilibrium time t can be calculated by

$$R_{ja} = \frac{T_{a,0} + \Delta V \times KC - T_{a,t}}{q} \qquad (6.1.1)$$

where $T_{a,0}$ is the initial ambient air temperature before heating power is applied

$T_{a,t}$ = final ambient air temperature when steady state is reached
ΔV = the voltage change in the sensing diode reading (V)
KC = the proportionality coefficient of temperature rise over the voltage change (°C/V)

Here, the small increase in the ambient temperature due to package heating, which is in the range of a few degrees Celsius, has been taken into account in Eq. (6.1.1) to obtain the accurate thermal resistance.

There is no steady-state JEDEC measurement method for the junction-to-case thermal resistance or R_{jc}. The cold plate testing based on the method 1012.1 in MIL-STD-883H can be used for thermal testing, provided that the cold plate thermal resistance itself is much smaller than R_{jc}. Fig. 6.1.5 shows the cold plate test jig for a molded package, for which the test die was mounted on a silicon interposer with through-silicon vias. The measured R_{jc} is in the range of 10°C/W. Nonetheless, such a thermal resistance is actually the junction-to-sink thermal resistance including, besides R_{jc}, the thermal interface resistance and heat sink thermal resistance.

Figure 6.1.5 Junction-to-case thermal resistance test with a liquid-cooled cold plate assembled on the top of the test jig.

The junction-to-board thermal resistance R_{jb} is a figure of merit for comparing the thermal performance of surface mount packages mounted on a standard board. The standard test boards should have two internal copper planes conforming to JEDEC 51-8 standard. The double ring cold plate, made of copper, is clamped onto the board, normally solder mask—covered traces, at a minimum of 5 mm clearance from the package. The double ring cold plate is a fluid cooled cold plate that clamps both sides of the test board such that the heat flows radially from the package along the plane of the test board. The double-ring cold plate test is illustrated in Fig. 6.1.6.

The cold plate clamping must have a minimum of 4 mm width. The measurement results are not sensitive to clamping force. Thermal interface materials (TIM) are not required between the clamp on the board surface after confirmation with experimental test. The openings in the top and bottom of the cold plate corresponding to the package footprint are to be insulated as shown in Fig. 6.1.6. The top insulation material with a conductivity less than $0.1 \text{ W} \cdot \text{m}^{-1} \cdot \text{K}^{-1}$ is faced with a low emissivity aluminized plastic film. There should be a 1—5 mm air gap between the package and the insulation and between the bottom of the test board and the insulation on the bottom of the fixture to minimize the heat loss to the insulation. Furthermore, in the latest version, the thermocouple with the wire diameter of 40 gauge instead of 30 gauge is suggested to solder to the board for more accurate temperature measurement. The power levels shall be chosen such that the junction temperature rise during testing is between 15 and 30°C to minimize the heat loss to the ambient other than the cold plate.

In practice, this junction-to-board thermal resistance is not often used, partially due to the fact that the test PCB is never realistic for a specific application. However, such a configuration is useful in the development of compact board thermal model for a

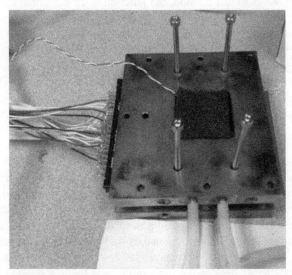

Figure 6.1.6 The test configuration for junction-to-board thermal resistance JB with double-ring cold plate test jig.

realistic PCB. The double-ring cold plate serves as the ideal boundary condition to determine the compact board thermal model of a package with equivalent junction-to-board thermal resistance to save the computational time and efforts. The thermal model with the boundary condition of the double-ring cold plate can be implemented in CFD software, such as Mentor FloTHERM.

6.1.3 Transient thermal test

A transient thermal test of the junction-to-case thermal resistance, known as TDI measurement, has been documented in JEDEC 51-14 [4]. In this method, the transient temperature responses are obtained for a package-heat sink assembly both with and without TIM during the cooling-down process. The structure functions related to the geometry and material thermophysical properties are obtained to obtain the cumulative thermal resistance and thermal capacitance, based on the analysis of the Cauer RC network. By identifying the separation point in the two curves for the structure functions, the junction-to-case thermal resistance can be determined.

Assume the chip temperature curves without and with thermal grease have been measured. The following steps shall be followed to compute the junction-to-case thermal resistance. Step 1: Convert the temperature curves R_{jc1} and R_{jc2} to the corresponding structure functions C_{th1} and C_{th2} using some dedicated software. Step 2: Interpolate both structure functions on a common R_{jc} scale and compute the difference $C_{th1}-C_{th2}$. Step 3: The junction-to-case thermal resistance R_{jc} is the point where the difference of $C_{th1}-C_{th2}$ starts to rise. If spurious peaks make the determination of R_{jc} impossible, the R_{jc} measurement shall be repeated with higher power dissipation to achieve a higher signal-to-noise ratio.

Based on the standard test method, the junction-to-case thermal resistance can be determined for a 2.5D package more accurately than the steady-state given in Ref. [2]. A case study was conducted for characterizing a molded 2.5D package with built-in heater and diode. The test started with the cooling-down process for the 2.5D package after the power is switched off. A commercial transient thermal tester branded T3Ster from Mentor was used for fast temperature recording. The test setup is shown in Fig. 6.1.7, which mainly comprises the tester with a power booster, test section, and software for postprocessing the temperature data to obtain the structure functions.

Both the cumulative and differential structure functions can be obtained to determine the junction-to-case thermal resistance, as indicated in Figs. 6.1.8 and 6.1.9, respectively. In comparison, the differential structure function curve has more abrupt turning in the curve and thus provides a more clear separation point to determine the thermal resistance. The transient thermal resistances at different heating power are shown in Table 6.1.1 at different heating power. It is seen that the results are smaller than the thermal resistance values obtained in the steady-state measurement. As has been explained, the steady-state thermal resistances comprises the heat sink and thermal interface resistances besides the junction-to-case thermal resistance, which may overestimate the thermal resistance significantly. A detailed discussion on the pros

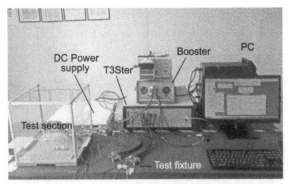

Figure 6.1.7 The T3Ster setup for thermal resistance measurement consisting of thermal transient tester, booster, thermostat for device under test (DUT), natural convection chamber, and data postprocessing software.

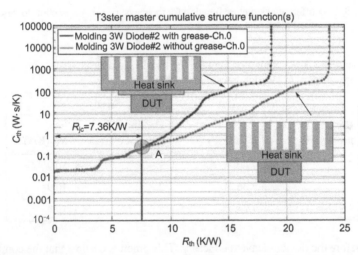

Figure 6.1.8 The cumulative structure function for the measurement of junction-to-case thermal resistance for a molded 2.5D package.

and cons of the TDI method is given in Ref. [5]. A comparison of the transient measurement is conducted with the steady-state method [6].

6.1.4 Burn-in test of electronic packaging

To weed out the weak parts and avoid early life failures, the packages are put under accelerated temperature and voltage stress conditions for a certain period of time. This is called burn-in. Burn-in can be understood as the preconditioning of assemblies and the accelerated power-on tests performed subjected to temperature, voltage, and the aging of components after fabrication but before operational use to eliminate

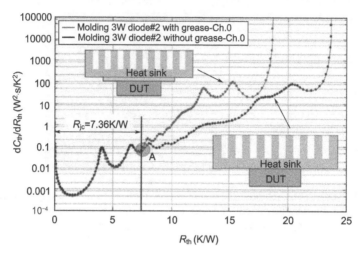

Figure 6.1.9 The differential structure function for the measurement of junction-to-case thermal resistance for a molded 2.5D package.

Table 6.1.1 A comparison between the transient and steady-state measurement of junction-to-case thermal resistances.

T3Ster		Steady-state	
Heating power (W)	R_{jc} (K/W)	**Heating power (W)**	R_{jc} (K/W)
2	7.40	1	10.9
3	7.36	2	10.9
4	7.34	4	10.8

failures before the devices leave the factory. This practice ensures that the components that get to the user have a minimum number of detectable failures and minimize field repair costs.

According to Refs. [7,8], the infant–mortality period of the life cycle results from failure is a weak subpopulation of a bimodal component lifetime distribution. The percentage of weak subpopulation, usually two percent or smaller, varies with component type and manufacturing lot, even for the same manufacturer. The period of infant mortality is about 1 year, after which the package becomes more reliable as shown in Fig. 6.1.10. Because of high failure rates, it would be expensive to use these components in an operational system during the early stage of their life cycle. The industry has tried to eliminate the infant mortality exhibited in industrial products with burn-in screenings to accelerate aging.

This kind of screening procedure includes static burn-in, dynamic burn-in, and system burn-in running with various programs and software. Test patterns are applied to the inputs and the outputs are monitored. The consensus in the manufacturers has been

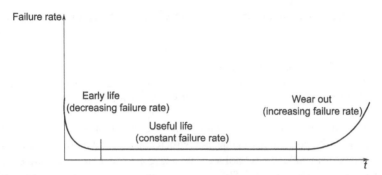

Figure 6.1.10 Bathtub curve describing the behavior of the hazard rate of semiconductor devices over time.

that certain components and assemblies that are marginal can be removed and replaced at a lower cost during manufacturing screening operations rather than after shipment to the customer.

Industry generally accepts the method 1012.10 from MIL-STD-883H (2010) [3] as the basis for burn-in conditioning by manufacturers of industrial electronic equipment. A list of the military standards has been applied with references. There are similar documents covering resistors, capacitors, and discrete semiconductors. Some company inspects its incoming IC's and generally subjects every IC to electrical and burn-in testing.

Because stringent thermal boundary conditions are imposed in the chip during the burn-in test, the thermal design of the test equipment is crucial in the test performance. Fig. 6.1.11 illustrates the dynamic system testing in a burn-in test platform at room

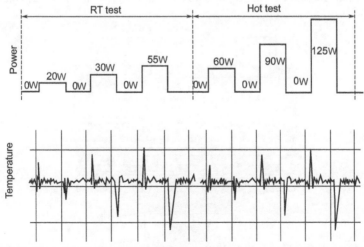

Figure 6.1.11 The system testing at room temperature and hot temperature in a burn-in test platform.

temperature and high temperature of $\sim 90°C$. The temperature fluctuations are controlled within a low range less than 5°C even at a high-power surge. A high-performance thermal control unit with fast thermal response and low inertia is required to meet the thermal design challenge.

Because of the increasing trend in DUT cores and integration with homogeneous and heterogeneous chips, the thermal design difficulty has been elevated to meet the testing challenges.

A practical design of the thermal control unit is elaborated here. Fig. 6.1.12 shows the schematic of the present thermal design, which consists of a thermoelectric module, a liquid-cooled heat sink, a TEC holder also acting as a heat spreader, and solid TIM [9]. The photograh and image of TEC are shown in Fig. 6.1.13. The purpose of thermoelectric module is to provide refrigeration cooling capacity to the DUT so that high-power testing can be conducted at ambient or subambient temperatures. Nonetheless, the selection of a TEC depends on not only the nominal cooling power but also the thermal load at device side. An analytical solution incorporating the TEC module parameters as well as thermal resistance factors that have been developed in Chapter 3 is employed for the selection of optimal TECs for the present tester design.

Three TEC coolers ranging from 187 W to 500 W in nominal cooling power are first assessed following the analytical method given in Section 3.7. TECs of high cooling capacities are considered and their details are listed in Table 6.1.2. The calculated cooling powers q_c at specified $T_j = 25°C$ are shown in Fig. 6.1.14 for both the ideal case of $R_{jc} = R_{hs} = 0$ K/W and the actual case with nonzero R_{jc} and R_{hs}. TEC 2 gives a maximum cooling power of 302 W at a feasible current of 10 A, near to the nominal power of 330 W. Although a much higher nominal power is claimed for TEC 3, ~ 452 W at a large current $I \sim 40$ A, it can only dissipate 217 W at the same feasible current of 10 A. In practice, it is the realistic power dissipations with thermal loads that are of the most interest. The actual case with device and heat sink thermal resistances of $R_{jc} = 0.6$ K/W and $R_{hs} = 0.04$ K/W is plotted in

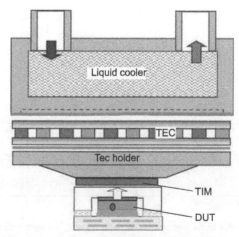

Figure 6.1.12 Schematic of the thermal design configuration for system burn-in testing.

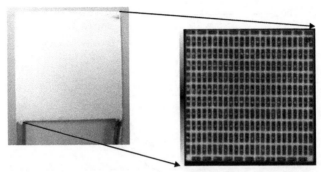

Figure 6.1.13 Photograph of TEC 1 and corresponding C-mode acoustic image at 35Mhz.

Table 6.1.2 List of thermoelectric cooler (TEC) details for system burn-in test.

Parameters	TEC 1	TEC 2	TEC 3
I_{max}	13.4	19	60.6
q_{max}	187	330	452
V_{max}	21.4	32	13.6
ΔT_{max}	58	64.5	49.6
T_{h0}	300.15	298.15	298.15
Z	0.00198	0.00236	0.00161
S_m	0.071	0.107	0.046
K_m	1.99	3.69	6.93
R_m	1.29	1.32	0.19

Fig. 6.1.14b. By considering the thermal resistances at both the cold side and hot side, TEC 2 provides the best thermal performance and the optimal current is around $I = 10$ A. For actual experimental implementation, TEC 2 with the highest actual cooling power and figure of merit is more suitable for high-power dissipations and selected for tests. As indicated in Fig. 6.1.13, TEC 2 has a size of 50 mm × 50 mm with thermoelectric pellets of 127 pairs.

There exists electrical contact resistance for a commercial TEC, which may have an adverse effect on TEC performance, especially for the TEC operated at high currents. Additional effort is made to obtain the module electrical resistance R_m at high current around 10 A, which is around 1.7Ω, 29% higher than the nominal value of 1.32Ω. A corrected calculation for TEC 2 is also conducted incorporating the effect of the increase in electrical resistance, which is illustrated in Fig. 6.1.14. As indicated in the corrected Rm curves, an increase in the electrical resistance will degrade the thermal performance, especially when the thermal resistances at the hot side and cold side of the TEC are nonzero.

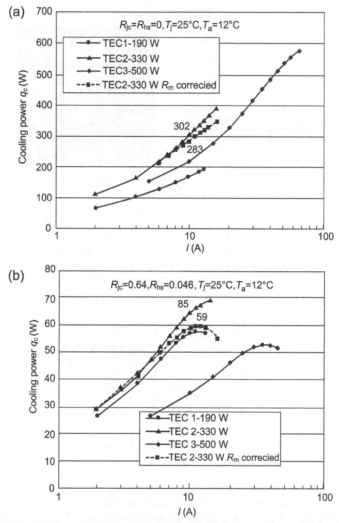

Figure 6.1.14 Cooling capacities for various TECs at given device temperatures of $T_j = 25°C$: (a) without thermal resistances and (b) with thermal resistances.

To remove the waste heat from the TEC hot side, a liquid cooler with alternatively stacked honeycomb microstructure is utilized and discussed in this section. Fig. 6.1.15 illustrates the heat sink structure. In the meantime, the CFD is conducted for this type of heat sink structure. Water is used as coolant and incompressible fluid is assumed. Because of the flow tortuosity, algebraic turbulent model is used. Half of the heat sink is incorporated in the model due to symmetricity. The computed flow field, pressure drop, and temperature profiles are illustrated in Fig. 6.1.16.

The application of the present TEC 2 model in processor burn-in test was also conducted based on a burn-in test platform. The test platform consists of a mother board to

Thermal and mechanical tests for packages and materials 305

Figure 6.1.15 Images of a honeycomb heat sink and its microstructure, with a unit cell represented in dashed line.

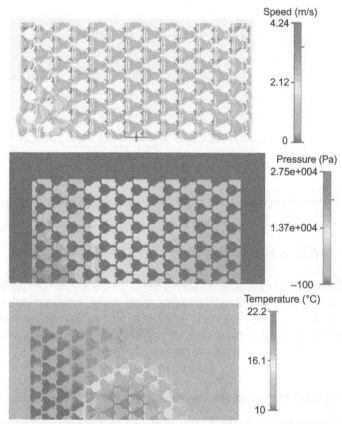

Figure 6.1.16 Simulated fluid flow, pressure drop, and temperature profiles at the bottom layer for the honeycomb heat sink.

provide electrical connections to the DUT. Loaded on the DUT, TEC 2 was sandwiched between a liquid-cooled heat sink and a heater plate with thermal grease filled at the thermal interfaces, as indicated in Fig. 6.1.12. A high-power lidded processor was used for the DUT with built-in thermal diode to record the junction temperature

T_j. In each test, the TEC current was fixed and the processor power q_c was varied to obtain T_j when the steady-state was reached. The plot of q_c versus T_j was attained, which appeared linear in the test range and, thus, q_c at specified T_j can be extracted. Fig. 6.1.17 shows a comparison of measured q_c results, together with the analysis results at $T_j = 25°C$ based on the method given in this section. Good agreement is achieved between measurements and analysis, in spite of some minor deviation observed in the experimental test.

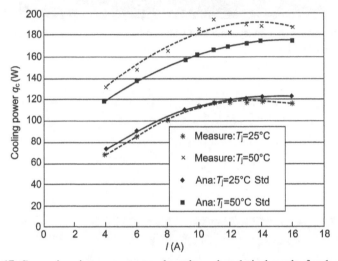

Figure 6.1.17 Comparison between measured results and analytical results for the processor burn-in test.

6.2 Material mechanical test and characterization

Material mechanical test refers to the testing to obtain mechanical properties such as stress—strain relationship, Young's modulus, yield stress, creep properties, fatigue behavior, viscoplastic and viscoelastic properties, etc. In this section, solder joint material and epoxy molding compound are selected as two typically advanced electronic packaging materials to represent metal and polymer materials, respectively.

6.2.1 Material test and characterization for solder alloys

For solder material testing, different types of testing methods and sample preparation are available. Tensile or shear test is usually used to obtain mechanical properties. Universal tester (e.g. Instron tensile tester), dynamic mechanical analysis (DMA), or other machines providing horizontal or vertical tension and compression loading can be used to conduct testing for solder material.

In the past decades, lead-free solders have been investigated widely. Among them, the Sn-Ag-Cu (SAC) solder becomes a promising candidate to replace Pb-based solders, and it is increasingly implemented in electronics industry. Nonetheless, drop reliability issue is encountered in the portable electronic devices with the use of the SAC solder due to its rigidity compared with the eutectic Sn-Pb solder [10,11]. It is shown that lowering Ag content in SAC solder results in better drop reliability due to ductile and compliant solder properties [12]. However, the low Ag content SAC solder leads to low thermal fatigue life due to its low fatigue resistance [13]. To enhance drop reliability and thermal fatigue life, Ni, Fe, Co, Zn, or rare-earth elements are added in the SAC solder to refine microstructures and decrease growth of intermetallic compound (IMC) [14–20]. Among additives, Ni is the commonly used doped material for SAC solder because it could enhances microstructure and decreases IMC growth, which improves the drop reliability of electronic package [14–16,21,22]. Therefore, we selected one low Ag content and Ni-doped solder, Sn-1.0Ag-0.5Cu-0.02Ni (or SAC105Ni0.02) solder, as an example to demonstrate material test and characterization for solder material.

6.2.1.1 Tensile test and analysis

Fig. 6.2.1 shows the geometry and dimensions of the bulk solder specimen with flat dog-bone shape used for tensile test. In sample preparation, solder was melted and maintained for 20 min at 100°C above the melting point of the solder, and then the melted solder was cast into the designed mold. Finally, solder specimens were naturally cooled down to the ambient temperature. Solder specimens were annealed at 100°C for 2 h to reduce residual stress before conducting test.

A universal tester equipped with a thermal chamber was used for the tensile test. The chamber can provide temperature from −35 to 250°C. The specimen was fixed at the ends of the solder specimen by testing jigs during the test. The applied force onto the solder specimen was measured by a load cell for calculating stress. A gauge length of 10 mm was used for measuring strain by an extensometer. Tensile force added on the specimen was measured by a load cell for stress calculation. The stress–strain curve can be obtained from the measurement data by extensometer and load cell. The strain rate can be controlled by adjusting loading speed. For testing conditions at

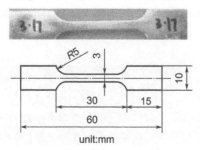

Figure 6.2.1 Solder specimen and its geometry for tensile test.

high and low temperatures, a thermal chamber was used to enclose the tested specimen to provide the designed temperature condition. Solder's material properties are very sensitive to temperature and strain rate. Tensile test was conducted at $-35°C$, $25°C$, $75°C$, and $125°C$, respectively, and with different strain rates of 10^{-2}, 10^{-3}, 10^{-4}, and 10^{-5} 1/s to investigate the effects of temperature and strain rate on the material properties including elastic modulus and yield stress. Five samples were chosen for testing at the same condition and the averaged result was used for analysis. Temperature- and strain rate–dependent material models of the SAC105Ni0.02 solder were established from testing results.

Fig. 6.2.2 shows the typical ductile failure mode of solder. Necking and surface coarsening phenomena were observed for the tested solder before complete failure. The stress–strain curves of solder shown in Fig. 6.2.3 reveals that strain rate affects mechanical properties significantly. The solder exhibits significant ductility with large elongation and plastic deformation before fracture. The yield stress, ultimate tensile stress (UTS), and elongation increase with the increase in strain rate. Fig. 6.2.4 shows the stress–strain curves of the solder at different testing temperatures. The solder becomes softer, subject to broken failure at higher testing temperature. Usually, solder creep is one dominant deformation mechanism at high-temperature tensile test, which makes necking easily happen during tensile test and thus elongation at fracture becomes lower. Elongation data show significant variation compared with elastic modulus and yield stress from tensile testing data.

Elastic modulus, E, is defined as the initial slope of the stress–strain curve. Yield stress, σ_y, is determined as the stress value corresponding to 0.2% plastic strain. Figs. 6.2.5–6.2.7 show the effect of temperature on the elastic modulus, yield stress, and UTS at various strain rates. It can be seen that the elastic modulus, yield stress, and UTS decrease with increasing temperature. The effect of temperature on the mechanical properties of solder at a large strain rate is more prominent compared with such effect at a low strain rate. Figs. 6.2.8–6.2.10 show the effect of strain rate on the elastic modulus, yield stress, and UTS at different test temperatures. It can be seen that the elastic modulus, yield stress, and UTS increase with increasing strain rate. Large strain rate makes the solder become harder, while high temperature makes the solder become softer.

The relationship of material properties and strain rate or temperature can be established through curve fitting for testing data. Usually, elastic modulus, yield stress, and UTS have linear function with temperature but logarithmic function with strain rate from experimental results. Strain rate – and temperature-dependent material models

Figure 6.2.2 Typical ductile failure mode of solder bar.

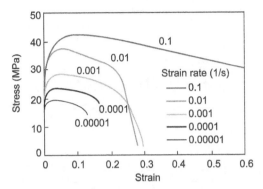

Figure 6.2.3 Effect of strain rate on the stress–strain curve of solder (test at 25°C).

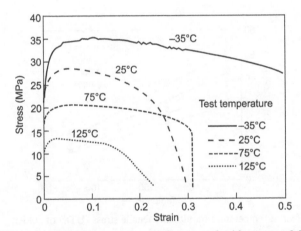

Figure 6.2.4 Effect of temperature on stress–strain curve of solder (test at 0.001 s^{-1}).

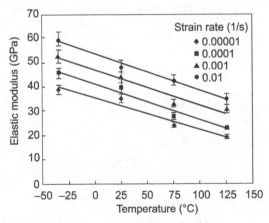

Figure 6.2.5 Effect of temperature on elastic modulus of solder.

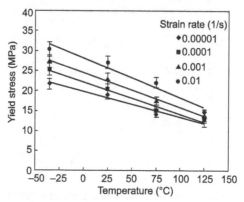

Figure 6.2.6 Effect of temperature on yield stress of solder.

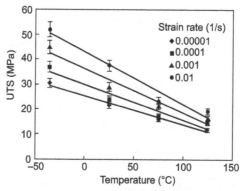

Figure 6.2.7 Effect of temperature on ultimate tensile stress (UTS) of solder.

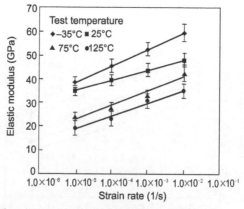

Figure 6.2.8 Effect of strain rate on elastic modulus of solder.

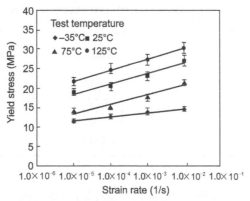

Figure 6.2.9 Effect of strain rate on yield stress of solder.

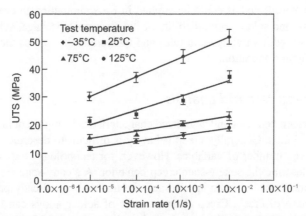

Figure 6.2.10 Effect of strain rate on ultimate tensile stress (UTS) of solder.

of elastic modulus, yield stress, and UTS are determined by following equations for the SAC105Ni0.02 solder based on experimental results:

$$E(T,\dot{\varepsilon}) = (a_1 T + a_2)\lg(\dot{\varepsilon}) + (a_3 T + a_4) \tag{6.2.1}$$

$$\sigma_y(T,\dot{\varepsilon}) = (b_1 T + b_2)(\dot{\varepsilon})^{(b_3 T + b_4)} \tag{6.2.2}$$

$$\mathrm{UTS}(T,\dot{\varepsilon}) = (c_1 T + c_2)(\dot{\varepsilon})^{(c_3 T + c_4)} \tag{6.2.3}$$

where T is temperature in °C, and $\dot{\varepsilon}$ is strain rate in s^{-1}. The material constants of a_1, a_2, a_3, and a_4 in the elastic modulus model; b_1, b_2, b_3, and b_4 in the yield stress model; c_1, c_2, c_3 and c_4 in the UTS model are listed in Table 6.2.1. The error analysis was conducted to evaluate the accuracy of models. Error analysis shows that the proposed models are accurate with error less than 10%. The calculated values and errors of the

Table 6.2.1 Material constants in mechanical property models of solder.

Elastic modulus model		Yield stress model		UTS model	
Variable	Value	Variable	Value	Variable	Value
a_1	−0.00592	b_1	−0.132	c_1	−0.243
a_2	6.07	b_2	35.57	c_2	59.76
a_3	−0.164	b_3	−4.80 × 10^{-5}	c_3	2.02 × 10^{-4}
a_4	65.59	b_4	0.0513	c_4	0.0745

mechanical properties are listed in Table 6.2.2. It can be seen that the curve-fitting models are accurate with the errors within 10.5% for the predicted elastic modulus, 9.3% for the predicted yield stress, and 14.4% for the predicted UTS. The above constitutive material models could be applied in FEA simulation to evaluate stress–strain response and solder joint reliability of the electronic package, subjected to the external loading with varying strain rates and temperatures, such as thermal cycling, cyclic bending, and vibration.

6.2.1.2 Creep test and analysis

The study of creep behavior and associated characteristics is important in the reliability tests of solder joints. Generally, creep behavior is difficult to describe because it depends on a large number of variables. However, for technological applications, it is sufficient to describe the steady-state creep behavior. A steady-state creep is reached after a constant stress has been applied and the material has already passed through a transient phase of creep. Creep characteristics of solder alloys can be studied by different experimental techniques such as lap shear creep [23,24], uniaxial tensile [24,25], stress relaxation [26], and indentation testing methods [27]. In this section, the tensile creep behavior of the bulk solder alloy of SAC105Ni0.02 was investigated at different temperatures and stress levels.

The same specimen used in tensile test was also applied in creep test. Creep tests were carried out at four isothermal conditions including −35°C, 25°C, 75°C, and 125°C. For each testing temperature, several stress levels were designed from 5 to 45 MPa. For testing conditions at high and low temperatures, a thermal chamber was used to enclose the tested specimen to provide the designed temperature condition. The specimens were held at the designed temperature for 10 min before loading. At the beginning, tensile force was applied and increased gradually to the designed value. The designed stress is held to the end of the test, and creep strain is recorded during the whole test. For the creep test at a high stress level or high temperature, the creep test continued to be sample broken. While for the creep test at a low stress level or low temperature, the test was stopped for saving time when the constant creep strain rate was observed. Tests were repeated for three samples at the same condition and the averaged results are used.

Table 6.2.2 Predicted mechanical property data and errors based on the material models.

Strain rate/ Temp.	Calculated values			Errors		
10^{-5}/s	E (GPa)	Yield (MPa)	UTS (MPa)	E	Yield	UTS
−35°C	39.97	22.62	31.42	3.4%	4.5%	3.4%
25°C	31.89	18.09	21.49	−8.9%	−5.2%	−0.5%
75°C	25.15	14.34	14.80	3.9%	0.3%	−7.7%
125°C	18.42	10.62	9.33	−4.5%	−9.3%	14.4%
10^{-4}/s						
−35°C	46.24	25.45	36.70	1.0%	0.8%	−0.7%
25°C	37.81	20.36	25.81	−4.7%	−0.5%	7.3%
75°C	30.77	16.14	18.20	10.5%	6.6%	6.2%
125°C	23.74	11.95	11.73	2.1%	−7.4%	−13.3%
10^{-3}/s						
−35°C	52.52	28.60	42.86	−0.5%	4.7%	−4.4%
25°C	43.73	22.88	31.00	−0.5%	−0.6%	7.1%
75°C	36.40	18.15	22.37	10.0%	3.5%	6.1%
125°C	29.07	13.44	14.76	−5.3%	−3.9%	−9.9%
10^{-2}/s						
−35°C	58.79	32.09	50.05	−1.7%	5.5%	−3.4%
25°C	49.64	25.69	37.23	2.6%	−5.2%	−0.7%
75°C	42.02	20.38	27.50	−1.9%	−7.4%	14.2%
125°C	34.40	15.10	18.57	−2.5%	3.7%	−3.4%

Fig. 6.2.11 shows a typical creep strain curve of the solder. Creep has three stages including primary, secondary, and tertiary stages. Primary creep happens at the beginning of the test and very short. The secondary creep stage, also called steady-state creep, is a long creep stage, and it occupies most of the creep lifetime. The tertiary creep stage occurs before creep rupture, which has fast creep strain rate. Creep strain rate is calculated by taking the derivative of creep strain with time. The minimum rate was taken as the creep strain rate of the steady-state creep stage. Creep strain rate at steady-state creep stage is an important parameter that usually determines lifetime of solder joints. Stress and testing temperature have a significant effect on creep strain rate.

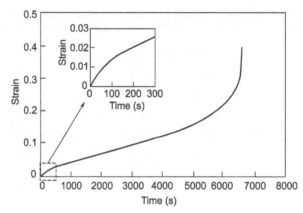

Figure 6.2.11 Typical tensile creep curve of solder (test conditions: 25°C and 20 MPa load).

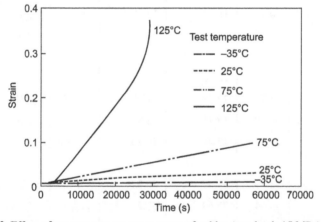

Figure 6.2.12 Effect of temperature on creep curve of solder (test load: 15 MPa).

Fig. 6.2.12 shows the effect of temperature on creep curve of solder. The creep strain rate increases and creep rupture time decreases significantly with increasing temperature at a certain stress level. Fig. 6.2.13 shows the effect of stress on creep curve of solder. The creep strain rate increases and creep rupture time decreases sharply when increasing the applied stress level. The creep rupture time as a function of stress at different temperatures is summarized in Fig. 6.2.14. It can be seen that the creep rupture time is sensitive to the stress level and temperature. Both high stress level and temperature accelerate creep rupture of solder due to the corresponding high creep strain rate.

Fig. 6.2.15 shows the relationship between creep stress and steady-state creep strain rate at test temperature of 25°C. The creep strain rate changes from 10^{-9} to 10^{-2} s^{-1} when stress level changes from 5 to 30 MPa. The creep strain rate is less dependent on stress in the low stress range, while it becomes more dependent on stress in the high

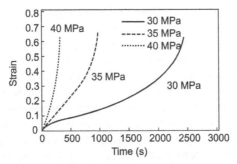

Figure 6.2.13 Effect of stress level on creep strain of solder (test at −35°C).

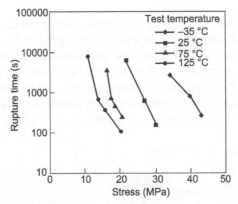

Figure 6.2.14 Relationship between creep rupture time and stress at different temperatures.

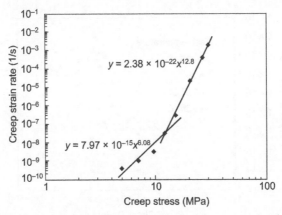

Figure 6.2.15 Relationship between creep stress and steady-state creep strain rate.

stress range. The power law constitutive model can be used to describe the relationship between creep strain rate and stress. The stress exponent n is larger in the high stress range than that in the low stress range. This phenomenon is referred as power law creep behavior breakdown. This power law breakdown behavior can be found at any given test temperature as shown in Fig. 6.2.16. The stress exponents in the low and high stress ranges can be determined based on creep strain rate data in this figure.

Fig. 6.2.17 shows the stress exponents separately for high and low stress ranges at different temperatures. The stress exponent decreases with increasing temperature for both high and low stress ranges. Stress exponent can vary from 2 to 13 for different solders due to different creep mechanisms. At low stress, creep deformation is controlled mainly by grain boundary sliding in which stress exponent is 2–4 [28]. At high stress, creep deformation is mainly due to dislocation climb and glide with high stress exponent [24]. At high temperature, diffusion-controlled creep is a dominant mechanism.

From the analysis mentioned above, the stress exponent and creep activation energy in the creep models are stress level and temperature dependent. The averaged values of stress exponent and creep activation energy can be determined by fitting all the creep experimental results under different stresses and temperatures. We conducted the curve fitting for the creep test data through iterative multivariable nonlinear regression to develop the creep constitutive models based on testing data shown in Fig. 6.2.16. The different types of creep constitutive models for SAC105Ni0.02 solder are presented below.

Norton's power law model (refer to Eq. (4.1.26)):

$$\dot{\varepsilon}_{cr} = 2.39 \times 10^{-7} \sigma^{8.90} \exp\left(-\frac{6282.7}{T}\right) \tag{6.2.4}$$

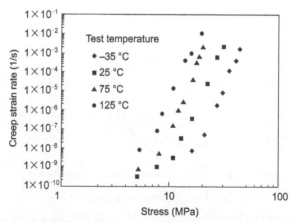

Figure 6.2.16 Steady-state creep strain rate at various load conditions.

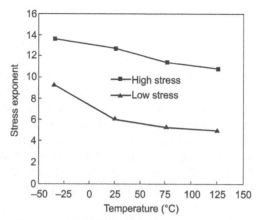

Figure 6.2.17 Effect of temperature and stress level on stress exponent.

Double power law model (refer to Eq. (4.1.27)):

$$\dot{\varepsilon}_{cr} = 9.87 \times 10^{-7} \sigma^7 \exp\left(-\frac{4883.1}{T}\right) + 5.01 \times 10^{-10} \sigma^{13} \exp\left(-\frac{8949.2}{T}\right) \quad (6.2.5)$$

Garofalo hyperbolic sine model (refer to Eq. (4.1.28)):

$$\dot{\varepsilon}_{cr} = 3.94 \times 10^4 [\sinh(0.0607\sigma)]^{6.32} \exp\left(-\frac{7024.2}{T}\right) \quad (6.2.6)$$

Exponential model (refer to Eq. (4.1.29)):

$$\dot{\varepsilon}_{cr} = 277.4 \exp\left(\frac{\sigma}{2.11}\right) \exp\left(-\frac{7744.3}{T}\right) \quad (6.2.7)$$

These four creep constitutive models have been applied in FEA simulation for fine pitch ball grid ball package subjected to thermal cycling load. FEA simulation results showed that different models lead to similar creep strain and creep strain energy density, and similar fatigue life was predicted based on results from these four creep models [29]. Among steady-state creep models, hyperbolic sine model is commonly implemented for electronic solder materials. The creep activation energy is an important parameter used for measuring the creep resistance of the solder alloy. The larger the creep activation energy is, the higher the creep resistance is. For the purpose of using the creep model in FEA simulation, Eq. (4.1.28) is usually rewritten as the following format:

$$\dot{\varepsilon}_{cr} = C_1 [\sinh(C_2 \sigma)]^{C_3} \exp\left(\frac{-C_4}{T}\right) \quad (6.2.8)$$

where C_1, C_2, C_3, C_4 are material constants, which are determined from tensile creep test results.

Table 6.2.3 lists material constants in Eq. (6.2.8) for SAC105Ni0.02 solder and other solder alloys for comparison. Creep activation energy of the SAC105Ni0.02 solder is comparable with that of high Ag content Sn-3.9Ag-0.6Cu solder, which indicates that SAC105Ni0.02 solder has similar creep resistance as the high Ag content SAC solder under TC condition. Ni addition in solder increases creep resistance and creep rupture lifetime due to refinement of IMC particles [20,24]. Compared with Sn-1.0Ag-0.5Cu solder as listed in Table 6.2.3 [30], adding 0.02%wt Ni improves creep resistance of the SAC105 solder. From other researchers' studies, SAC105Ni0.02 solder has similar mechanical properties as SAC105 solder, but the SAC105Ni0.02 solder improves drop performance compared with SAC105 solder [33,34].

6.2.2 Material test and characterization for polymers

Dynamic mechanical analysis (DMA) is a technique used to study the viscoelastic behavior of polymers. A small oscillation stress is applied on the material as a function of time and the strain response is measured, allowing one to determine the complex modulus. The variation of the temperature and frequency of the load causes the complex modulus to vary, and the glass transition temperature, T_g, can be evaluated through this approach. Suppose a sinusoidal force is applied to the material, the time-varying stress (σ) and strain (ε) are written as

$$\sigma = \sigma_0 \sin(\omega t + \delta) \qquad (6.2.9)$$

$$\varepsilon = \varepsilon_0 \sin(\omega t) \qquad (6.2.10)$$

where ω is frequency of strain oscillation, t is time, and δ is phase lag between stress and strain. σ_0 and ε_0 are the maximum amplitude of stress and strain, respectively. The complex modulus of the material $E^* = \sigma/\varepsilon$ is defined as the ratio of stress to the strain of the material. It consists of storage modulus E' (real part) and loss modulus E'' (imaginary part), which can be expressed as

$$E^* = E' + iE'' \qquad (6.2.11)$$

where

$$E' = \frac{\sigma_0}{\varepsilon_0} \cos \delta$$

$$E'' = \frac{\sigma_0}{\varepsilon_0} \sin \delta$$

Table 6.2.3 Material constants of hyperbolic sine creep model for different solders.

Solder alloys	C_1	C_1(1/MPa)	C_2	C_4(K)	Q (kJ/mol)	References
Sn-1.0Ag-0.5Cu-0.02Ni	3.94×10^4	0.0607	6.32	7024.2	58.4	This study
Sn-1.0Ag-0.5Cu	2.8×10^{-4}	0.0581	7.1	5292.3	44	[30]
Sn-2.0Ag-0.5Cu	3.7×10^{-4}	0.0659	6.8	9165.3	76.2	[14]
Sn-2.0Ag-0.5Cu-0.05Ni	2.8×10^{-4}	0.0677	7.9	10,175.6	84.6	[14]
Sn-1.0Ag-0.3Cu	3.1×10^{-4}	0.0576	5.1	5412.6	45	[17]
Sn-3.9Ag-0.6Cu	0.184	0.221	2.89	7457	62.0	[31]
63Sn-37Pb	10	0.2	2.0	5400	44.9	[32]
Sn-3.5Ag	9×10^5	0.0653	5.5	8690	72.2	[35]

The phase angle δ is calculated as

$$\tan \delta = \frac{E''}{E'} \quad (6.2.12)$$

Epoxy molding compound (EMC) material was tested by DMA to obtain temperature-dependent modulus, T_g, and viscoelastic model. The tested sample had a dimension of 40 mm (length) by 10 mm (width) by 1 mm (thickness). The specimen was clamped on the fixture at one end and the load applied at the other end. The geometry deformation was set for single cantilever bending mode in strain control with amplitude of 20 μm.

Temperature sweep with fixed frequency method was used to obtain the temperature-dependent modulus and glass transition temperature of the EMC specimen. The oscillation frequency was fixed at 1 Hz while the temperature was ramping from 30 to 250°C at a heating rate of 3°C/min. There are mainly two methods to evaluate the glass transition temperature of a material by DMA spectra. A direct and simple way is to determine the transition temperature at which the peak of tan δ or loss modulus occurs. Another method is called the evaluation of modulus step. In this method, the glass transition temperature is measured as the temperature of the midpoint of inflectional tangent. The onset temperature is determined as the intersection of the inflectional tangent with the tangent extrapolated from temperatures below the glass transition temperature. The T_g determined by the first method using the peak tan δ is always higher than that by the modulus step method. In this study, the peak of tan δ was adopted for evaluating the T_g of the EMC specimen, as it is the most straightforward method, and its results had least variation on the DMA spectra.

Temperature step with frequency sweep method was used to construct the master curves and to determine the corresponding shift factors. The testing temperature increased from 30°C to 210°C with a step of 10°C. The storage modulus (E'), loss modulus (E''), and tan δ were measured at a series of frequencies (10−0.1 Hz) while the temperature was kept as constant at each temperature step. The selected frequencies (Hz) were 10.00, 6.30, 3.00, 2.50, 1.60, 1.00, 0.63, 0.40, 0.25, 0.16, and 0.10. Time−temperature superposition principle (TTSP) was developed to describe the equivalence of time and temperature effects on linear viscoelastic polymers. The correspondence between time and temperature can be consequently achieved by time−temperature shifting with suitable shift factors. Utilizing this principle, polymer's behavior in elevated temperatures can be used to generate a master curve to predict the polymer behavior in a much longer time range than testing time under reference temperature. The key idea of using TTSP is to accelerate the polymer resin test by increasing temperature.

Three specimens were tested for the EMC material, and the DMA curve is shown in Fig. 6.2.18. It clearly shows that modulus (unit in MPa in figure) is temperature dependent and has significant change from below T_g to above T_g. The value of T_g can be determined based on tan δ curve.

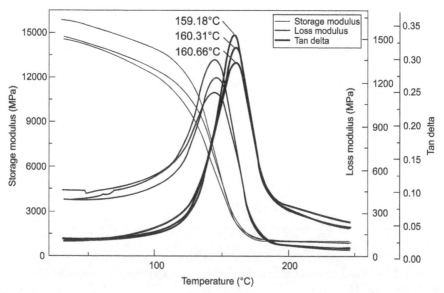

Figure 6.2.18 DMA curves for the epoxy molding compound (EMC) specimens.

The master curve of the storage modulus can be expressed as a Prony series:

$$E(t) = E_\infty + (E_0 - E_\infty) \sum_{i=1}^{N} C_i \exp\left(-\frac{t}{\tau_i}\right) \quad (6.2.13)$$

where E_0 is the elastic modulus at $t = 0$, E_∞ is the elastic modulus as $t = \infty$, and $i = 10$ is selected in all the curve fitting in this study. The Williams–Landel–Ferry Equation (or WLF Equation) is empirically obtained and associated with time and temperature superposition for almost all amorphous polymers above the glass transition temperature:

$$\lg a_T = \frac{-C_1(T - T_r)}{C_2 + (T - T_r)} \quad (6.2.14)$$

where the two constants, C_1 and C_2, in the WLF Equation can be determined by fitting the discrete values of shift factor a_T.

Fig. 6.2.19 shows the master curves of three EMC specimens at reference temperature of 150°C. The repeatability of testing data is quite well. The parameters in Prony series can be determined through curve fitting. Fig. 6.2.20 shows the shift factors fitted by WLF Equation for EMC specimens at all temperatures with $T_0 = 150°C$. The fitted parameters in WLF Equation also can be obtained through curve fitting. Such parameters can be implemented in FEA modeling and simulation to investigate effect of viscoelastic behavior of EMC material on stress and reliability of electronic package.

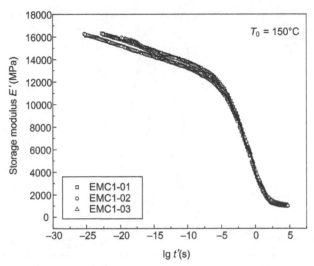

Figure 6.2.19 Master curves of the storage modulus for epoxy molding compound (EMC) specimens at 150°C.

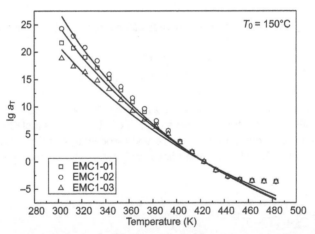

Figure 6.2.20 Relationship between lg a_T and temperature for epoxy molding compound (EMC) with $T_0 = 150°C$ (Williams–Landel–Ferry [WLF] model).

References

[1] JEDEC JC-15 websites. http://www.jedec.org.
[2] Thermal Characteristics, MIL-STD-883H: Test Method Standard, Microchips, Method 1012.1, US Dept. Defense, 2010.
[3] Burn-In, MIL-STD-883H: Test Method Standard, Microchips, Method 1015.10, US Dept. Defense, 2010.
[4] JEDEC 51-14. http://www.jedec.org.

[5] H.Y. Zhang, Y. Sui, T. Lin, H. Liu, Transient thermal characterization of junction to case thermal resistance for 2.5D packages, in: ICEPT Conference, 2018, pp. 115–120.

[6] H.Y. Zhang, X.W. Zhang, B.L. Lau, S. Lim, L. Ding, M.B. Yu, Thermal characterization of both bare die and overmolded 2.5-D packages on through silicon interposers, IEEE Trans. Compon. Packag. Manuf. Technol. 4 (5) (2014) 807–816.

[7] A. Vassighi, et al., CMOS IC technology scaling and its impact on burn-in, IEEE Trans. Device Mater. Reliab. 4 (2) (2004).

[8] W. Kuo, Y. Kuo, Facing the headaches of early failures: a state-of-the-art review of burn-in decisions, Proc. IEEE 71 (11) (1983).

[9] H.Y. Zhang, M. Tarin, R.K. Kumar, Y.C. Mui, Thermal design considerations for system level testing of electronic packages, in: ICEPT Conference 2010, 2010.

[10] K.J. Puttlitz, G.T. Galyon, Impact of the ROHS. Directive on high-performance electronic systems, J. Mater. Sci. Mater. Electron. 18 (2007) 331–346.

[11] D. Frear, Issues related to the implementation of Pb-free electronic solders in consumer electronics, J. Mater. Sci. Mater. Electron. 18 (2007) 319–330.

[12] D.A.A. Shnawah, S.B.M. Said, M.F.M. Sabri, I.A. Badruddin, F.X. Che, High-reliability low-Ag-content Sn-Ag-Cu solder joints for electronics applications, J. Electron. Mater. 41 (2012) 2631–2658.

[13] S. Terashima, Y. Kariya, T. Hosoi, M. Tanaka, Effect of silver content on thermal fatigue life of Sn-xAg-0.5Cu flip-chip interconnects, J. Electron. Mater. 32 (2003) 1527–1533.

[14] A.A. El-Daly, A.M. El-Taher, Evolution of thermal property and creep resistance of Ni and Zn-doped Sn-2.0Ag-0.5Cu lead-free solders, Mater. Des. 51 (2013) 789–796.

[15] I.E. Anderson, Development of Sn-Ag-Cu and Sn-Ag-Cu-X alloys for Pb-free electronic solder applications, J. Mater. Sci. Mater. Electron. 18 (2007) 55–76.

[16] T. Laurila, J. Hurtig, V. Vuorinen, J.K. Kivilahti, Effect of Ag, Fe, Au and Ni on the growth kinetics of Sn-Cu intermetallic compound layers, Microelectron. Reliab. 49 (2009) 242–247.

[17] A.A. El-Daly, A.E. Hammad, G.S. Al-Ganainy, M. Raga, Influence of Zn addition on the microstructure, melt properties and creep behavior of low Ag-content Sn-Ag-Cu lead-free solders, Mater. Sci. Eng. A 608 (2014) 130–138.

[18] D.A.A. Shnawah, S.B.M. Said, M.F.M. Sabri, I.A. Badruddin, F.X. Che, Microstructure, mechanical, and thermal properties of the Sn-1Ag-0.5Cu solder alloy bearing Fe for electronics applications, Mater. Sci. Eng. A 551 (2012) 160–168.

[19] J.X. Wang, S.B. Xue, Z.J. Han, S.L. Yu, Y. Chen, Y.P. Shi, et al., Effects of rare earth Ce on microstructures, solderability of Sn-Ag-Cu and Sn-Cu-Ni solders as well as mechanical properties of soldered joints, J. Alloy. Comp. 467 (2009) 219–226.

[20] W. Xiao, Y. Shi, G. Xu, R. Ren, F. Guo, Z. Xia, Y. Lei, Effect of rare earth on mechanical creep-fatigue property of Sn-Ag-Cu solder joint, J. Alloy. Comp. 472 (2009) 198–202.

[21] Y.W. Wang, Y.W. Lin, C.T. Tu, C.R. Kao, Effects of minor Fe, Co, and Ni additions on the reaction between Sn-Ag-Cu solder and Cu, J. Alloy. Comp. 478 (2009) 121–127.

[22] P. Liu, P. Yao, J. Liu, Evolutions of the interface and shear strength between SnAgCu-xNi solder and Cu substrate during isothermal aging at 150 °C, J. Alloy. Comp. 486 (2009) 474–479.

[23] S.W. Shin, J. Yu, Creep deformation of Sn-3.5Ag-xCu and Sn-3.5Ag-xBi solder joints, J. Electron. Mater. 34 (2005) 188–195.

[24] M. He, S.N. Ekpenuma, V.L. Acoff, Microstructure and creep deformation of Sn-Ag-Cu-Bi/Cu solder joints, J. Electron. Mater. 37 (2008) 300–306.

[25] M. Amagai, M. Watanabe, M. Omiya, K. Kishimoto, T. Shibuya, Mechanical characterization of Sn-Ag-based lead-free solders, Microelectron. Reliab. 42 (2002) 951–966.

[26] S.G. Jadhav, T.R. Bieler, K.N. Subramanian, J.P. Lucas, Stress relaxation behavior of composite and eutectic Sn-Ag solder joints, J. Electron. Mater. 30 (2001) 1197.

[27] Y.C. Liu, J.W.R. Teo, S.K. Tung, K.H. Lam, J. Alloy. Comp. 448 (2008) 340.

[28] G.S. Zhang, H.Y. Jing, L.Y. Xu, J. Wei, Y.D. Han, Creep behavior of eutectic 80Au/20Sn solder alloy, J. Alloy. Comp. 476 (2009) 138.

[29] F.X. Che, W.H. Zhu, S.W. Edith, Poh, X.R. Zhang, X.W. Zhang, T.C. Chai, et al., Creep properties of Sn-1.0Ag-0.5Cu lead-free solder with the addition of Ni, J. Electron. Mater. 40 (3) (2011) 344−354.

[30] A.A. El-Daly, G.S. Al-Ganainy, A. Fawzy, M.J. Younis, Structural characterization and creep resistance of nano-silicon carbide reinforced Sn-1.0Ag-0.5Cu lead-free solder alloy, Mater. Des. 55 (2014) 837−845.

[31] Q. Xiao, W.D. Armstrong, Tensile creep and microstructural characterization of bulk Sn3.9Ag0.6Cu lead-free solder, J. Electron. Mater. 34 (2) (2005) 196−211.

[32] S. Wiese, E. Meusel, Characterization of lead-free solders in flip chip joints, J. Electron. Packag. 125 (2003) 531−538.

[33] F.X. Che, W.H. Zhu, E.S.W. Poh, X.W. Zhang, X.R. Zhang, The study of mechanical properties of Sn-Ag-Cu lead-free solders with different Ag contents and Ni doping under different strain rates and temperatures, J. Alloy. Comp. 507 (2010) 215−224.

[34] M. Amagai, Y. Toyoda, T. Tajima, High solder joint reliability with lead free solders, in: Proc. 53rd Electron. Comp. Technol. Conf., 2003, pp. 317−322.

[35] S. Ridout, M. Dusek, C. Bailey, C. Hunt, Assessing the performance of crack detection tests for solder joint, Microelectron. Reliab. 46 (2006) 2122−2130.

System-level modeling, analysis, and design

7.1 System-level thermal modeling and design

7.1.1 Introduction to system-level thermal modeling and design

7.1.1.1 Description of the system-level thermal modeling

The system-level thermal modeling and design deal with the cooling and temperature control of more than one chip or components. The terminology system can be ambiguous considering the different scenarios. Professor Rao Tummala [1] from Georgia Institute of Technology first defined the system to be all the electronic products. As such, the system involves not only active components such as central processing unit (CPU), memory, graphic processing unit, RF, MEMS sensors, and actuators but also a myriad of passives such as capacitors, resistors, and inductors. All the components are to be packaged, not necessarily in one package, and assembled on one or more printed circuit boards (PCBs). More broadly speaking, a number of electronic boards are to be configured in a tray, a number of trays are to be stacked in a rack, and a number of racks are to be stored in a room to fulfill the functions. The last case may represent the system for a modern data center.

Because of the different levels of the electronic systems, the thermal solutions vary depending on various factors such as the power, heat flux dissipation, technology feasibility, and implementation cost [2]. Fig. 7.1.1 illustrates the hierarchy of the system-level thermal modeling and designs at different levels. The tray-level model in a burn-in test system is displayed in Fig. 7.1.2, which includes three full sets of computer board subsystems. Each board is mounted with high heat flux components such as dual processors with the corresponding thermal heads for cooling purpose and other components such as memory modules, graphic card, hard disk, voltage regulating modules (VRMs), capacitors and resistors, and so on.

To facilitate the operation, the thermal head can be made of air-cooled heat sink, liquid-cooled heat sink, or even refrigeration cooling block with in-built expansion valve to allow heat dissipation in the order of several hundred Watts. Besides the thermal head apparatus dedicated to the cooling of the processors on the boards, three axial fans are installed at the inlet of the tray to force the indoor air into the tray to cool the rest of the components. The vertical stacking of a number of trays with thermal heads forms a rack system.

7.1.1.2 Chip-level cooling

Chip-level cooling could be the most expensive thermal solution as it is to be integrated during the device fabrication stage before the packaging. The chip-level cooling

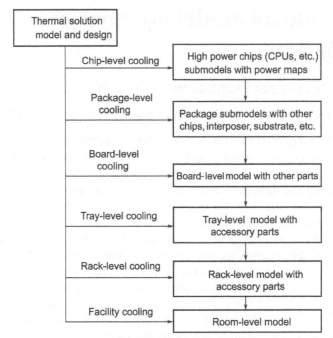

Figure 7.1.1 Description of system-level thermal modeling and design of electronic system at different levels.

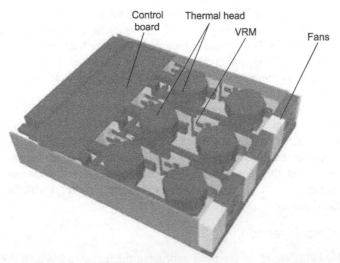

Figure 7.1.2 The thermal model for a tray with three printed circuit boards (PCBs) and three inlet fans, each board mounted with a set of computing subsystem including dual processors with their respective thermal heads, hard disk, memory modules, and voltage regulating power modules.

could interfere with the fabrication process and thus cause incompatibility in process and implementation. IBM [3] investigated the dielectric cooling with R1234z as the direct cooling medium as funded by DARPA; the pin-fin microstructure was fabricated on the backside of the silicon die for impingement cooling. To facilitate the central impingement flow, the central holes were made on the manifold chip so that the backside cooling could be achieved. This additional fabrication of microstructure might affect the process for the chip circuits, and the whole structure might cause premature failures due to the damage of the mechanical integrity of silicon material. Although this is a new cooling method for chip-level cooling, the implementation could be costly considering the fabrication process and the assembly and integration of the evaporative coolant.

Nonetheless, novel chip-level cooling can be gradually implemented by introducing new materials. One example is the introduction of carbon nanotubes and graphene on the chip [4]. The chip process is being improved with time so that the incorporation of new designs and materials in the next generation of chips can be achieved.

7.1.1.3 Package-level cooling

Most of the air cooling and liquid cooling techniques can be implemented at the package level. The most prevalent method is to mount heat sink and heat spreader on the package. The enlarged fin surface can remove the heat directly from the chip and package. In general, air cooling provides a simple yet effective cooling solution. However, with the increase in heat flux dissipation, the air cooling technology is not sufficient anymore. According to Ref. [5,6], the upper limit of the power dissipation heat flux for dedicated air cooling solutions is around 50 W/cm^2. The use of vapor chamber heat sink could extend the cooling power up to 100 W per a 12 mm × 12 mm chip. When the heat flux goes up to and even beyond 100 W/cm^2, which is the case for high performance computing chips, air cooling methods have become inadequate for most applications [6—8]. As such, liquid cooling technology with higher power dissipation capability for high performance microelectronic devices is required.

Basically, there are two major operational modes of liquid cooling, namely the single-phase cooling and two-phase cooling. A number of researchers have explored the advantages of using liquid cooling to mitigate the present thermal management problems [9—12]. Madhour et al. [9] offered a parametric study with single- and two-phase cooling technologies to address the cooling requirements for the vertically integrated microprocessors. Mohapatra and Loikits [10] provided a study on what attributes are more effective in selecting a working fluid for a microchannel application. Wei and Joshi [11] presented a thermal resistance modeling approach to identify the potential beneficial and deleterious effects associated with stacked microchannels.

Considering the higher pressure drop and more complexity in a two-phase liquid cooling system, utilizing the single-phase liquid cooling technology for high heat flux microprocessors is considered a more viable option [12—20]. For a single-phase liquid cooling technology, both microchannel and microjet heat sinks can dissipate high heat fluxes anticipated in high-power electronic devices [15—17]. A review of the high heat flux cooling technologies is given in Ref. [12], which pointed out the chip cooling power limitation of 300 W/cm² based on the microchannel structure under straight flow configuration. An overview of the thermal performance of the heat sinks with miniaturized structure is displayed in Fig. 7.1.3.

Colgan et al. [17] described a practical implementation of a single-phase silicon microchannel cooler designed for cooling very high-power chips, and the cooler of this design was able to cool a single chip with the average power densities of 300 W/cm² by bonding this chip to the cooler using the silver epoxy or solder. Brunschwiler et al. [20] demonstrated that a microjet array impinging cooling method can provide a high heat transfer coefficient up to $8.7 \times 10^4 \, W \cdot m^{-2} \cdot K^{-1}$. Compared with the impinging microjets on the plain backside of chip, microchannel cooling has a lower heat transfer coefficient, but, with the enlarged fin areas, the overall heat transfer can be enhanced. Recently, the hybrid microcooler, combining the merits of both microjet array impingement and microchannel flow, has been developed for cooling the high-power devices [21,22].

It is noted that all of the above studies are focused on the development of the cooling methods with enhancement structure such as microchannels/microjet array to maximize their heat dissipation capability. The thermal design for the external heat exchanger, as part of the thermal system, is not well addressed and discussed. In many cases, however, the heat exchanger could be a limiting factor in the application scenarios such as computers, data center, and transportation systems, where both weight and size become of a practical concern. Miniaturized heat exchanger with low pressure drop and high heat exchange efficiency is a major design objective for cooling system designers. Available studies on a single miniaturized heat exchanger are reported in Refs.[23,24], which, nonetheless, did not consider the hydraulic performance and thermal efficiency of the full system, including the chip cooler for integrated circuits cooling.

To better understand the problem of a heat exchanger, a computer casing integrated with a microchannel cooler for the microprocessor and a heat exchanger was implemented in the laboratory without optimization of the heat exchanger, as illustrated in Fig. 7.1.4. The fan was mounted on the top of the casing for sucking air out of

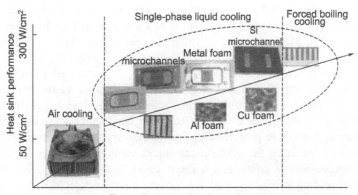

Figure 7.1.3 An overview of the heat sink performance for the high-power chips.

System-level modeling, analysis, and design

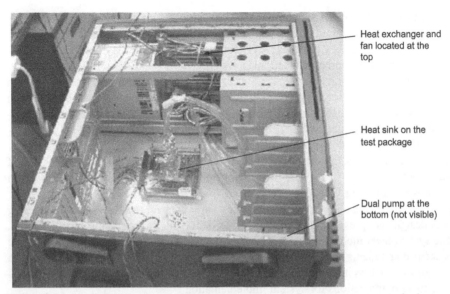

Figure 7.1.4 A computer casing integrated with a cooler for the microprocessor and a heat exchanger is demonstrated by removing the side casing.

the computer casing. An aluminum microchannel heat sink was assembled on the high-power chip for thermal management. The test chip was built in with both thermal diode for temperature monitoring and heating resistors for chip heating. The measured chip junction temperature was 55°C based on 60 W power input at an ambient temperature of 25°C, indicating a junction to the ambient thermal resistance of 0.50°C/W. Nonetheless, the thermal resistance for junction to the heat sink itself was 0.30°C/W, indicating a temperature rise of 18°C for chip over the heat sink inlet coolant. A fast analysis at the temperature rise indicates that the temperature difference over the heat exchanger amounts to around 12°C, which is nontrivial and should be addressed in the thermal design at the system level.

7.1.1.4 System-level modeling at the tray, rack, and room levels

The thermal system cooling at the tray, rack, and room levels may occur at the processor manufacturing site and data center for data communication and processing. A numerical procedure to model individual servers within each rack was developed by Rambo and Joshi [25]. A commercial finite volume method was developed for the multiscale model of air-cooled data centers. The multiscale model of a representative data center consists of ∼ 1 500 000 grid cells and needs more than 2400 iterations to obtain a converged solution. This model took about 8 h to converge on a 2.8 GHz Xeon with 2 GB memory. Even so, this model is still a significant departure from reality because it does not include finer details at the server and chip level. Unfortunately, a fully detailed multiscale CFD model of operational data centers, which may contain hundreds and thousands of racks, seems infeasible due to available computing capacities. As such, more compact submodels with greatly reduced details and thus grid cells should be developed, which could run much faster by keeping the influences of all the key parameters. Such modeling reduction is essential, especially for optimization of complex electronics systems.

The air flow in a rack with seven layers of trays was investigated in Ref. [26], in which each tray consists of three sets of computer subsystems, and each subsystem consists of mother boards, memory cards, hard disk drives, and so on. A comprehensive review of literature on data center numerical modeling with a study on the necessity of compact air flow/thermal modeling for data centers is available in Ref. [27]. A more detailed discussion on the thermal analysis and design of thermal cooling of data center from the tray to the room level can be found in the monograph as given by Joshi and Kumar [2].

In the following parts of this section, the case studies for the system-level modeling and designs are presented and discussed. As defined in the beginning of this section, the system-level modeling consists of at least two components of different types. The system components can be either electronic components or thermal components to demonstrate the system design methodologies, together with engineering considerations and optimizations for system implementation.

7.1.2 Modeling and design of cooling system with cooler and heat exchanger

A typical liquid cooling system for microelectronic chip is shown in Fig. 7.1.5. It mainly consists of a microcooler to remove the heat from the high-performance chip, a pump to transport the heat through the liquid flow, and an external heat exchanger, where the heat is transferred to the secondary fluid to reject the heat to the ambient.

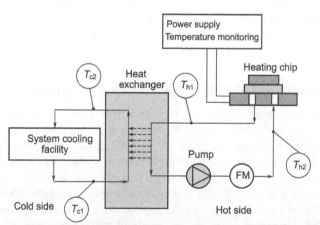

Figure 7.1.5 Typical liquid cooling heat exchanger to transport heat from heat-generating chip (the hot side) to the cold side. The fluids are separated at the exchanger, whereas the heat is exchanged from hot side to cold side. T_{c1}, T_{c2}, T_{h1}, and T_{h2} represent temperatures of cold inlet, cold outlet, hot inlet, and hot outlet, respectively, with FM indicating flow meter.

A conventional fin-tube heat exchanger requires fans to remove the heat from extended fins attached to the peripheral of the cooling tube. Such a configuration could be oversized in space-constrained electronic systems. The air flow, if not well guided, would also overheat the indoor air or even the peripheral electronic parts, accompanied with other design limitations such as annoying fan noise. Instead, a liquid-to-liquid heat exchanger has the advantage of higher heat exchange efficiency, smaller size, and centralized management of the secondary liquid flow through facility cooling such as the coolant distributed unit at the site.

The design target for this thermal system is to have a heat dissipation capability of 175 W from a chip of 7 mm × 7 mm, which corresponds to a heat flux of 350 W/cm². The heat exchanger works with the coolant inlet temperature of 25°C. The allowable junction temperature is 85°C. The design target of the junction-to-ambient thermal resistance for this system is about 0.343°C/W. The following design criteria should be taken into account in the design of the heat exchangers:

① The flow rate in the hot side chip cooling loop should be maintained at a low level. In the present study, the flow rate is fixed to 0.4 L per minute with water as coolant. This avoids large pressure drop and also minimizes the negative consequences in case of aqueous liquid leakage, which is detrimental to the electronic components. The pump can also be miniaturized to save space. The thermal performance shall not be affected with the dedicated design of the microcooling devices.

② The heat exchanger should have compact size with high heat exchange efficiency, or heat exchange density as termed in this section, and should not be located too far away from the electronic chip to minimize pressure loss along the piping and tubing. Reduction in size and weight has been one of the key design consideration.

③ It is required to maintain the thermal resistance and pressure drop of the heat exchanger at a level obviously lower than the heat sink with the task of chip cooling. Because of the chip fragility, the design burden for the chip cooling should be mitigated by minimizing the overall pressure drop and thermal resistance.

④ Design methodology for the heat exchanger is required. The thermal performance of the heat exchanger shall be obtained, which is to be done through both numerical analysis and experimental characterization. However, in the thermal system with a heat sink and a heat exchanger, the thermal design budget for the heat exchanger should be set in connection with the heat sink performance. Namely,

$$R_{ja} = R_{js} + R_{sa} \tag{7.1.1}$$

Considering the chip cooler inlet temperature of 40°C, as defined by the ambient temperature in the ITRS 2012 roadmap, the thermal resistance for the cooler should be maintained not more than 0.25°C/W. Thus, the heat exchanger thermal resistance shall not exceed 0.093°C/W.

In designing and selecting a heat exchanger, the inlet temperature difference of the cold flow and hot flow, $T_{h1} - T_{c1}$, is known for determining heat exchanger thermal performance and heat exchanged. In the chip cooling case, the selection criteria is to maintain T_{h2} at preset $T_s = 40°C$. Sine T_{h1} is not known a priori, and T_{h1} can be correlated as follows given the heat dissipation q and hot side flow rate VF.

$$T_{h1} = T_{h2} + q/(\rho c_p \text{VF}). \tag{7.1.2}$$

where ρc_p is the heat capacitance of coolant. In the present case, $T_{h1} = 46.25°C$ based on flow rate of 0.4 L/min with water as coolant. The designs of chip cooler and heat exchanger are to be discussed in what follows.

7.1.2.1 Silicon-based microcooler for chip cooling

A silicon-based microcooler, featuring both microchannels and jet structure, was developed to dissipate the heat flux for the high-power IC chip. Hybrid multiple jets and microchannels, shortened as MJMCs, have been designed inside the cooler [21].

Fig. 7.1.6 shows the geometrical design of the hybrid MJMC silicon microcooler, including the heat generating chip, the microchannel plate, and the microjet plate with the draining trenches. As shown in Fig. 7.1.6c, there are two adjacent circular nozzles in blue color for the incoming flow and two draining trenches in red color adjacent to the two nozzles in the proposed MJMC design. Besides the flow speed, the thermal performance is the microcooler depending on geometrical factors including nozzle diameter D, the nozzle length L, and the microchannel height H.

During operation, the jet flows from the two micronozzles impinge on the hot wall underneath the heat generating chip, turn around along the microchannel, and then exit through the adjacent draining trenches, as indicated by the arrows in Fig. 7.1.6b–7.1.6c. In the conventional design with a single jet at the center and multiple draining drenches at the far ends of the microchannels, the forced flow would be overheated and the heat dissipation is not efficient. In the present design, the adjacent drainage design would allow faster removal of heated fluid without being overheated when passing through the long channels and thus more efficient heat transfer could be achieved. As explained in the previous work [21], there is no accumulation of the cross-flow to cause detrimental influence on the adjacent jet impingement with this MJMCs configuration. Therefore, the impingement jet flow can be effectively drained for each individual nozzle, which enables more uniform and higher cooling capability for the chip.

The computational fluid dynamics (CFD)−based modeling and simulation is conducted using COMSOL and ANSYS FLUENT solver for the microcooler design optimization. For the targeted chip size of 7×7 mm^2, the microjet array is designed to cover an area of 8×8 mm^2 to fully cover the chip-heating area. Considering the fabrication capability and process ability, the nozzle diameter D is fixed at $D = 100$ μm. The nozzle length (l), microchannel width (w), and jet to wall distance (H) are varied in the modeling to evaluate their effects on the thermal performance of the microcooler. After optimization, the geometrical parameters in the numerical simulation are listed as the followings. Nine draining trenches with a width of 150 μm are configured in the microcooler. The number of the nozzles along the microchannel flow direction in x direction is 16 (with a pitch of 250 μm in y direction between two trenches and

550 μm in x direction across the trenches). The number of the nozzles along the draining trench direction is 21 in y direction with a pitch of 350 μm.

Deionized water is used as the coolant with an inlet temperature of 40°C. Heat source of 175 W is set for the heat-generating chip, and a gage pressure of 20 kPa is set at inlet of the microcooler. A mesh of 1 million is generated, and the grid independence study is conducted. The mesh number is sufficient to obtain accuracy within around 1% for the hydraulic and thermal performance.

The effects of the nozzle length, microchannel width, and jet to wall distance on the thermal performance of the microcooler are respectively presented in Figs. 7.1.7—7.1.9. Fig. 7.1.7 shows that the chip junction-to-microcooler thermal resistance increases as the nozzle length increases slightly. This is because the pressure loss in the nozzle increases as the nozzle length increases, thus flow rate decreases and the jet thermal resistance increases. However, it can be found that the increment of thermal resistance is insignificant within 5%, when the nozzle length varies from 300 to 600 μm based on the feasible fabrication range.

Fig. 7.1.8 shows that the thermal resistance first decreases and then increases with the microchannel width increasing from 150 to 300 μm, while the thermal resistance increases again as the microchannel width increases further to 300 μm. This can be attributed to the reduction of heat conduction capability through the thinned microchannel fins due to the increase in microchannel width. Fig. 7.1.9 shows that the optimal thermal performance can be achieved when the jet to wall distance is 250 μm. Based on above numerical analysis, the microcooler with a nozzle diameter of 100 μm, a nozzle length of 400 μm, a jet to wall distance of 250 μm, and a microchannel width of 250 μm was fabricated for the experimental study. The optimal cooler thermal resistance of 0.164°C/W also meets the design budget less than 0.25°C/W.

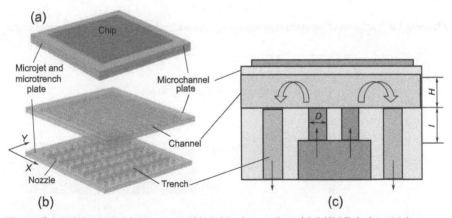

Figure 7.1.6 Schematic of the proposed hybrid microcooler with MJMC design: (a) heat-generating chip assembled with the microchannel plate; (b) isometric view of the microchannel plate and microjet plate; and (c) side view of the microjet flow configuration [21].

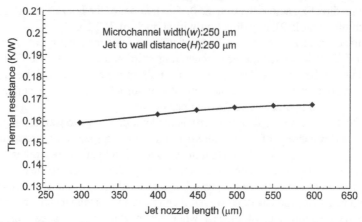

Figure 7.1.7 Variation of junction-to-microcooler thermal resistance with Jet nozzle length [6].

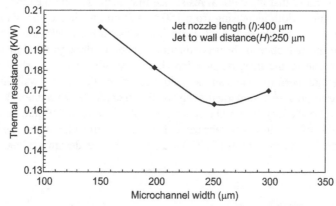

Figure 7.1.8 Variation of junction-to-microcooler thermal resistance with microchannel width [6].

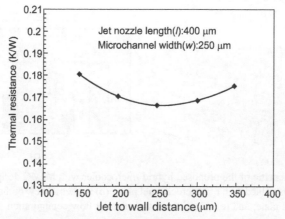

Figure 7.1.9 Variation of junction-to-microcooler thermal resistance with jet to wall distance [6].

The hybrid microcooler is designed with two patterned silicon layers, which is bonded together through the tin—gold alloy to form the cooler. The first layer consists of the patterned microchannels and fins, and the second layer has double-side pattern layer, which consists of the nozzles and jet pattern on one side and inlet flow manifolds and draining trenches on the other side. The process layers for the bonding of two silicon substrates are schematically shown in Fig. 7.1.10 [22].

The fabrication process steps of the microchannel and nozzle wafers are illustrated in Fig. 7.1.11. The process starts with an 8-inch wafer. First, a layer of SiO_2 with a 3 μm thick is deposited by plasma-enhanced chemical vapor deposition (PECVD) process in a Novellus PECVD system, and then the SiO_2 layer is patterned and etched as hard mask with microchannel patterns and micronozzle patterns. During the fabrication of microstructure, the deep reactive ion etching (DRIE) technique was utilized to make the microchannels and nozzles with uniform size.

The microchannels are further etched using the DRIE process to reach a depth of 250 μm. After that, the backside of the wafer is grinded to 350 μm thickness. At last, the metal layers of Titanium 1000 A, Platinum 2000 A, gold 2 μm, tin 2 μm, and gold 500 A are deposited consequently, which were used for the AuSn eutectic bonding with the heat-generating chip.

7.1.2.2 Modeling of heat exchanger HEX-A

The heat exchanger should be designed along with the microcooler development in the system design. Because of the size and weight constraints, it is not appropriate to use a conventional fan tube—fin heat exchanger, which has a much larger size. Instead, a liquid-to-liquid heat exchanger is examined here for the integration with the microcooler. Two different liquid-to-liquid heat exchanger configurations, HEX-A and HEX-B, are studied in the present case.

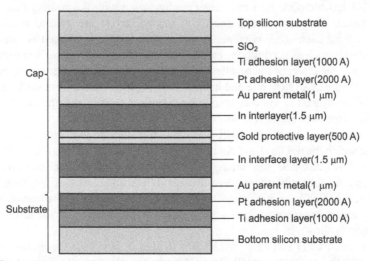

Figure 7.1.10 Schematic of various bonding layers with the gold—indium composites process for the bonding of silicon microcooler to the silicon substrate.

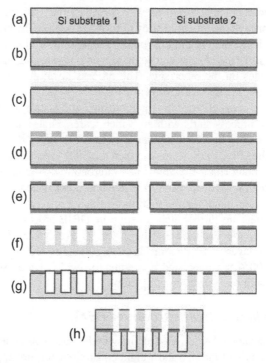

Figure 7.1.11 Fabrication process of the microcooler with the similar steps for both the channel wafer and the nozzle wafer: (a) original wafer, (b) SiO_2 deposition, (c) photoresist deposition, (d) channel and nozzle pattern, (e) SiO_2 etch, (h) DRIE etching, (g) back grinding and metallization, and (h) alignment and bonding.

HEX-A has the plate heat exchanger configuration with the counter flow. In this configuration, the hot fluid and cold fluid channel layers are stacked alternatively, separated from each other with plates made of metals such as stainless steel (SS). Micropatterns such as herringbone can be fabricated based on vendor's expertization. Nonetheless, the effect of micropattern on the thermal performance may not be significant due to the small flow rate and laminar flow regime in the present study. Bonding of the different metal layers was done through brazing tooling and process at vendor's site to form the heat exchanger without fluid leakage.

The second heat exchanger HEX-B is customized design with cross-flow configuration, in which the hot fluid and cold fluid layers are stacked alternatively, but the two types of flow channels are arranged orthogonally to form the cross-flow heat exchanger. As this is at the heat exchanger design stage, fast prototyping can be done through CNC machining without resorting to the brazing process. Brazing process can be expensive due to tooling setup. A comparison of the different heat exchangers is shown in Fig. 7.1.12.

Commercial CFD software FloTHERM is used to construct the thermal models. Because of the relatively small Reynolds number, the laminar flow is assumed. Both the inlet and outlet are modeled as 10 mm × 10 mm rectangular holes. For the

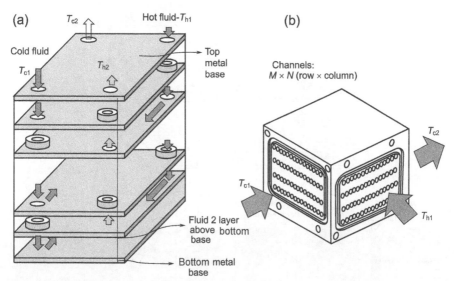

Figure 7.1.12 Comparison of different heat exchangers: (a) HEX-A with parallel plates and (b) HEX-B with cross-flow.

cold side, the inlet flow rate is varied from 0.4 to 2 L/min at fixed temperature of 25°C, whereas the hot-side inlet is set to 0.4 L/min. The hot inlet temperature is set to a fixed temperature of either 46.25 or 40°C for case study. A mesh of 134K is generated, and the grid independence study is conducted. The mesh number is sufficient to obtain accuracy within around 1% for thermal performance.

Fig. 7.1.13 shows velocity profiles and temperature profiles at different fluid layers for the HEX-A configuration. It is seen that relatively similar temperature profile and velocity profile are obtained for the fluid layers. The heat exchanged can be derived from difference of enthalpy for either the hot fluid or cold fluid, namely,

$$q = \text{EH}_{h1} - \text{EH}_{h2} \tag{7.1.3a}$$

$$q = \text{EH}_{c2} - \text{EH}_{c1} \tag{7.1.3b}$$

where EH_{h1}, EH_{h2}, EH_{c1}, and EH_{c2} are the enthalpies of the inlet and outlet fluids at the hot side and the inlet and outlet fluids at the cold side, respectively. The discrepancy of heat amount for the two fluids is kept within 1% as part of computing convergence criteria. The thermal resistance is calculated as

$$R_{sa} = (T_{h1} - T_{c1})/q \tag{7.1.4}$$

Fig. 7.1.14 shows the heat exchanged with varying cold fluid flow rate. Firstly, the heat exchanged could increase with flow rate increase at both cold and hot fluids. The copper heat exchanger is better than SS at 1 Lpm or larger. At fixed hot flow case (0.4 L/min), however, the heat exchange levels off with the increase in the cold

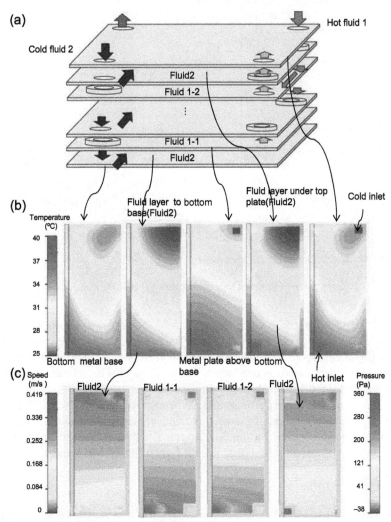

Figure 7.1.13 Computed profiles at different fluid layers for the HE-A configuration (a), with temperature profile (b), and velocity and pressure profiles (c).

flow. The exchanged heat at 1.89 L/min (0.5Gpm) is around 842 W, which is close to the heat exchange capacity as is indicated in the supplier's datasheet. It is noted that the supplier's data are only available for the high flow rates, which may not guide the present design with low flow rate suitable for chip cooling. The simulation cases are also conducted with different inlet temperatures. The inlet temperature change at the cold side does not have obvious effect in thermal resistance.

7.1.2.3 Modeling and design of heat exchanger HEX-B

For the HEX-B with cross-flow configuration, the thermal model is first built with 4 × 9 channels in each fluid side with a heat exchanger core size of 20 mm. At least

System-level modeling, analysis, and design

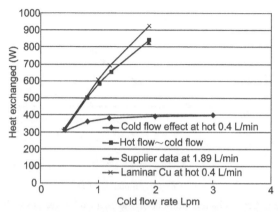

Figure 7.1.14 Computed results for the HEX-A with counter flow configuration: heat exchanged for SS heat exchanger with the hot flow 0.4 L/min, hot flow same as cold flow, and copper heat exchanger with hot flow 0.4 L/min.

seven cells are assigned in each channel cross section to ensure the computational accuracy. The resulting mesh is 1.5 M for the 15-channel case. Computation shows that such configuration only provides 110 W heat exchanged at $T_{h2} = 40°C$, which is too low for system design. Therefore, the design is further optimized to increase the power dissipation. The optimized heat exchanger core has a footprint of 32 mm × 32 mm with channel number of 15 in each row for both fluids. Fig. 7.1.15 shows the fluid flow and temperature profile of HEX-B. A slightly nonuniform

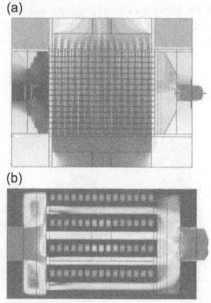

Figure 7.1.15 The computed results for HEX-B with 15 channels in each layer: (a) plain view of the temperature and velocity profile and (b) side view of velocity profile.

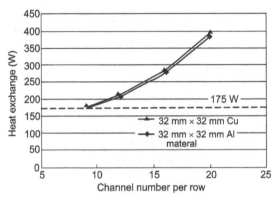

Figure 7.1.16 The effect of channel number on the heat exchanger power based on design constraint of $T_{h2} = 40°C$.

velocity profile may be present, which nonetheless does not affect the heat exchange performance in the present two-fluid case.

Fig. 7.1.16 shows the computed effect of channel number on thermal performance based on the design constraint $T_{h2} = 40°C$ and hot flow of 0.4 L/min, respectively. It is seen that the thermal performance increases as the channel number increases, and the heat exchanged also exceeds the design power of 175 W for channel number around or larger than 15 for both conditions, meeting the design target.

The effect of heat exchanger material type is shown in Fig. 7.1.17. SS, with thermal conductivity of $16 \text{ W} \cdot \text{m}^{-1} \cdot \text{K}^{-1}$, would have unfavorable thermal performance and is not a preferred material for cross-flow configuration. On the other hand, aluminum heat exchanger has similar performance as the copper heat exchanger. The aluminum heat sink is suggested for practical fabrication due to the light weight.

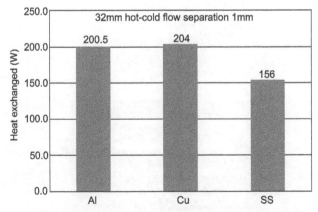

Figure 7.1.17 The effect of heat sink material effect for HEX-B with 20 channels in each layer at $T_{h2} = 46.2°C$.

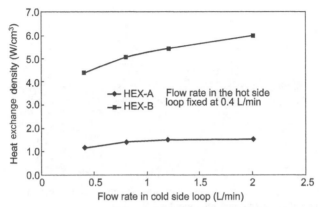

Figure 7.1.18 Comparison of heat exchange density for the HEX-A and HEX-B heat exchangers at different cold side flow rates.

A comparison of the heat exchange density for the commercial HEX-A and the newly developed HEX-B heat exchangers is illustrated in Fig. 7.1.18. The heat exchange density is defined by the exchanged heat over the heat exchanger volume without counting in the connecting joints and parts. It is seen from Fig. 7.1.8 that the HEX-B has a much higher heat exchange density in comparison due to the more compact design.

The photo of the fabricated HEX-B, together with HEX-A and an air-to-liquid heat exchanger, is shown in Fig. 7.1.19. It is seen that the present heat exchanger (HEX-B) has the advantage of small size over the other two types of heat exchangers.

An experimental setup was established to evaluate the hydraulic and thermal performance of the proposed liquid cooling system. Before the setup of the

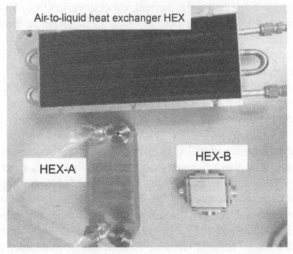

Figure 7.1.19 Comparison of different heat exchangers: HEX-A, HEX-B, air-to-liquid heat exchanger (HEX).

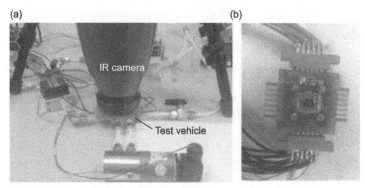

Figure 7.1.20 Photo of the experimental apparatus: (a) chip temperature measurement with infrared camera and (b) assembled microcooler TV.

experimental system, the microcooler test vehicle (TV) was first assembled in the following way. The bottom side of the microcooler was attached on the PCB using epoxy, and the thermal test chip on the top surface of the microcooler was wire bonded to the PCB for electrical connection. Fig. 7.1.20 shows the experimental measurement apparatus, in which the infrared (IR) camera was used for chip temperature measurement. Both the commercial heat exchanger (HEX-A) and the designed HEX-B were used in the flow loop test.

A micropump forced the water through the filter and a flow meter before it entered the microcooler. The water temperature at the inlet of the microcooler and ambient temperature were around 25°C. The differential pressure transducers were attached to a system to measure the pressure drop. The test chip temperature at steady state, which usually took 30 min, was measured and recorded using the IR camera.

The thermal resistances were measured to authenticate the thermal system design. In the tests, the power applied to the thermal test chip varied from 50 to 200 W. The flow rate remains 0.4 L/min in the hot flow side and 2 L/min in the cool flow side during the test.

The measured thermal resistances of the microcooler and the heat exchanger, together with their summation corresponding to the overall thermal resistance of the system, are illustrated in Fig. 7.1.21. Note the hot side flow is fixed to be 0.4 L/min. It can be seen that the thermal resistance decreases as the cold flow rate increases causing the decrease in the overall system thermal resistance.

Taking a close examination on Fig. 7.1.21, it can be found that the microcooler is the main contributor to the overall system thermal resistance, accounting for around 80%–90% of the total thermal resistance, while the heat exchanger only accounts for about 10%–20% of the overall system thermal resistance. In comparison, it can be seen that the thermal resistance of the customized HEX-B is slightly larger than that for the commercial HEX-A, while as mentioned in the previous context, HEX-B is much smaller and lighter than HEX-A. The pressure drop in HEX-B is only half of the pressure drop in HEX-A. Thus, it can be concluded that HEX-B is an effective design in the trend toward to miniaturization.

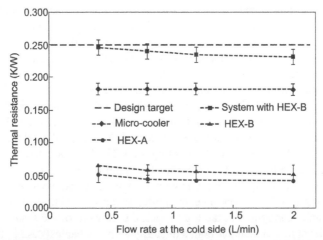

Figure 7.1.21 Measured thermal resistances for the microcooler and heat exchangers with varying flow rate at the cold side.

7.1.3 Modeling of burn-in rack system with multiple trays

7.1.3.1 Model development for the rack-level test system

A burn-in test is referred to the accelerated test of semiconductor devices under stress conditions of high-power and high temperature. Such a testing is known to weed out weak components and to ensure that highly reliable processors are shipped to market without premature fails. In the past, the burn-in testing was mostly conducted for devices arranged in an oven environment without sufficient functional and power tests. The productivity was also low.

Recently, the functional burn-in systems have been demanded to handle advanced testing with increased functionality and test efficiency, which is accompanied with higher power dissipations in a shrinking space from thermal viewpoints. To enhance productivity, the burn-in trays are to be stacked in a rack to enable integrated testing. This leads to a rack-level power dissipation of 10–20 kW or even higher under controlled temperature levels. More efficient yet compact flow and thermal management techniques are to be examined and optimized for the burn-in boards, trays, and racks with processor chips and densely packed modules and cards. Enhanced thermal head techniques, such as active heat sinks, indirect liquid cooling, or refrigeration cooling, can be used for the cooling of the microprocessor chips [28].

On the other hand, the densely packed modules and cards such as graphic modules, VRMs, and Dram II memory cards are to be cooled by either direct forced air convection or on-card conduction associated with secondary convection to the air flows. Failures on the air-cooled cards are becoming one of major factors affecting the productivity. For this purpose, plenty of fans with high capacity are employed to provide sufficient flow rates into the burn-in system.

However, partially increasing the fan rotation speed and capacity may induce undesirable noise, which is harmful to human audibility in a manufacturing environment. Therefore, upfront simulation and modeling on the functional burn-in system are required to provide insightful understanding of air flow patterns and to optimize fan number and configuration. As one would expect, the flow can be maximized, the noise level can be minimized, or both can be simultaneously achieved.

For this purpose, the CFD model is established for the numerical modeling. In the tray-level model, the duct tray is configured with three computer modules including mother boards, and dual processors are considered. The thermal head trays are incorporated on top of the duct tray with a cooling loop. The fan characteristic curves from suppliers are used in the model to determine the system operation points. For comparison study, two tray configurations, one with three-fan tray and the other with four-fan tray are investigated. In the rack-level modeling, seven stacked trays are included, and an exhaust fan is configured at the top to suck the flow out of the system.

The tray-level model includes both a duct tray with mechanical casing, PCBs, components and a thermal tray with thermal head, tubings, and insulation. The duct tray model is based on the original CAD solid geometrical model generated from the SOLIDWORKS software, which has been incorporated with both mechanical parts and electronic parts. Because of overwhelmingly large quantity of parts and complexity of part geometry, a direct geometrical model conversion from the CAD model to the simulation model is virtually not feasible on an office workstation, not to mention the CFD simulation with the detailed geometry.

To reduce the modeling complexity, the CAD model of the duct tray is dissembled into a number of subassemblies. Each subassembly is further simplified through removing aerodynamically and thermally insignificant parts, such as screws, small through-holes (less or around 1 mm), low-profile resistors, and thin wires. The electrolytic capacitors are usually significantly large (in the range of 10—20 mm) and thus represented as rectangular blocks of same volumes in the geometry model. The geometrically simplified subassemblies and parts are then input to the simulation model and reassembled into a tray model for mesh generation and fine-tuning. Because there is no CAD model available for the thermal tray, the geometrical data from available thermal tray hardware were measured based on a practical thermal tray used for the model construction. The study methodology is illustrated in Fig. 7.1.22.

The CPU packages under testing are typical computer processors in flip-chip pin grid array format. Although microprocessors attain the highest power dissipation in the testing, the processor temperature could be well controlled by the directly attaching thermal head without being affected by the surrounding air flow. Thus, the processor is modeled with simplified blocks with the same outer dimensions instead of detailed package models. Attached on each graphics card is a plate fin heat sink for the cooling purpose. To reduce the effort in the tray-level modeling, the detailed heat sink is replaced with a volume flow resistor of the same flow resistance level to facilitate the simulation of air flow at the rack level. The volume flow resistor acts as a porous medium, so that the pressure drop could be obtained without excessive modeling of details.

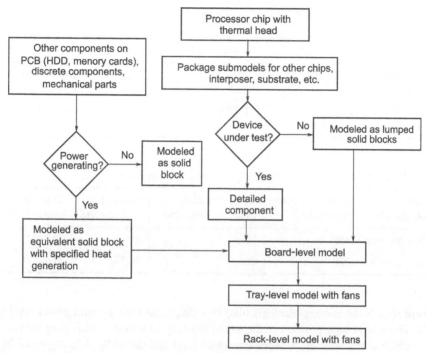

Figure 7.1.22 The flowchart for the modeling from the board- to the rack-level model.

7.1.3.2 Model results and discussion

Zhang et al. [26] numerically simulated the air flow in a system test hardware consisting of seven trays in a rack, each tray consisting of three PCB boards with each board mounted with dual processors, memory cards, hard disk driver and power supply units, and other accessory parts. The tray has been shown in Fig. 7.1.3. The cooling is implemented by direct cooling of the processors with refrigeration thermal head, as well as upstream fan blowing. In addition, a rack system exhaust fan is mounted at the top of the rack to guide the flow exiting from all the trays.

In the tray-level modeling, only one tray is modeled without the exhaust fan effect. The specification for the two types of axial fans in the trays is tabulated in Table 7.1.1. Based on the tray-level model, the computed fan operation points are listed in Table 7.1.2. It is noted that the fan noise N_{db} with the sound power of SP for each tray is calculated based on the following equation.

$$N_{db} = N_{db0} + 10 \lg\left(\frac{SP}{SP_0}\right) \tag{7.1.5}$$

Table 7.1.1 Fan specifications for the different tray configuration.

Model	Noise at 1 m (dB)	Nominal pressure drop (Pa)	Flow rate at 0 pressure (CFM)
Fan 1 (three-fan tray), 24 W	59	178	190, powerful and noisy
Fan 2 (four-fan tray), 4.3 W	42.5	75.8	74, moderate and quiet

Table 7.1.2 The computed fan operation parameters at the tray level.

Tray configuration	Tray operation flow (CFM)	Tray operation pressure drop (Pa)	Noise at $r = 1$ m (dB)	Total fan power (W)
Three-fan tray	420	90	64	72
Four-fan tray	220	26	48.5	17.2

where N_{db0} is the nominal decibels (dB) of a single fan with a sound power level of SP_0. The spatial superposition of the sound intensity of n fans would cause increased decibels. The relationship between the sound level and the distance is expressed by

$$\frac{SP}{SP_0} = n\left(\frac{r_0}{r}\right)^2 \tag{7.1.6}$$

with r_0 is the standard distance of 1 m and r the evaluation distance in practice. It is seen that, even though the four-fan tray had one more fan, the overall fan noise was only 48.5 dB, which is smaller than 64 dB for the three-fan tray configuration. This can be attributed to the reason that the four-fan tray could operate at the much lower flow rate and thus reduced noise level. In addition, the operational power for the four-tray configuration was lower, only 17.2 W as against 72 W for the three-fan configuration, which is preferred in the real implementation.

The computed velocity profiles for the two-tray configurations are displayed in Fig. 7.1.23, shown with the cross-sectional planes at the heights of $z = 10$ and 50 mm above the boards. The maximum velocity is reduced, from 14.98 m/s for the three-fan tray to the 8.09 m/s for the four-fan tray. In addition, the velocity profiles are also more uniform for the four-tray configuration, as marked in the encircled dashed lines.

The rack-level model is also developed by stacking the trays vertically. In addition, an exhaust fan is installed at the top of the rack to suck the air flow from the exiting ports of the trays. The developed rack-level model and the computed velocity and pressure profiles are shown in Fig. 7.1.24 for the three-fan tray case and Fig. 7.1.25 for the four-fan tray, respectively. It is seen that relatively uniform air flow could be obtained for the trays located at different heights in the rack. An improved air flow uniformity is

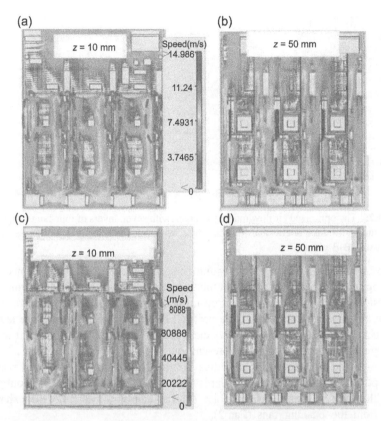

Figure 7.1.23 Computed velocity profile: (a) $z = 10$ mm for three-fan tray, (b) $z = 50$ mm for three-fan tray, (c) $z = 10$ mm for four-fan tray, and (d) $z = 50$ mm for four-fan tray.

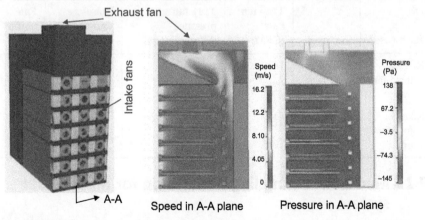

Figure 7.1.24 The air flow model for a rack with seven layers of three-fan trays, each tray configured with three boards and three inlet fans, with velocity magnitude and pressure levels shown in A-A plane.

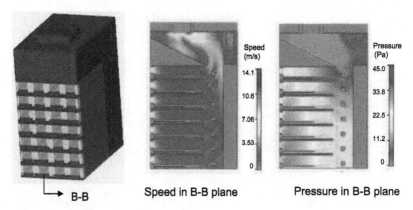

Figure 7.1.25 The optimized air flow model for a rack with seven layers of four-fan trays, each tray configured with three boards and four inlet fans, with velocity magnitude and pressure levels shown in B-B plane [26].

attained by arranging the four-fan configuration so that the thermal heads connecting with the top of the processors would not block the direct blowing of the fans from the inlets. Besides, the resulting pressure drop in the four-fan tray is also reduced, with the maximum pressure drop of 29.5 Pa versus 114 Pa with reduced pressure levels are shown in Fig. 7.1.25. Not only was the fan noise reduced, the fan power level was also greatly reduced based on the present four-fan tray configuration. Table 7.1.3 lists the computed fan parameters including the flow rate, pressure drop, and noise-level fan power. Since the fan noise level is minimized for the four-fan tray configuration with smaller pushing fans, the noise source mainly comes from the exhaust fan, not from the pushing fans (Fan 2).

Table 7.1.3 The computed fan operation parameters at the rack level.

Rack cases	Fan used	Tray fan flow (CFM)	Tray fan pressure difference (Pa)	Noise source at 3 m (dB)	Fan power (W)
R1 rack: three-fan tray	21 Fan 1, exhaust fan	1830	114	Tray fan (63)	504
R2 rack: four-fan tray	28 Fan 2, exhaust fan	1416	29.5	Exhaust fan (55)	240

7.2 Mechanical modeling and design for microcooler system

As the electronics industry continues the trend toward smaller form factor, greater functionality, and faster processing speed, the high-power dissipation across a smaller chip area can result in hot spots. These are usually high heat flux concentrated in small areas, which could lead to extremely high junction temperature and drastic breakdown of integrated circuits devices. As chip power densities are increasing beyond air cooling limits, the liquid cooling technology is around the corner. Because of high heat transfer coefficient associated with it, microchannel cooling is an attractive approach, which is widely investigated by researchers [13,29]. Singh et al. [30] fabricated on-chip microchannels on the reverse side of the device to improve the cooling performance and reported the in situ impact analysis of very high heat flux transient on the operation and mitigation of electronic devices through simultaneous thermal, fluidic, pressure, and performance measurement. It is not feasible to form the microchannels directly on the backside surface of the chip to achieve the chip-level thermal management due to the potential yield loss. Instead, a separate microchannel cooler is bonded to the backside of the chip to achieve the package-level thermal management.

Die attach (DA) material is a critical material affecting package reliability, especially for high temperature power electronics application. Studies by Sabbah et al. [31] showed that AuGe alloy shows good performance under temperature loading compared with the silver sintered joint for high temperature power electronics. With the advantages of high thermal conductivity and thin bonding layer, AuSn eutectic bonding can achieve the best thermal and mechanical properties among the three tested bonding materials including AuSn, silver paste, and SnAgCu305 solder paste [32]. Studies by Quintero et al. [33] showed that the AuSn SLID (solid-liquid interdiffusion) system proves to be a reliable alternative for high temperature applications compared with nanosilver paste and high Pb solder. Yoon et al. [34] investigated two eutectic compositions for Au-Sn: Au-rich Au-20 wt.% Sn and Sn-rich Au-90 wt.% Sn. The results showed that the Sn-rich, AuSn/Ni flip-chip joint was mechanically much weaker than the Au-rich, AuSn/Ni flip-chip joint. Indium (In) solder strength can be greatly enhanced by forming an intermetallic bonding of AuIn2 [35]. Indium-based solder was used to join two surface under low temperature ($<200°C$) bonding process, and the AuIn2 intermetallic compound (IMC) remelting temperature is more than 495°C [36]. Different DA materials of AuSn, AuGe, SnAg, and AuIn are to be evaluated numerically to provide design and process guidelines for the design of high reliable package solution.

In this section, the silicon microchannel cooler or SMC as described in Section 7.1.2 is to be examined from mechanical and process viewpoints fabricated from silicon. A thermal test chip with eight heating spots making hot spot heat flux reaching up to 10 kW/cm^2 is first mounted onto a synthetic diamond heat spreader and then mounted onto the SMC cooler through thermocompression bonding (TCB) process. Furthermore, mechanical reliability of the cooler system is still one of the important concerns considering manufacturing process condition and application condition. It is the aim of this mechanical study to investigate reliability of the cooler system through thermomechanical modeling and characterization. Finite element analysis (FEA) is implemented for mechanical reliability assessment of the SMC cooler. Parametric studies are conducted to optimize the cooler system design and material evaluation.

7.2.1 Silicon microcooler system

Based on analyses for simulation results and experimental data, the final TV is fabricated as shown in Fig. 7.2.1. A 7 mm × 7 mm × 0.1 mm thermal chip is mounted on a 9 mm × 9 mm × 0.4 mm diamond heat spreader using AuSn DA material with 10 μm thick bonding layer. Then the diamond heat spreader is bonded onto a 21 mm × 12 mm × 0.75 mm SMC cooler, which includes Si microchannels and Si nozzles. The SMC cooler system is finally mounted onto 1.6 mm-thick PCB using epoxy glue with the bond line thickness (BLT) of 70 μm. The designed outlet and inlet on PCB were used to connect with the manifold. The coolant water enters the SMC cooler through inlet, flows over microchannels, and then flows back to manifold through two outlet trenches.

The thermal test chip, as shown in Fig.7.2.2, is designed with eight hot spots located at the center line of the chip. The size of hotspot is 450 × 300 μm. The highly doped n-type resistors were built on the Si test chip as hotspot heaters. The total heating power for the thermal test chip is around 110 W, which makes hotspot heat flux reaching up to 10 kW/cm^2 by dividing the heating power by hotspot area. The Si thermal test chip with eight tiny resistors is customized to mimic heat generation in the GaN-on-Si device. The maximum operation temperature of this type of device can be as high as 200°C. Both GaN and silicon materials have similar thermal conductivity [37]. Therefore, using silicon thermal chip to mimic GaN-on-Si device can provide similar temperature distribution.

TCB is used to assemble SMC cooler. Fig. 7.2.3 shows the schematic of TCB bonding process. The thermal chip surface is metallized with Ti/Pt/Au/Sn/Au layers for bonding. The hybrid microcooler is fabricated by bonding two Si plates together, one is with 18 microchannels and the other is with a 14 × 18 microjet array. The internal dimensions of the microcooler have been optimized, considering the combined effect of heat convection and conduction. As shown in Fig. 7.2.3 b, the nozzle diameter is 100 μm, depth is 400 μm, and pitch is 350 μm. The channel width, fin width, and depth are 250, 100, and 250 μm, respectively. The channel length is 5 mm. The size of the outlet trench is 6.5 × 0.5 mm^2. The microchannels or microjets are etched on different wafers with DRIE process. After etching, the wafer is metalized with 4 μm

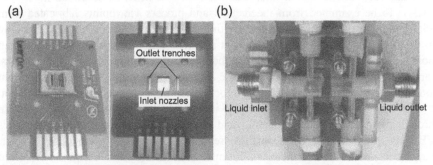

Figure 7.2.1 Assembled microcooler: (a) top and bottom view of SMC and (b) SMC connected to manifold.

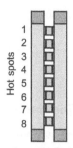

Figure 7.2.2 Layout of hot spots designed on the thermal test chip.

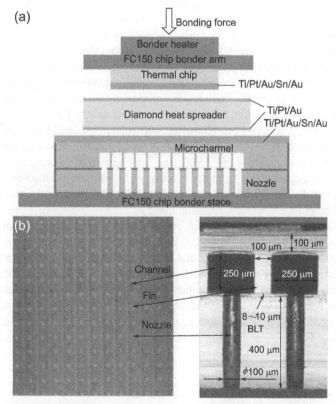

Figure 7.2.3 Microcooler fabrication: (a) schematic of bonding process and (b) image of bonded SMC microcooler.

eutectic AuSn solder alloy by evaporation. The 21×12 mm^2 microchannel plate and the microjet plate are bonded together through the TCB process with bonding temperature of 280°C and the bonding force of 49 N. After the assembly of the microcooler, the outside surfaces of the bonded microcooler are then metalized for the following bonding process. All the prepared components, including the chip, heat spreader, and microcooler, are bonded in one step through the TCB process. Thicknesses of the thermal chip, DA, diamond heat spreader, and glue are optimized through experiment and numerical simulation.

7.2.2 Finite element modeling and simulation

FEA modeling and simulation was implemented for mechanical reliability assessment of the SMC cooler. Five types of FEA models were conducted considering process flow and application conditions, including shear test modeling analysis (model 1), bonding the thermal chip to the heat spreader (model 2), entire cooler structure assembly (model 3), thermomechanical coupling analysis considering hot spot heating (model 4), and thermal cycling (TC) reliability test model (model 5).

To investigate bonding strength when using different DA materials, a type of simple sample was built to conduct shear test at room temperature. The interface area is 2×2 mm^2. To extend the understanding of bonding strength using shear test, shear test FEA modeling as shown in Fig. 7.2.4 (model 1) was conducted for the shear test configuration. A 2×2 mm^2 chip or diamond was attached onto a 5×5 mm^2 silicon substrate with different DAs. The effects of shear height and DA layer thickness on stress distribution were simulated. The shear height is defined as a gap between the shear head and the top surface of the substrate. The bottom of the substrate was fixed. Shear head was modeled as a rigid body. Shear force was added onto shear head to a certain value in the horizontal direction (x) gradually. DA material was modeled as a bilinear elastic-plastic behavior, which is listed in Table 7.2.1 for different DAs.

Fig. 7.2.5 shows a 3D quarter FEA model of bonding a thermal chip to a heat spreader (model 2). Symmetric boundary condition (BC) was applied along two symmetric planes. Different DA materials of AuSn, AuGe, AuIn, and SnAg were modeled for comparison. Bonding temperature of each DA as listed in Table 7.2.1 was considered as a stress-free temperature. Heat spreader thickness, bonding materials, BLT, and thermal chip thickness were optimized through FEA modeling and simulation.

The optimized parameters by model 2 were used in model 3 as shown in Fig. 7.2.6 of 3D quarter FEA model for the cooler assembly due to symmetric geometry to

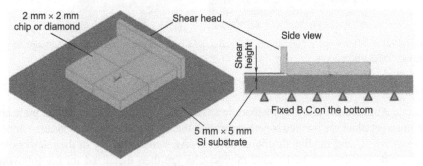

Figure 7.2.4 Shear test simulation model (model 1).

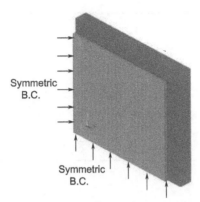

Figure 7.2.5 Quarter finite element analysis (FEA) model of thermal chip bonded with heat spreader (model 2).

evaluate the reliability of entire cooler structure after assembly process. In the model 3, epoxy glue BLT between the cooler and the PCB was optimized to reduce warpage and stress by considering PCB assembly process condition as a temperature loading. Thermomechanical coupling simulation (model 4) was conducted for the same cooler system by considering temperature nonuniform distribution due to hot spot and cooling effect when the thermal power was applied onto the thermal chip. Temperature distribution and stress were analyzed based on simulation results of model 4 to evaluate cooler system under application condition. The final TV was tested under TC condition with temperature range from −40 to 125°C to assess TC reliability performance. In the design stage, TC simulation was conducted for cooler system to optimize design and improve TC performance. FEA geometry model for TC simulation (model 5) is the same as models 3 and 4 as shown in Fig. 7.2.6.

Table 7.2.1 lists materials and material properties used in different FEA models. DA materials were modeled with an elastic-plastic or creep behavior. Temperature-dependent material properties were modeled for epoxy glue, DA, silicon, and diamond. PCB material has the orthotropic material properties, where there are different properties for in-plane (x, y) and out-of-plane (z) directions.

7.2.3 Results and discussion

7.2.3.1 Shear test and simulation of model 1

Shear test sample was made by bonding 2×2 mm^2 Si die or diamond onto silicon substrate using different DA materials through TCB process. Shear test was conducted using a Dage 4000 Pull/Shear Tester. Si substrate was fixed onto a XY stage, which was used for tested sample alignment. Shear test head pulled the sample in the horizontal direction with shear speed of 500 μm/s and shear height (H) of 15 or 250 μm. For each test sample, an average shear force was obtained based on four tested samples. Shear strength was calculated based on the maximum shear force and interface area.

Table 7.2.1 Material properties used in finite element models.

Materials	Elastic modulus (GPa)	Poisson's ratio	CTE (ppm/°C)	Bonding temperature (°C)	Yield stress (MPa)	Tangent modulus (MPa)	Thermal conductivity (W·m^{-1}·K^{-1})
Silicon	130	0.27	2.8				$152 \times (273/T)^{1.334}$ (T in K)
Diamond	1000	0.1	1				1500@25°C, 1400@500°C
AuSn	68	0.4	16	300	200	680	57
AuIn	69.25@25°C 62.7@150°C 57.5@250°C	0.32	13	380	896	1980	44
AuIn	40	0.4	18	280	120	500	66
AuGe	47.5@25°C 33.1@110°C 13.8@200°C	0.4	22	250	30	1000	78
Glue	5.42 below T_g 0.4 above T_g	0.3	31 below T_g 158 above T_g	$T_g = 80°C$			0.2
PCB	26(x,y), 11(z)	0.39, 0.19	15(x,y), 31(z)				18.9(x,y), 0.4(z)

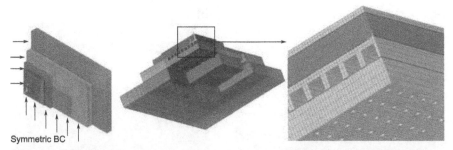

Figure 7.2.6 Quarter finite element analysis (FEA) model of SMC cooler system used for assembly process modeling (model 3), thermomechanical coupling analysis (model 4), and TC reliability assessment (model 5).

Fig.7.2.7 shows the maximum shear force from shear tests for different samples and shear height. When using AuSn-bonding material, shear strength for Si - AuSn - Si and diamond - AuSn - Si structures are similar (27 MPa vs. 25 MPa). Shear test with 15 μm shear height leads to higher shear force by 4% than shear test with 250 μm shear height. AuIn-bonded sample has higher shear strength by 6% than AuSn-bonded sample.

Fig. 7.2.8 shows scanning electron microscope images of bonding interfaces after TCB process for Si - AuSn - Si and Si - AuIn - Si-bonded samples. Good bondability and microstructure were achieved for both AuSn and AuIn-bonding materials. The formed IMCs have higher remelting temperature, so the bonded samples have good performance under high-temperature condition.

Fig. 7.2.9 shows shear stress simulation result for AuSn-bonding layer under 107 N shear force and 15 μm shear height. Nonuniform shear stress distribution is obtained. Shear stress of AuSn layer is higher at shear head side and then decreases gradually from the shear head side to the opposite side. Both shear and bending effects exist

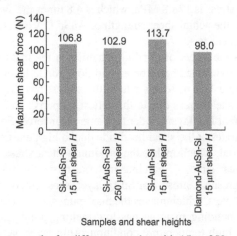

Figure 7.2.7 Shear force results for different samples with 15 or 250 μm shear height.

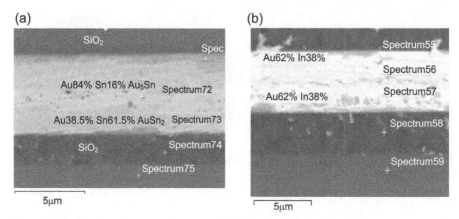

Figure 7.2.8 Scanning electron microscope (SEM) images of bonding interfaces: (a) Si-AuSn-Si and (b) Si-AuIn-Si.

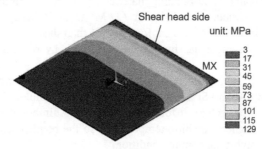

Figure 7.2.9 Shear stress distribution for AuSn layer from shear test simulation.

on the DA layer during shear test. The lateral edge closing to shear head has the maximum shear stress due to stress concentration and boundary effect at this area. The maximum shear stress is 128.8 MPa, which is 4.8 times the average shear strength (27 MPa). Therefore, the actual shear strength of AuSn DA material is larger than the average value from shear test.

Fig. 7.2.10 shows the effect of bonding material on the maximum shear stress of bonding layer under the same shear force condition. AuGe layer leads to the largest shear stress among four DA materials due to the stiffest property, while SnAg layer leads to the smallest shear stress due to the softest property. Fig. 7.2.11 shows stress distribution for different DA layers under the same shear force. For AuGe layer, shear stress highly concentrates at the shear head side due to high elastic modulus and yield stress of AuGe material, which makes the maximum shear stress very high.

The very high shear stress of bonding material raises the issue that failure may migrate to chip at high shear stress. For SnAg layer, shear stress distributes more uniform than other DA layers, which makes the maximum shear force much lower. However, the soft SnAg-bonding material has low melting point and is prone to creep fatigue failure under high temperature condition. AuIn layer has similar shear stress

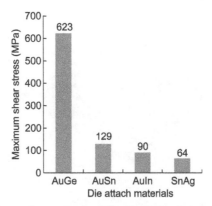

Figure 7.2.10 The maximum shear stress for different die attach materials under the same shear force.

distribution as AuSn layer. AuSn-bonding material was finally selected for microcooler system assembly due to its high shear strength, good processability, and high melting point.

Fig. 7.2.12 shows the effect of AuSn layer thickness on the maximum shear stress. The maximum shear stress decreases with the increase in bonding layer thickness, which means that the increase in bonding layer thickness can improve shear performance. For example, shear stress reduces by 7% when DA thickness increases from 5 to 12 μm. On the other hand, thick DA layer helps improve thermal fatigue life of bonding layer when subjected to TC loading.

The different shear heights, including 15, 50, 150, and 250 μm, were simulated to study the effect of shear height on shear stress of AuSn layer. Fig. 7.2.13 shows the effect of shear height on the maximum shear stress. The maximum shear stress decreases with the increase in shear height under the same shear force, which is consistent with shear test results. Shear stress could reduce by 12% when shear height changes from 15 to 250 μm. Larger bending moment exerts on bonding layer for a larger shear height. To make sure the bonding layer subject to pure shear loading condition, lower shear height should be used.

7.2.3.2 Simulation results of model 2: chip bonded with heat spreader

Selection of bonding pressure is critical for the successful TCB process. However, bonding pressure—induced stress is very small compared with coefficient of thermal expansion (CTE) mismatch—induced stress in TCB process simulation [38]. Therefore, only bonding temperature—induced stress was simulated. The bonding temperature(slightly higher than melting point of each DA)listed in Table 7.2.1 was defined as a stress-free temperature when a certain DA bonding material was used. The reference model as shown in Fig. 7.2.5 has a 7 mm × 7 mm × 0.1 mm thermal chip, a 9 mm × 9 mm × 0.4 mm diamond heat spreader, and a 10 μm AuSn DA layer.

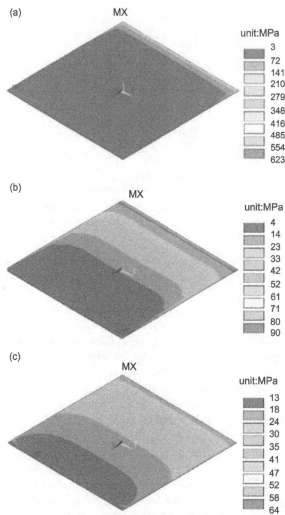

Figure 7.2.11 Shear test induced shear stress distribution for die attach materials: (a) AuGe, (b) AuIn, and (c) SnAg.

Fig. 7.2.14 shows the effect of diamond heat spreader thickness on the maximum first principal stress (S1) of thermal chip, DA, and diamond heat spreader. DA has larger stress than thermal chip and heat spreader. With the increase in heat spreader thickness, stress in thermal chip and DA increases but stress in heat spreader decreases. When heat spreader thickness increases from 200 to 400 μm, stress increases by 20% for thermal chip and by only 2% for DA, but stress reduces by 20% for heat spreader. Fig. 7.2.15 shows the effect of diamond heat spreader thickness on package warpage. Warpage decreases with the increase in heat spreader thickness due to stiffer structure. Warpage reduces by 70% when heat spreader thickness increases from 200 to 400 μm.

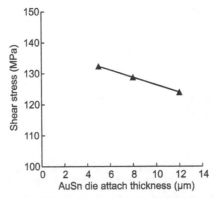

Figure 7.2.12 Effect of AuSn layer thickness on the maximum shear stress.

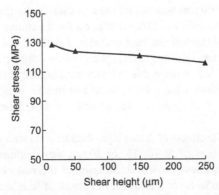

Figure 7.2.13 Effect of shear height on the maximum shear stress of AuSn layer.

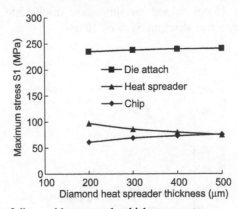

Figure 7.2.14 Effect of diamond heat spreader thickness on stress.

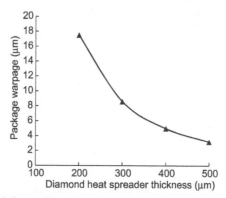

Figure 7.2.15 Effect of diamond heat spreader thickness on package warpage.

By considering low stress and warpage, and good thermal performance, diamond heat spreader thickness of 400 μm was optimized and used in the final TV.

Fig. 7.2.16 shows the effect of DA material on the maximum stress. DA has significant effect on stress of thermal die, heat spreader, and itself. AuGe induces the highest stress among four DA materials due to its highest bonding temperature and yield stress. AuIn and SnAg lead to lower stress due to lower bonding temperature and yield stress. Based on mechanical simulation results, shear test results, bondability, and intermetallic reliability study [32,33], AuSn was selected as DA material for the final TV fabrication.

To optimize BLT, the effects of AuSn layer thickness on stress were simulated. The AuSn layer with thickness of 5, 10, 15, or 20 μm was modeled. Fig. 7.2.17 shows the effect of AuSn thickness on stress. Stresses of thermal chip and DA decrease with the increase in AuSn layer thickness, but stress of heat spreader increases with the increase in AuSn layer thickness. For example, stress decreases by 15% for both DA and thermal chip and increases by 8% for heat spreader when AuSn layer thickness increases from 5 to 15 μm. Based on simulation results and processability, the thickness of AuSn layer was determined to be 10 μm.

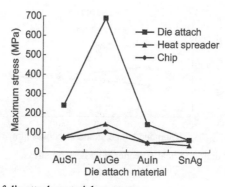

Figure 7.2.16 Effect of die attach material on stress.

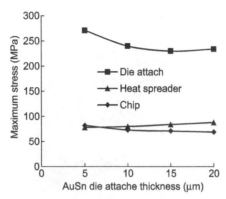

Figure 7.2.17 Effect of AuSn die attach thickness on stress.

7.2.3.3 Simulation results of model 3: cooler system after assembly

In assembly process simulation, the model 4 as shown in Fig. 7.2.6 has a 9 mm × 9 mm × 0.4 mm diamond heat spreader, a 7 mm × 7 mm × 0.1 mm thermal chip, and 10 μm-thick AuSn DA, 40 μm thick glue, and 1.6 mm thick PCB. The glue curing temperature of 150°C was used as a stress-free temperature for glue and PCB materials in the microcooler system assembly modeling. Fig. 7.2.18 shows the stress results for the microcooler system after assembly process. The maximum principal stress (S1) occurs at the chip/diamond interface (encircled chip corner in Fig. 7.2.18a). The maximum shear stress (S_{xz}) occurs at the cooler/PCB interface (circled cooler corner in Fig. 7.2.18b). Glue thickness is important in the cooler system assembly, and its influence was simulated with varying thicknesses 20, 40, 70, or 100 μm. Fig. 7.2.19 show the effects of glue thickness on stress of glue and warpage of microcooler system. Thicker glue helps reduce stress of glue material. When glue thickness is more than 70 μm, this effect is not significant. When glue thickness increases from 20 to 70 μm, stress reduces by 20%. The effect of glue thickness on the stresses of other materials is not significant. Warpage decreases with the increase in glue thickness. The glue thickness of 70 μm was determined for the final TV.

7.2.3.4 Simulation results of model 4: thermomechanical coupling

The FEA model used in thermomechanical coupling simulation is the same as that used in cooler system assembly as shown in Fig. 7.2.6. In thermal modeling, heat transfer coefficient of 2.3×10^5 W·m^{-2}·K^{-1}, which was calculated from thermal cooling simulation, was applied to the cooling channel surfaces. Heat flux of 10 kW/cm^2 was applied onto four hot spot areas in the quarter model according to the designed areas. Fig. 7.2.20 shows the temperature distribution induced by hot spots and cooling effects. The maximum temperature occurs at the hot spot areas of the

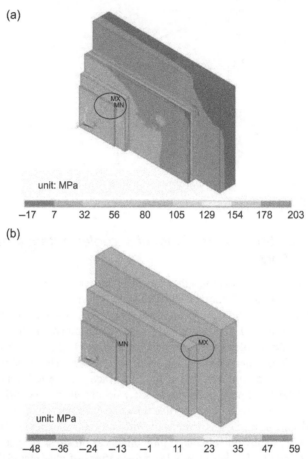

Figure 7.2.18 Stress contour of microcooler system after assembly: (a) principal stress S1 and (b) shear stress S_{xz}.

thermal chip. Nonuniform temperature distribution would raise thermal stress that was simulated by subsequent mechanical stress modeling.

To characterize the thermal performance and verify the thermomechanical coupling simulation results, experiment was conducted for the physical microcooler. The DC power was supplied to the hot spots with dissipation of 1×10^4 kW/cm² hot spot heat flux, and the flow rate of 0.4 L/min for the coolant water was provided by a microgear pump. The heated water from the microcooler was cooled down by a water bath and then flowed back to the cooling system. The inlet water ambient temperature was around 25°C. The temperature at the surface of the thermal chip was measured using an IR camera at steady-state condition. In the test, the steady state (temperature variation ±0.1°C) was usually reached within 30min. Before the measurement, the chip was coated with a thin layer of matt black paint, which was used to reduce reflection, provide high emissivity factor, and ensure accurate thermographic testing using

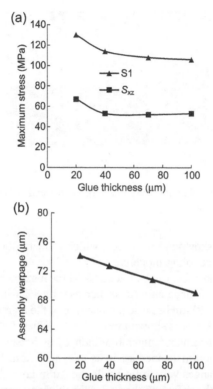

Figure 7.2.19 Effect of glue thickness on (a) stress of glue and (b) warpage of microcooler system.

Figure 7.2.20 Temperature distribution of cooler system from simulation result.

IR camera. The emissivity of the coating was estimated at 0.9. The maximum chip temperature measured by IR camera occurred at the hot spots located near the test chip center, as illustrated in Fig. 7.2.21. The maximum chip temperature can be kept at around 157°C. For the GaN-on-Si device, the maximum operation temperature can be as high as 200°C. By increasing the pumping power, further cooling effect can be achieved. Simulation result as shown in Fig. 7.2.20 is consistent with temperature

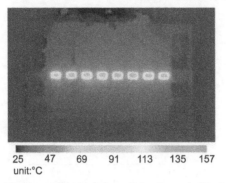

Figure 7.2.21 Infrared (IR) thermography image of the thermal chip for 1×10^4 kW/cm^2 hot spot heat flux dissipation.

measurement result as shown in Fig. 7.2.21, which validates the inputs and settings in the thermomechanical coupling modeling.

Fig. 7.2.22 shows the effect of DA thickness on the maximum stress of each part in the microcooler system. Silicon chip stress increases by 8% with the increase in AuSn DA thickness from 5 to 20 μm because temperature of the thermal chip increases with the increase in DA thickness as shown in Fig. 7.2.23. When AuSn thickness increases from 5 to 20 μm, the maximum temperature induced by hotspot increases by 7%. DA stress decreases by 50% with the increase in its thickness from 5 to 20 μm. The stress of the heat spreader changes very little while the AuSn layer thickness increases.

Fig. 7.2.24 shows the effect of thermal chip thickness on the maximum stress. Stresses increase with the increase in thermal chip thickness because the maximum temperature of the microcooler system increases significantly with the increase in thermal chip thickness as shown in Fig. 7.2.25. When thermal chip thickness increases from 100 to 300 μm, the maximum temperature induced by hotspots increases by 28%, which induces high stress due to significant temperature nonuniformity.

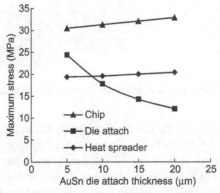

Figure 7.2.22 Effect of AuSn layer thickness on the maximum stress induced by hot spots effect.

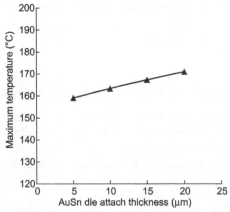

Figure 7.2.23 Effect of AuSn layer thickness on the maximum temperature of thermal chip induced by hot spots.

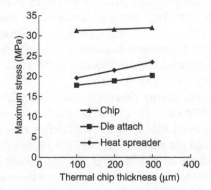

Figure 7.2.24 Effect of thermal chip thickness on the maximum stress induced by hot spots.

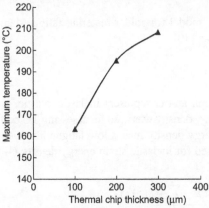

Figure 7.2.25 Effect of thermal chip thickness on the maximum temperature of thermal chip induced by hot spots.

However, the stress in each part of the microcooler system is not as high as bonding- and assembly process—induced stress mentioned above, which demonstrates that the microcooler system design is successful in terms of mechanical stress and maximum temperature.

7.2.3.5 Simulation results of model 5: thermal cycling simulation

Solder DA layer is a weak structure of the electronic assembly when subjected to TC test due to creep fatigue failure mechanism of the solder alloy. Solder DA exhibits creep and plastic behavior at high-temperature condition. The typical creep strain rate for solder is usually expressed by a hyperbolic sine equation below:

$$\dot{\varepsilon}_{cr} = C_1 [\sinh(C_2 \sigma)]^{C_3} \exp\left(\frac{-C_4}{T}\right) \qquad (7.2.1)$$

where C_1, C_2, C_3, and C_4 are material constants, which can be obtained from the creep test. Zhang et al. [39] published material constants for the AuSn solder alloy: $C_1 = 9 \times 10^5$, $C_2 = 0.0653$, $C_3 = 5.5$, and $C_4 = 8690$ (K). In the FEA modeling, the elastic-plastic-creep (EPC) constitutive model was simulated for AuSn DA material. The EPC model can be realized in ANSYS using combined model of bilinear kinematic hardening plasticity with implicit creep [40].

Fig. 7.2.26 shows strain energy density results from FEA simulation for three different DA materials after three TC cycles. DA1 refers to AuSn layer between the thermal chip and the diamond heat spreader; DA2 refers to AuSn layer between the diamond heat spreader and the microcooler; and DA3 refers to AuSn layer between the Si microchannel and the Si nozzle. DA1 has the highest strain energy density due to CTE mismatch between Si and diamond. DA3 has the lowest strain energy density results because both sides of DA are the same material of Si. Plastic deformation of AuSn layer is dominant compared with creep deformation because AuSn material has a high creep resistance.

When using the EPC model for solder material, fatigue damage parameter can be selected and expressed as [40]

$$W_{in} = W_{pl} + W_{cr} \qquad (7.2.2)$$

where the subscripts in, pl, and cr represent inelastic, plastic, and creep, respectively. Total inelastic strain energy density was used for assessing TC performance of the solder. Large value of strain energy density means low fatigue life of the solder. The volume-averaged method was used for inelastic strain energy density (W) calculation [41]:

$$\Delta W_{ave} = \frac{\sum_i^n W_{i,2} \cdot V_{i,2}}{\sum_i^n V_{i,2}} - \frac{\sum_i^n W_{i,1} \cdot V_{i,1}}{\sum_i^n V_{i,1}} \qquad (7.2.3)$$

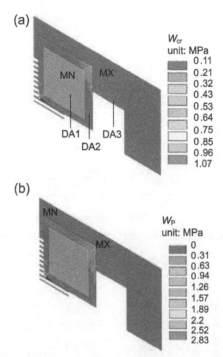

Figure 7.2.26 Strain energy density simulation results for AuSn layers: (a) creep strain energy density W_{cr} and (b) plastic strain energy density W_P.

where $W_{i,2}$ and $W_{i,1}$ are the total accumulated strain energy density in one element at the end point and the starting point of one thermal cycle, respectively, $V_{i,2}$ and $V_{i,1}$ are the element volume at the end point and start point of one cycle, respectively, and n is the number of selected elements to calculate averaged strain energy density. In this study, whole layer elements on each AuSn layer were selected for averaging. Three TC cycles were simulated to obtain converged accumulation results per cycle.

Fig. 7.2.27 shows volume-averaged strain energy density for different DA layers. Potential fatigue failure of DA layer from high to low follows DA1, DA2, and DA3. Fig. 7.2.28 shows the effect of AuSn layer thickness on strain energy density results. Thicker AuSn layer helps to reduce stain energy density accumulation, which means that thicker AuSn layer improves its TC life, but this effect is not significant. AuSn layer with 10 μm thickness after bonding process was adopted in the final TV. TC test results showed that all AuSn layers have good reliability without any failure after 1000 cycles.

7.2.3.6 Reliability test and hydraulic leak test

With the help of simulation analysis and shear test results, the final TV was fabricated as shown in Fig. 7.2.1. The designed outlet and inlet on PCB were used to connect with the manifold for hydraulic leak test. The as-assembled SMC cooler system had a good

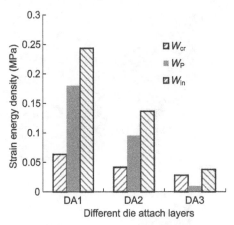

Figure 7.2.27 Volume-averaged strain energy density accumulation per cycle for different die attach (DA) layers.

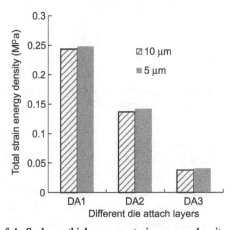

Figure 7.2.28 Effect of AuSn layer thickness on strain energy density.

bonding and sealing condition without leak. The microcooler system was then subjected to the TC test with temperature range from −40 to 125°C and cycle duration of 1 h per cycle. After 1000 TC cycles, failure analysis for the SMC cooler system was conducted using microscope and X-ray. There was no failure, such as delamination and cracking. Then hydraulic leak test was conducted for the microcooler system after 1000 TC cycles. A microgear pump forced the coolant water through a 15 μm filter and a flow meter before entering the microcooler. A differential pressure transmitter was attached to the manifold to measure pressure drop. Pressure drop increases with the increase in flow rate as shown in Fig. 7.2.29. To maintain 400 mL/min flow rate, the required pressure drop is around 30 kPa. No leakage was observed during

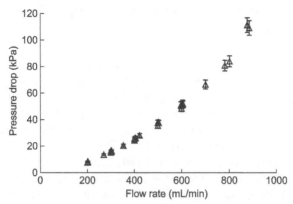

Figure 7.2.29 Pressure drop as a function of the flow rate from experiment.

the hydraulic test even when the flow rate increased to 900 mL/min. The desirable pressure drop validates that the microcooler system has a good bonding and sealing quality even after long-term TC reliability test.

7.2.4 Summary

Mechanical and thermomechanical modeling and characterization for silicon microcooler system design were carried out. Five different FEA models were used to study the shear test results, bonding process, assembly process, thermal–mechanical coupling induced stress, and TC performance. Through simulation and shear test characterization results, AuSn was finalized as a desirable DA material, and its optimal thickness was determined as 10 µm. Diamond thickness was optimized as 400 µm to reduce process-induced stress and warpage and also to achieve a good thermal performance. Thermal chip thickness was designed with 100 µm to improve mechanical reliability and thermal performance during process and field application.

Shear test simulation provided the understanding on shear strength and factors affecting shear test results. The maximum shear stress from simulation is much higher than the average shear stress calculated from shear test data due to nonuniform stress across the bonding layer. Lower shear height is recommended for shear test to make sure shear stress distribution much even across the bonding layer. TC simulation results showed that DA layer between the thermal chip and the diamond heat spreader is the critical layer than other DA layers. AuSn DA demonstrates a good TC performance.

Through numerical modeling and characterization, the final TV of the microcooler system was successfully fabricated. Hydraulic test and TC reliability test results and failure analysis verified that the designed microcooler system has a good reliability in terms of both thermal and mechanical performances.

7.3 Codesign modeling and analysis for advanced packages

Through-silicon interposer (TSI) is a successful application of through-silicon via (TSV) technology [42,43]. The TSI technology realizes that the heterogeneous integration of IC chips and vias offers vertical connection from the chips with fine pitch I/Os to the substrate with medium pitch I/Os [43]. However, TSV fabrication is facing many challenges [44—50], such as Cu protrusion, temporary bonding/debonding process, wafer warpage control, and long-term reliability. TSI is not a cost-effective technology due to wafer processes, material cost, and yield loss. New technologies have been explored to provide the similar function as TSI interposer but without TSV formation, such as Non-TSV Interposer [51] and TSV-Free Interposer (TFI) technology [52]. In TFI technology, the back end of line (BEOL) layers are formed based on a Si wafer, followed by chip-to-wafer bonding, wafer-level compression molding and back grinding (BG), carrier wafer removal, solder ball attachment, and singulation. Therefore, TFI technology eliminates TSV fabrication and reduces manufacturing and material cost.

TFI technology still faces several challenges needed to be addressed, such as wafer warpage during wafer fabrication process, assembly-induced package warpage, and package/board-level solder joint reliability. FEA is a powerful and useful tool and has been widely used in the packaging design stage for virtual prototyping [46,47,53,54]. In this study, a SMART (structure-material-assembly-reliability-thermal) codesign methodology with FEA simulation has been established and applied for TFI technology to optimize structure design, wafer process, assembly process, material selection, and thermal performance. Codesign modeling methodology for TFI technology includes TFI wafer process simulation to reduce wafer warpage, TFI package assembly process simulation to minimize package warpage, package-level TC simulation to improve C4 (controlled collapse chip connection) solder joint reliability, board-level temperature cycling (TCOB) simulation to improve board-level solder joint reliability, and thermal simulation to evaluate and enhance thermal performance of TFI package. Through codesign modeling, successful wafer and assembly processes and reliable and high-performance package solution can be achieved for advanced package with TFI technology.

7.3.1 Test vehicle and experimental procedure

Fig.7.3.1 shows a schematic of TFI package layout, in which one 15 mm × 15 mm × 0.56 mm larger graphics processing unit (GPU) chip and two 5.5 mm × 7 mm × 0.49 mm smaller High Bandwidth Memory (HBM) chips are mounted onto a 25 mm × 18 mm TFI interposer. The final TFI package excluding C4 joint and stiffener has a thickness of 0.62 mm. To achieve such package, following process flow for TV fabrication is carried out. The TFI interposer with three levels of 0.4 μm/0.4 μm fine line/space redistribution layer (RDL) is first fabricated on a Si carrier wafer using BEOL process. Then chip-on-wafer bonding process is conducted

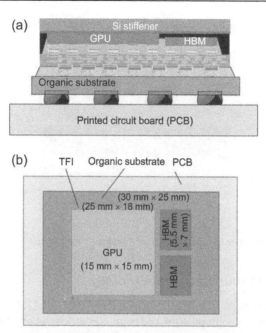

Figure 7.3.1 Test vehicle with TSV-Free Interposer (TFI) technology: (a) schematic of cross section of TFI package assembly and (b) layout of IC chips in TFI package.

to mount GPU and HBM chips onto the wafer and followed by wafer-level underfilling. After wafer-level compression molding and BG, GPU chips are exposed and HBM chips are embedded with 100 μm overmold. Si stiffener with 500 μm thickness is then attached onto the top of the molded wafer to help reduce wafer warpage. After carrier wafer removal, under bump metallization (UBM) and C4 bumping with 300 μm pitch are conducted. After singulation process, the obtained TFI package with the stiffener has a total thickness of 1.15 mm. The TFI package is then mounted onto a 30 mm × 25 mm organic substrate with C4 solder joints and finally attached onto a PCB through ball-grid array (BGA) solder joints. Underfill is used to protect C4 solder joints but not used for BGA solder joints. The final TV design and material selection are determined based on codesign modeling results for achieving successful TFI wafer process, package assembly process, long-term package/board-level solder joint reliability and acceptable thermal performance.

Experiment is conducted by using a simple bilayer structure as shown in Fig. 7.3.2 to validate the wafer warpage modeling method and results and also help to select suitable epoxy molding compound (EMC) material. Liquid- or granular-type EMC is put onto a 12″ Si carrier with the calculated weight of the EMC material. Molding temperature for each EMC is applied accordingly. After wafer-level compression modeling, the molded wafer has 400 μm thick mold compound and 770 μm thick Si carrier. Wafer warpage after post mold cure is measured by the IR measurement equipment. The measured warpage data are used for correlating with FEA simulation results.

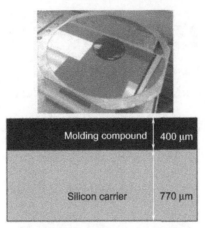

Figure 7.3.2 Bilayer wafer structure for simulation result validation.

7.3.2 Codesign modeling methodology

Fig. 7.3.3 shows contents and categories included in the SMART codesign modeling methodology. The main purpose of codesign method is to optimize structure and geometry design, wafer process, assembly process, material selection, and thermal performance. Codesign modeling method for TFI technology in this study includes wafer-level warpage, package warpage, package/board-level solder joint reliability, and thermal performance simulation as shown in Fig. 7.3.4.

Wafer process simulation using a 3D quarter model as shown in Fig. 7.3.5a has been established to simulate compression molding, BG, stiffener attach, carrier wafer removal, and back-side RDL processes. For small feature such as RDL, vias, UBM, and microbumps, thin layer is simplified with considering layer thickness and majority material (relative high volume) in wafer-level simulation. Through wafer-level modeling, suitable EMC materials and Si stiffener are recommended to control wafer warpage less than 2 mm for 12 inch wafer, which is a requirement of warpage handling by lithography equipment. Effect of organic substrate CTE and stiffener thickness on assembly induced package warpage is simulated by a 3D half model of the package as shown in Fig. 7.3.5b to reduce package warpage. Long-term package- and board-level

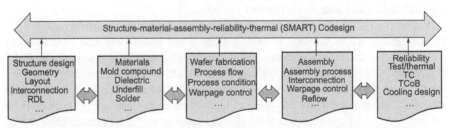

Figure 7.3.3 Structure-material-assembly-reliability-thermal (SMART) codesign modeling contents and categories.

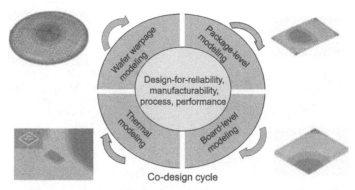

Figure 7.3.4 Codesign modeling for TSV-Free Interposer (TFI) package technology.

TC reliability is simulated to predict C4 and BGA solder joint thermal fatigue lives. Comparison between 3D half model in Fig. 7.3.5b and 3D slice model in Fig. 7.3.5c is carried out for C4 joint reliability. To reduce solving time and computing resource, a 3D slice model as shown in Fig. 7.3.5d is also used for investigating board-level solder joint reliability. In the 3D slice model, detailed structure and small feature size are included. Proper BCs are applied for each model as shown in Fig. 7.3.5. Table 7.3.1 lists material properties used in each model. Stress-free temperature is chosen for materials according to their active (processing) condition in the actual processes. Temperature range from −40 to 125°C and cycle duration of 1 h are modeled for TC loading. The recommended materials and structure design based on reliability are aligned with that from wafer and package warpage simulation results. The optimized TV structure and materials are finalized based on codesign modeling results.

7.3.3 Results and discussion

7.3.3.1 Wafer warpage simulation and validation

First of all, the wafer warpage modeling method is validated by experimental results. Fig. 7.3.6 shows correlation between experimental and simulation results (based on structure in Fig. 7.3.2). Simulation result has a good agreement with warpage measurement data. EMC3 leads to the lowest warpage among three EMCs due to the lowest CTE mismatch between EMC3 and Si wafer. The validated modeling method can be implemented in the real TV fabrication process to analyze and optimize wafer warpage.

To identify wafer warpage direction during wafer process, warpage direction and sign are defined in Fig. 7.3.7. Process-dependent modeling is conducted for wafer warpage simulation using element birth and death technique provided by software. Fig. 7.3.8 shows process-dependent wafer warpage data at room temperature (25°C) for the reference model. In the reference model, original molding thickness is 720 μm, 100 μm thick molding compound (EMC1) is removed after BG, and

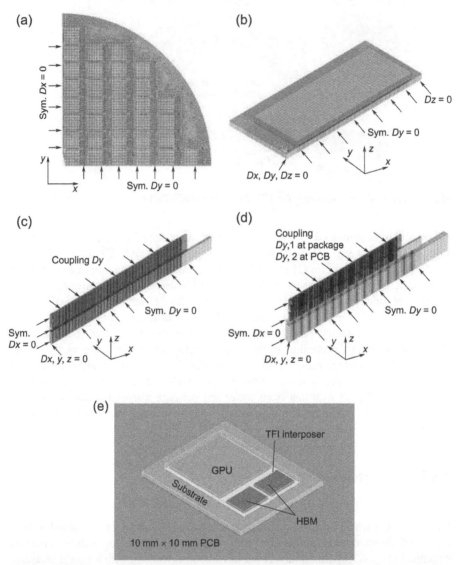

Figure 7.3.5 Finite element analysis (FEA) models for different modeling categories: (a) 3D quarter model for wafer warpage simulation, (b) 3D half model for package-level simulation, (c) 3D slice model for package-level simulation, (d) 3D slice model for board-level simulation, and (e) thermal model.

770 μm thick Si carrier and 500 μm thick Si stiffener are modeled. Two critical processes inducing large warpage are identified by simulation results, i.e., molding process and carrier wafer removal process. Concave shape warpage (>1 mm as shown in Fig. 7.3.9a) occurs after wafer-level compression molding and reduces slightly after BG process. The molded wafer becomes flat after stiffener attach. However, convex

Table 7.3.1 Material properties used in finite element models.

Materials	Modulus (GPa)	CTE (ppm/°C)	Poisson's ratio	T_g (°C)	Remark	Thermal conductivity ($W \cdot m^{-1} \cdot °C^{-1}$)
EMC1-reference	18($<T_g$), 1($<T_g$)	7($<T_g$), 22($>T_g$)	0.3	160	Molding 125°C	0.8
EMC2	34($<T_g$), 0.45($>T_g$)	6($<T_g$), 22($>T_g$)	0.3	180	Molding 135°C	
EMC3	10($<T_g$), 0.5($>T_g$)	4($<T_g$), 21($>T_g$)	0.3	170	Molding 125°C	
Underfill	13($<T_g$), 0.5($>T_g$)	23($<T_g$), 80($>T_g$)	0.3	120	Curing 125°C	0.8
Silicon	131	2.8	0.28			130
Passivation	1.6	62	0.3	>250	Curing 180°C	0.25
Bonding material	2($<T_g$), 0.2($>T_g$)	60($<T_g$), 200($>T_g$)	0.35	220		0.2
SiO_2	73	0.5	0.17			1.5
Organic substrate/PCB	25 (in plane), 11 (out of plane)	15 (in plane), 46 (out of plane)	0.39 (in plane), 0.11 (out of plane)		Orthotropic	18.9 (in plane), 0.4 (out of plane)
Solder—SnAgCu	60.1 (−40°C), 41.7 (25°C), 28.3 (125°C), 0.2 (220°C)	22	0.35		Creep model	58.7
Cu	117	17	0.34		Elastic-plastic model	386
Solder mask	2.4 ($<T_g$), 0.2($>T_g$)	50 ($<T_g$), 160($>T_g$)	0.3	101		0.2

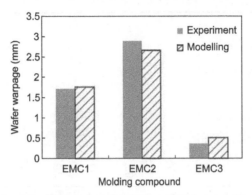

Figure 7.3.6 Validation of wafer warpage simulation results by experimental data.

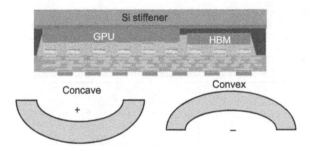

Figure 7.3.7 Definition and sign of wafer warpage.

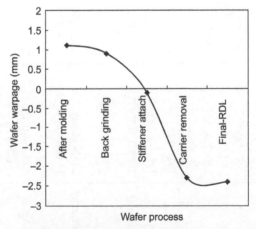

Figure 7.3.8 Wafer process–dependent warpage from finite element analysis (FEA) simulation.

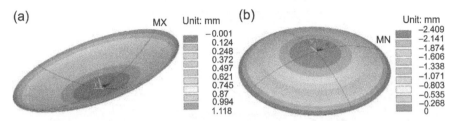

Figure 7.3.9 Wafer warpage simulation results: (a) after compression molding and (b) after final RDL.

shape warpage (>2 mm as shown in Fig. 7.3.9b) occurs after carrier removal and final RDL process, which arises a challenge for final RDL process. To control wafer warpage less than 2 mm through all processes, optimization needs to be conducted through FEA simulation.

7.3.3.2 Wafer warpage analysis and optimization

The effect of mold compound on wafer warpage is significant, as shown in Fig. 7.3.10, because wafer warpage is mainly induced by CTE mismatch between Si and EMC materials. EMC1 and EMC2 lead to similar wafer warpage, which is different from warpage results in Fig. 7.3.6 for the bilayered simple structure due to different structures and volume ratios of EMC to Si. EMC3 results in the lowest wafer warpage among three EMCs due to low CTE value of 4 ppm/K, which is one of the candidates to ensure wafer warpage less than 2 mm during wafer processes.

Fig. 7.3.11 shows the effect of Si stiffener thickness on wafer warpage considering two major processes after stiffener attach, i.e. carrier removal and final RDL process. Wafer warpage is more than 2 mm after carrier removal and final RDL process if the stiffener is not applied. Even though EMC3 has low CTE value, wafer with EMC3 still has large warpage when the stiffener is not used because EMC3 has low modulus,

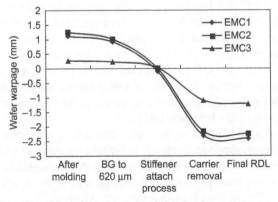

Figure 7.3.10 Effect of molding compound material on wafer warpage.

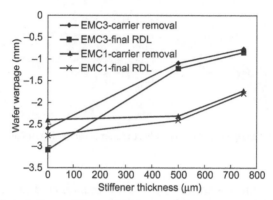

Figure 7.3.11 Effect of silicon stiffener thickness on wafer warpage.

which makes structure much softer and flexible. Stiffener makes structure stiffer and not easy to deform. Wafer warpage decreases with the increase in stiffener thickness, especially for wafer with soft EMC material (EMC3 in this study). For the purpose of wafer warpage reduction, 500 μm thick Si stiffener and EMC3 material are desirable choice for TFI package fabrication.

7.3.3.3 Package warpage analysis and optimization

Reflow condition with peak temperature of 250°C is modeled for package assembly to simulate C4 solder joint reflow process—induced package warpage. C4 joints have 300 μm pitch and 160 μm stand-off height after assembly, and organic substrate has 1 mm thickness. The warpage requirement is less than 100 μm to ensure the good joint quality of C4 joint after reflow process. After C4 joint reflow process, underfilling process is conducted to protect C4 joints. Fig. 7.3.12 shows package warpage at both room temperature (25°C) and solder melting temperature (220°C) at which solder starts to solidify and stress starts to build up in solder and high package warpage at this temperature will affect solder joint formation quality. Package warpage has a convex shape at 25°C and increases with the increase in substrate CTE. Package warpage has a different definition from wafer warpage. Stiffener helps to reduce package warpage by 40%−70%. Both EMC1 and EMC3 lead to similar package warpage at 25°C. However, package warpage has a concave shape at 220°C and also increases with the increase in substrate CTE. Effect of stiffener on package warpage becomes not significant at high temperature. Based on simulation results, organic substrate with low CTE ($\leqslant$10 ppm/K) and applying stiffener on the top of the package are recommended to control package warpage less than 100 μm.

Fig. 7.3.13 shows the effect of organic substrate thickness on package warpage. For package with stiffener, substrate thickness influences on warpage are not significant. Effect of organic substrate CTE on package warpage is more significant than its thickness effect. To control warpage of the TFI package with stiffener, selecting low CTE substrate is more efficient than changing substrate thickness.

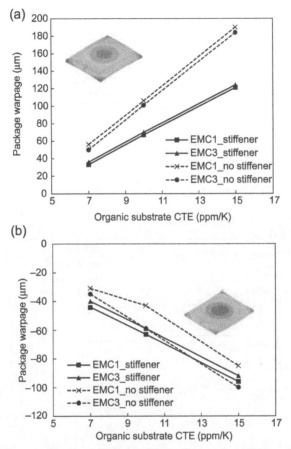

Figure 7.3.12 Effect of substrate coefficient of thermal expansion (CTE) and stiffener on package warpage at (a) 25°C and (b) 220°C.

7.3.3.4 C4 solder joint thermal cycling reliability

Package-level TC reliability is modeled for TFI package using a 3D half model (Fig. 7.3.5b) and a slice model (Fig. 7.3.5c) to investigate and predict C4 joint thermal fatigue life. The Sn-Ag-Cu lead-free solder is modeled for C4 joints using hyperbolic—sine creep constitutive model [55]. Temperature range from −40°C to 125°C and cycle duration of 1 h are modeled for TC loading. Fig. 7.3.14 shows creep strain energy density of C4 joints under TC loading for the TFI package with 500 μm thick stiffener based on simulation results from the half model. For TFI package without underfill, C4 joints at package edges (Fig. 7.3.14a) have large creep strain energy density value, which is corresponding to low solder joint life. Underfill is used to improve fatigue life of C4 joints at package corners and edges and makes C4 joints under Si die area have much uniform life (Fig. 7.3.14b). Volume-averaged creep strain energy density accumulation per cycle based on the interfacial layer of solder and

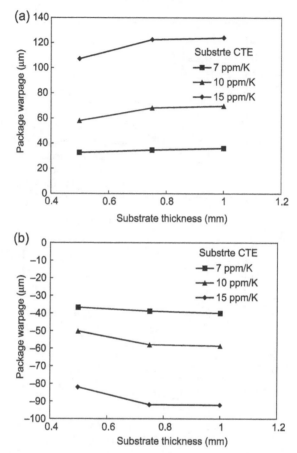

Figure 7.3.13 Effect of substrate thickness on package warpage at (a) 25°C and (b) 220°C.

package is used for fatigue life prediction. The energy based life model of Sn-Ag-Cu solder is expressed below:

$$N_f = 137.6 \times W_{cr}^{-1.112} \qquad (7.3.1)$$

where W_{cr} is creep strain energy density from simulation results, and then fatigue life, N_f, can be predicted.

Fig. 7.3.15 shows fatigue life comparison of C4 joints at the package edge (column C1 in Fig. 7.3.14b) from the 3D half modeling results (Fig. 7.3.5b). For TFI package with stiffener but without underfill, C4 joint life is much sensitive to joint location, and the corner joint has much lower fatigue life. For TFI package without underfill, EMC3 leads to lower C4 joint life than EMC1 due to higher CTE mismatch between EMC3 and organic substrate. C4 joints show much uniform and improved fatigue life when underfill is applied in the package. The effect of stiffener on C4 solder joint life is not

System-level modeling, analysis, and design 381

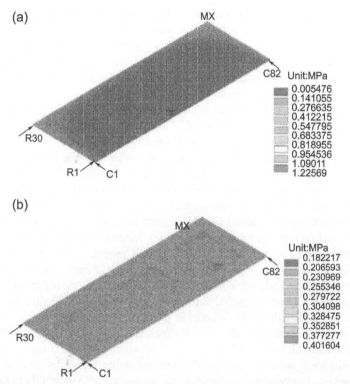

Figure 7.3.14 Creep strain energy density of C4 joints: (a) without underfill and (b) with underfill.

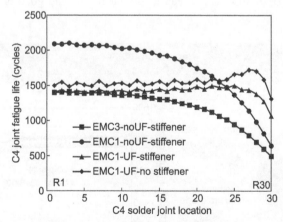

Figure 7.3.15 Life comparison of C4 solder joint based on simulation results from 3D half model.

significant when underfill is applied in package. Package with stiffener has slightly lower C4 joint fatigue life compared with package without stiffener because Si stiffener makes package not flexible and increases CTE mismatch between package and organic substrate. The main purpose of stiffener is to help to reduce wafer/package warpage significantly. Therefore, underfill and stiffener are all needed to improve C4 solder joint reliability and reduce package warpage.

3D half model is one very time-consuming model, which needs more than 50 h to solve one TC simulation for each case of TFI package. Simplified 3D slice model cutting along TFI package center to edge with one row C4 joints (Fig. 7.3.5c) is adopted for parametric study. In the 3D slice model, only large GPU die is modeled with the same distance from die edge to package edge as in the 3D half model. Fig. 7.3.16 shows simulation results and fatigue life of C4 joints from 3D slice model. C4 joints have similar life with helping from underfill. Life prediction based on bottom interfaces (organic substrate side) of C4 solder joint has slightly lower value compared with that based on top side interface (chip side). Simulation results from the 3D slice model are compared with those from the 3D half model, as shown in Fig. 7.3.17. Similar life prediction from 3D slice model and 3D half model indicates that slice model is one accurate and simple model that can be used to do more parametric studies effectively. In this TFI package TC simulation, solving time of 3D slice model is just one-tenth that of 3D half model. When TFI package with underfill, EMC1 and EMC3 leads to similar C4 joint life (Fig. 7.3.17), which is different from results of TFI package without underfill (Fig. 7.3.15).

Underfills are important to improve C4 joint reliability, especially for large package size. However, it is essential and critical to choose suitable underfills to protect C4 joints. Otherwise, underfills may decrease C4 joint reliability [56]. Table 7.3.2 lists 4 underfills and their properties. CTE and Young's modulus are two critical mechanical properties affecting C4 joint reliability. Fig. 7.3.18 shows the effect of underfills on C4 joint reliability. Underfills 1 and 2 lead to similar results and good protection on C4

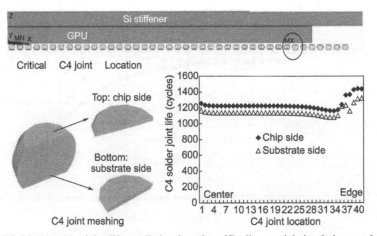

Figure 7.3.16 C4 solder joint life prediction based on 3D slice model simulation results.

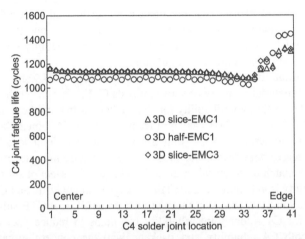

Figure 7.3.17 Comparison of C4 joint life prediction between 3D slice model and 3D half model.

Table 7.3.2 Mechanical Properties of different underfills.

Material	Modulus (GPa)	CTE (ppm/°C)	Poisson's ratio	T_g (°C)
Underfill 1	13/0.2	23/80	0.3	120
Underfill 2	12/0.5	24/75	0.3	115
Underfill 3	7/0.5	37/101	0.3	118
Underfill 4	6/0.5	37/125	0.3	150

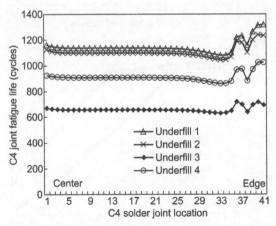

Figure 7.3.18 Effect of underfills on package-level C4 solder joint thermal cycling (TC) reliability.

joint reliability due to their similar mechanical properties and similar CTE value as lead-free solder. Underfill 3 leads to poor C4 joint reliability due to high CTE and low T_g. Underfill 1 is recommended to improve C4 joint reliability for TFI package, and it will be used in the following simulation.

Organic substrate is another important material affecting C4 joint reliability. The most important material property of substrate is its CTE. Fig. 7.3.19 shows the effect of substrate CTE on C4 joint reliability for TFI package with or without underfill using 3D slice model. Substrate CTE has significant effect on C4 joint reliability of TFI package without underfill, and C4 joint life is sensitive to joint location. C4 joint fatigue life increases with decreasing substrate CTE because low substrate CTE leads to low CTE mismatch between substrate and package. The effect of substrate CTE on C4 joint reliability is not significant for TFI package with underfill, and C4 joint life is not sensitive to joint location. For package with underfill, low CTE substrate has no improvement for C4 solder joint reliability but helps to reduce package warpage. Considering both C4 reliability and package warpage concern, organic substrate with CTE of 10 or 7 ppm/K is recommended.

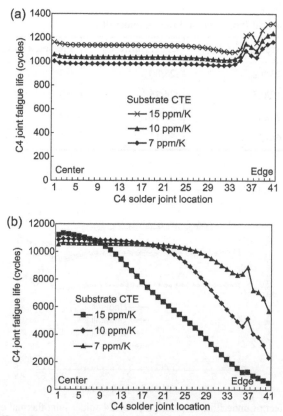

Figure 7.3.19 Effect of substrate coefficient of thermal expansion (CTE) on package-level C4 solder joint thermal cycling (TC) reliability: (a) with underfill and (b) without underfill.

7.3.3.5 Board-level solder joint reliability analysis

When TFI package is assembled onto the PCB board, TCOB reliability is another concern for BGA solder joint, which is simulated using a 3D slice model (Fig. 7.3.5d). In the 3D slice TCOB model, underfill 1 is applied for C4 joints but not for BGA joints, and EMC3 is used as molding compound. BGA joint has a pitch of 0.9 mm. Organic substrate has the same CTE of 15 ppm/K as the PCB board in TCOB simulation and substrate thickness is 1 mm and PCB thickness is 1.6 mm. PCB keeps a constant CTE of 15 ppm/K in TCOB simulation. C4 joints have slightly higher life under package-level TC test than that under board-level TCOB test for the same TC test condition due to stiffer structure of board-level assembly compared with package-level assembly. Fig. 7.3.20 shows that BGA solder joint reliability is sensitive to joint location due to DNP (distance to neutral point) effect as underfill is not used for BGA joints. The bottom interface of BGA joint (PCB board side) is critical compared with the top interface (package substrate side). The critical BGA joint is adjacent to Si stiffener edge, not package substrate edge. BGA solder joints show higher fatigue life compared with C4 joints.

Fig. 7.3.21 shows the effect of organic substrate CTE on critical C4 joint reliability and comparison between package TC and board-level TCOB conditions. When TFI package with underfill 1 for C4 joints, C4 joint reliability is not sensitive to organic substrate CTE in TCOB testing condition. Fig. 7.3.22 shows that substrate CTE affects BGA joint reliability significantly, which is different from its effect on C4 joint reliability. CTE mismatch between chip and substrate and between underfill and C4 joint

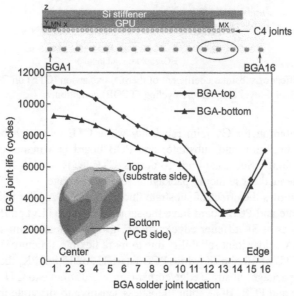

Figure 7.3.20 Effect of location on ball-grid array (BGA) solder joint thermal cycling (TC) reliability (substrate CTE = 15 ppm/K).

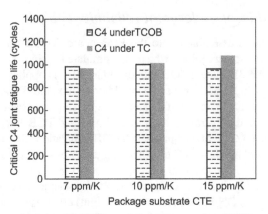

Figure 7.3.21 Effect of substrate coefficient of thermal expansion (CTE) on C4 joint reliability under thermal cycling (TC) and board-level temperature cycling (TCOB) conditions.

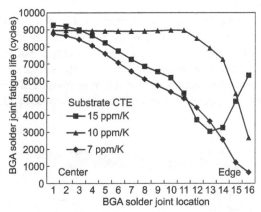

Figure 7.3.22 Effect of substrate coefficient of thermal expansion (CTE) on BGA solder joint reliability under board-level temperature cycling (TCOB) condition.

are main contribution for C4 joint reliability while CTE mismatch between whole package (including chip and substrate) and PCB board is a main contribution for BGA joint reliability. Substrate with 10 ppm/K CTE leads to high BGA life, which is also recommended for reducing package assembly warpage.

Fig. 7.3.23 shows the effect of substrate thickness on BGA joint reliability. When package substrate and PCB board have the same CTE value (15 ppm/K), the critical BGA joint locates at Si stiffener edge/corner (Fig. 7.3.23a), and thin substrate helps to improve BGA solder joint reliability due to more flexible structure. When package substrate has CTE of 10 ppm/K while PCB has CTE of 15 ppm/K, the critical BGA joint locates at package substrate edge/corner (Fig. 7.3.23b) due to CTE mismatch between substrate and PCB. BGA joint life is less sensitive to substrate thickness when package with 10 ppm/K CTE substrate compared with package with 15 ppm/K CTE

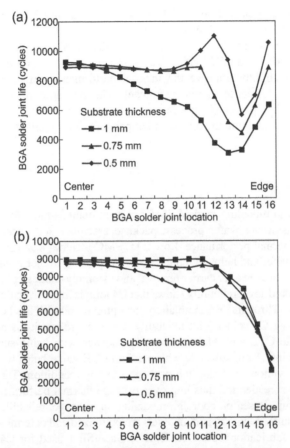

Figure 7.3.23 Effect of substrate thickness on BGA solder joint reliability: (a) substrate CTE of 15 ppm/K and (b) substrate CTE of 10 ppm/K.

substrate, which is also applicable for the effect of substrate thickness on the critical BGA joint life. BGA joint life decreases with the increase in substrate thickness. C4 joint reliability is not sensitive to substrate thickness and CTE under TCOB condition. Organic substrate with CTE of 10 or 15 ppm/K provides good BGA joint reliability. Considering package warpage, organic substrate with 10 ppm/K is finally recommended.

7.3.3.6 Thermal performance evaluation

3D simulation model as show in Fig. 7.3.5e has been constructed using a CFD software for thermal performance modeling. The equivalent thermal conductivity in the in-plane and out-of-plane direction is calculated based on volume-averaging method for GPU and HBM microbump joint layer, C4 joint and underfill layer, and TFI interposer layer by considering joint and bump layout design, materials, and their volume.

Thermal modeling is carried out for TFI package considering 2 power supply cases, i.e. 1 W for GPU chip and 1 W for each HBM chip and 2 W for GPU chip and 0.7 W for each HBM chip. Fig. 7.3.24 shows the effect of Si stiffener on the maximum temperature of GPU and HBM chips. The stiffener can work as a heat spreader for HBM chip and slightly improve the thermal performance. Effect of stiffener on GPU thermal performance is not significant. The package can dissipate around 3.5 W total heating power (Case2 in Fig. 7.3.24) with helping by 500 μm thick stiffener, while maintaining the operation temperature limit of HBM chip temperature less than 85°C.

7.3.4 Summary

SMART codesign modeling methodology has been demonstrated for TFI packaging technology considering wafer process, package assembly and package/board-level reliability and thermal performance. Low CTE EMC material has matched properties with Si carrier wafer and helps to reduce wafer warpage. Stiffener is an effective way to reduce wafer warpage after carrier removal and assembly-induced package warpage. Package- and board-level reliability shows that C4 joint is more critical than BGA joint for TFI package. Through FEA simulation, the optimal underfill has been recommended for improving C4 solder joint reliability. C4 joints reliability is not sensitive to organic substrate CTE and thickness for TFI package with using suitable underfill for C4 solder joints. Thin substrate with medium CTE value helps to improve BGA joint reliability without violating the package warpage condition. The stiffener also works as a heat spreader and thus improves package thermal performance, especially for the HBM chip. Based on codesign modeling on warpage, reliability, and thermal performance results, geometry and materials are optimized and determined for the final TV design and fabrication: low CTE EMC3, Underfill 1 used for C4 solder joints, 500 μm thick Si stiffener, and 0.75 mm thick substrate with 10 ppm/K CTE.

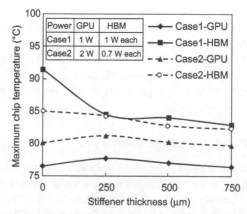

Figure 7.3.24 Effect of stiffener on the maximum chip temperature from thermal simulation results.

References

[1] R.R. Tummala, Fundamentals of Microsystems Packaging, McGraw-Hill, New York, 2001.
[2] Y. Joshi, P. Kumar, Energy Efficient Thermal Management of Data Centers, Springer Science+Business Media, LLC, 2012.
[3] T.J. Chainer, M.D. Schultz, P.R. Parida, M.A. Gaynes, Improving data center energy efficiency with advanced thermal management, IEEE Trans. Compon. Packag. Manuf. Technol. 7 (8) (2017) 1228–1239.
[4] C. Xu, H. li, K. Banerjee, Modeling, analysis, and design of graphene nano-ribbon interconnects, IEEE Trans. Electron Devices 56 (8) (2009) 1567–1578.
[5] H.Y. Zhang, D. Pinjala, P.-S. Teo, Thermal management of high power dissipation electronic packages on high density substrates: from air cooling to liquid cooling, in: EPTC Conference, 2003.
[6] G. Tang, Y. Han, B.L. Lau, et al., Development of a compact and efficient liquid cooling system with silicon microcooler for high-power microelectronic devices, IEEE Trans. Compon. Packag. Technol. 6 (5) (2016) 729–739.
[7] J.B. Marcinichen, J.A. Olivier, J.R. Thome, On-chip two-phase cooling of data centers: cooling system and energy recovery evaluation, Appl. Therm. Eng. 41 (2012) 36–51.
[8] J.F. Tullius, R. Vajtai, Y. Bayazitoglu, A review of cooling in microchannels, Heat Transf. Eng. 21 (7–8) (2011) 527–541.
[9] Y. Madhour, S. Zimmermann, J. Olivier, J. Thome, B. Michel, D. Poulikakos, Cooling of next generation computer chips: parametric study for single- and two-phase cooling, in: Proc. 17th Int. Workshop THERMINIC, 2011, pp. 1–7.
[10] S.C. Mohapatra, D. Loikits, Advances in liquid coolant technologies for electronics cooling, in: Proc. 21st Annu. IEEE Semiconductor Thermal Meas. Manage. Symp., 2005, pp. 354–360.
[11] X. Wei, Y. Joshi, Stacked microchannel heat sinks for liquid cooling of microelectronic components, J. Electron. Packag. 126 (1) (2004) 60–66.
[12] B. Agostini, M. Fabbri, J.E. Park, L. Wojtan, J.R. Thome, B. Michael, State of the art of high heat flux cooling technologies, Heat Transf. Eng. 28 (4) (2007) 258–281.
[13] H.Y. Zhang, D. Pinjala, T.N. Wong, Y.K. Joshi, Development of liquid cooling techniques for flip chip ball-grid array packages with high heat flux dissipations, IEEE Trans. Compon. Packag. Technol. 28 (1) (2005) 127–135.
[14] H. Bostanci, D. Van Ee, B.A. Saarloos, D.P. Rini, L.C. Chow, Thermal management of power inverter modules at high fluxes via two-phase spray cooling, IEEE Trans. Compon. Packag. Manuf. Technol. 2 (9) (2012) 1480–1485.
[15] M.K. Sung, I. Mudawar, Single-phase hybrid micro-channel/microjet impingement cooling, Int. J. Heat Mass Transf. 51 (17–18) (2008) 4342–4352.
[16] X. Song, et al., Surrogate-based analysis and optimization for the design of heat sinks with jet impingement, IEEE Trans. Compon. Packag. Manuf. Technol. 4 (3) (2014) 429–437.
[17] E.G. Colgan, et al., A practical implementation of silicon microchannel coolers for high power chips, IEEE Trans. Compon. Packag. Technol. 30 (2) (2007) 218–225.
[18] H.Y. Zhang, D. Pinjala, Y.K. Joshi, T.N. Wong, et al., Fluid Flow and heat transfer in liquid cooled foam heat sinks for electronic packages, IEEE Trans. Compon. Packag. Technol. 28 (2) (2005) 272–280.
[19] T. Chen, S.V. Garimella, Flow boiling heat transfer to a dielectric coolant in a microchannel heat sink, IEEE Trans. Compon. Packag. Technol. 30 (1) (2007) 24–31.

[20] F. Alfieri, M.K. Tiwari, I. Zinovik, D. Poulikakos, T. Brunschwiler, B. Michel, 3D integrated water cooling of a composite multilayer stack of chips, J. Heat Transf. 132 (12) (2010), 121402−1−121402−9.

[21] Y. Han, B.L. Lau, H. Zhang, X. Zhang, Package-level Si-based micro-jet impingement cooling solution with multiple draining microtrenches, in: Proc. 16th IEEE EPTC, 2014, pp. 330−334.

[22] B.L. Lau, Yong Han, H.Y. Zhang, L. Zhang, X.W. Zhang, Development of fluxless bonding using deposited gold-indium multi-layer composite for heterogeneous silicon micro-cooler stacking, in: Proc. 16th IEEE EPTC, 2014.

[23] L. Luo, Y. Fan, X. Yuan, N. Midoux, Integration of constructal distributors to a mini crossflow heat exchanger and their assembly configuration optimization, Chem. Eng. Sci. 62 (2007) 3605−3619.

[24] A. Lozano, F. Barreras, N. Fueyo, S. Santodomingo, The flow in an oil/water plate heat exchanger for the automotive industry, Appl. Therm. Eng. 28 (2008) 1109−1117.

[25] J. Rambo, Y. Joshi, Multi-scale modeling of high power density data centers, in: ASME InterPACK'03, 2003.

[26] H. Zhang, Y.C. Mui, M. Rathin, Air flow modeling and analysis for thermal management in functional burn-in systems, in: EPTC Conference, 2007, pp. 172−178.

[27] J. Rambo, Y. Joshi, Modeling of data center airflow and heat transfer: state of the art and future trends, Distrib Parallel Databases 21 (2007) 193−225.

[28] J.I. Tustaniwskyj, et al., High performance active temperature control of a DUT, in: 20th SEMI-THERM, 2004.

[29] E.G. Colgan, B. Furman, M. Gaynes, et al., High performance and sub-ambient silicon microchannel cooling, J. Heat Transf. 129 (8) (2007) 1046−1051.

[30] S.G. Singh, S.P. Duttagupta, A. Agrawa, In situ impact analysis of very high heat flux transients on nonlinear p-n diode characteristics and mitigation using on-chip single- and two-phase microfluidics, J. Microelectromech. Syst. 18 (6) (2009) 1208−1219.

[31] W. Sabbah, S. Azzopardi, C. Buttay, R. Meuret, E. Woirgard, Study of die attach technologies for high temperature power electronics: silver sintering and gold-germanium alloy, Microelectron. Reliab. 53 (2013) 1617−1621.

[32] T.-Y. Chung, J.-H. Jhang, J.-S. Chen, et al., A study of large area die bonding materials and their corresponding mechanical and thermal properties, Microelectron. Reliab. 52 (2012) 872−877.

[33] P. Quintero, T. Oberc, P. McCluskey, Reliability assessment of high temperature lead-free device attach technologies, in: Proc. 58th Electron. Comp. Technol. Conf., 2008, pp. 2131−2138.

[34] J.-W. Yoon, H.-S. Chun, B.-I. Noh, J.-M. Koo, J.-W. Kim, H.-J. Lee, S.-B. Jung, Mechanical reliability of Sn-rich Au-Sn/Ni flip chip solder joints fabricated by sequential electroplating method, Microelectron. Reliab. 48 (2008) 1857−1863.

[35] J. Lian, S.J.W. Chun, M.S. Goorsky, J. Wang, Mechanical behavior of Au-In intermetallics for low temperature solder diffusion bonding, J. Mater. Sci. 44 (2009) 6155−6161.

[36] Y.-C. Sohn, Q. Wang, S.-J. Ham, B.-G. Jeong, K.-D. Jung, M.-S. Choi, et al., Wafer-level low temperature bonding with Au-In system, in: Proc. 57th Electron. Comp. Technol. Conf., 2007, pp. 633−637.

[37] J.P. Calame, R.E. Myers, S.C. Binari, F.N. Wood, M. Garven, Experimental investigation of microchannel coolers for the high heat flux thermal management of GaN-on-SiC semiconductor devices, Int. J. Heat Mass Transf. 50 (2007) 4767−4779.

[38] F.X. Che, J.-K. Lin, K.Y. Au, H.-Y. Hsiao, X. Zhang, Stress analysis and design optimization for low-k chip with Cu pillar interconnection, IEEE Trans. Compon. Packag. Technol. 5 (9) (2015) 1273−1283.

[39] G.S. Zhang, H.Y. Jing, L.Y. Xu, J. Wei, Y.D. Han, Creep behavior of eutectic 80Au/20Sn solder alloy, J. Alloy. Comp. 476 (2009) 138–141.
[40] F.X. Che, J.H.L. Pang, Fatigue reliability analysis of Sn-Ag-Cu solder joints subject to thermal cycling, IEEE Trans. Device Mater. Reliab. 13 (1) (2013) 36–49.
[41] J.H.L. Pang, F.X. Che, Isothermal cyclic bend fatigue test method for lead-free solder joints, J. Electron. Packag. 129 (4) (2007) 496–503.
[42] M. Sunohara, T. Tokunaga, T. Kurihara, M. Higashi, Silicon interposer with TSVs (Through Silicon Vias) and fine multilayer wiring, in: Proc. 58th Electron. Compon. Technol. Conf., 2008, pp. 847–852.
[43] C.C. Lee, C.P. Hung, C. Cheung, P.F. Yang, C.L. Kao, D.L. Chen, et al., An overview of the development of a GPU with integrated HBM on silicon interposer, in: Proc. 66th Electron. Compon. Technol. Conf., 2016, pp. 1439–1444.
[44] F.X. Che, W.N. Putra, A. Heryanto, A.D. Trigg, X. Zhang, C.L. Gan, Study on Cu protrusion of through-silicon via (TSV), IEEE Trans. Compon. Packag. Manuf. Technol. 3 (5) (2013) 732–739.
[45] A. Heryanto, W.N. Putra, A.D. Trigg, S. Gao, W.S. Kwon, F.X. Che, et al., Effect of copper TSV annealing on the via protrusion for TSV wafer fabrication, J. Electron. Mater. 41 (2012) 2533–2542.
[46] F.X. Che, H.Y. Li, X. Zhang, S. Gao, K.H. Teo, Development of wafer-level warpage and stress modelling methodology and its application in process optimization for TSV wafers, IEEE Trans. Compon. Packag. Manuf. Technol. 2 (6) (2012) 944–955.
[47] F.X. Che, Dynamic stress modelling on wafer thinning process and reliability analysis for TSV wafer, IEEE Trans. Compon. Packag. Manuf. Technol. 4 (9) (2014) 1432–1439.
[48] K.N. Tu, Reliability challenges in 3D IC packaging technology, Microelectron. Reliab. 51 (2011) 517–523.
[49] S.K. Ryu, K.H. Lu, X. Zhang, J.H. Im, P.S. Ho, R. Huang, Impact of near-surface thermal stresses on interfacial reliability of through silicon vias for 3-D interconnects, IEEE Trans. Device Mater. Reliab. 11 (1) (2011) 35–43.
[50] E. Beyne, Reliable via-middle copper through-silicon via technology for 3-D integration, IEEE Trans. Compon. Packag. Manuf. Technol. 6 (7) (2016) 983–992.
[51] F.Y. Liang, H.H. Chang, W.T. Tseng, J.Y. Lai, S. Cheng, M. Ma, et al., Development of non-TSV interposer (NTI) for high electrical performance package, in: Proc. 66th Electron. Compon. Technol. Conf., 2016, pp. 31–36.
[52] F.X. Che, M. Kawano, M.Z. Ding, Y. Han, S. Bhattacharya, Study on low warpage and high reliability for large package using TSV-Free interposer technology through SMART Co-design modelling, IEEE Trans. Compon. Packag. Manuf. Technol. 7 (11) (2017) 1774–1785.
[53] F.X. Che, X. Zhang, J.K. Lin, Reliability study of 3D IC packaging based on through-silicon interposer (TSI) and silicon-less interconnection technology (SLIT) using finite element analysis, Microelectron. Reliab. 61 (2016) 64–70.
[54] L.J. Ladani, Numerical analysis of thermo-mechanical reliability of through silicon vias (TSVs) and solder interconnects in 3-dimensional integrated circuits, Microelectron. Eng. 87 (2010) 208–215.
[55] J.H.L. Pang, B.S. Xiong, Mechanical properties for 95.5Sn-3.8Ag-0.7Cu lead-free solder alloy, IEEE Trans. Compon. Packag. Technol. 28 (4) (2005) 830–840.
[56] F.X. Che, Study on board level solder joint reliability for extreme large fan-out WLP under temperature cycling, in: Proc. 18th Electron. Packag. Technol. Conf., 2016, pp. 207–215.

Appendix 1 Nomenclature

Chapter 2: Electrical modeling and design

Abbreviation
TSV   Through-silicon via

Nomenclature (units of measure)

$\vec{B}$	Magnetic flux density (Wb/m^2)
C	Interconnect capacitance per unit length (F/m)
C_{dep}	Depletion capacitance (F)
C_L	Load capacitance (F)
C_{ox}	Oxide capacitance (F)
C_{Si}	Silicon capacitance (F)
D_{cnt}	CNT diameter (m)
$\vec{E}$	Electric field (V/m)
f	Frequency (Hz)
f_{cnt}	CNT filling ratio
Fm	Ratio of metallic CNTs in the bundle
G	Interconnect conductance per unit length (S/m)
G_{Si}	Silicon conductance (S/m)
h	Planck's constant (J·s)
H	Substrate thickness (m)
H_{TSV}	TSV height (m)
I	Current (A)
$I_0(\cdot), I_1(\cdot)$	Zero- and first-order Bessel functions of the first type, respectively
$\vec{J}$	Current density (A/m^2)
k	Thermal conductivity (W·m^{-1}·K^{-1})
L	Interconnect inductance per unit length (H/m)
l	Interconnect length (m)
L_{eff}	Effective inductance of the TSV inductor (H)
L_{TSV}	TSV inductance (H)
N_a	p-type bulk doping concentration (1/m^3)
N_{ch}	Number of conducting channels
N_{cnt}	Number of CNTs in a bundle
N_i	Intrinsic carrier concentration (1/m^3)
N_n	Electron concentration (1/m^3)
N_p	Hole concentration (1/m^3)
P	Pitch between adjacent TSVs (m)
Q	Electrical charge (C)
q	Heating power (W)
R	Interconnect resistance per unit length (Ω/m)

r	Radius (m)
R_L	Load resistance (Ω)
R_{mc}	Imperfect contact resistance (Ω)
R_{on}	on resistance (Ω)
R_{TSV}	TSV resistance (Ω)
R_{via}	TSV radius (m)
S	Spacing between adjacent interconnects (m)
S_{11}, S_{21}	Scattering parameters and insertion loss
T	Temperature (K or °C)
t	Interconnect thickness (m); time(s)
t_{ox}	Oxide layer thickness (m)
V_{dd}	Power supply voltage (V)
v_F	Fermi velocity (m/s)
V_{FB}	Flatband voltage (V)
v_p	Phase velocity (m/s)
V_{Th}	Threshold voltage (V)
w	Interconnect width (m)
w_{dep}	Depletion layer width (m)
Y	Admittance (S)
Z	Impedance (Ω)
Z_0	Characteristic impedance (Ω)

Greek letters

γ	Propagation constant (1/m)
ε	Permittivity (F/m)
λ	Wavelength (m)
λ_{eff}	Effective mean free path (m)
μ	Permeability (H/m)
ρ	Electric charge density (C/m^3)
σ	Conductivity (S/m)
τ	Time constant (s)
$\tau_{50\%}$	Time delay (s)
ω	Angular frequency (rad/s)

Chapter 3: Thermal modeling, analysis, and design

Abbreviation

BGA	Ball-grid array
LGA	Land-grid array
PCB	Printed circuit board
PGA	Pin-grid array
RDL	Redistribution layer
TEC	Thermoelectric cooler
TIM	Thermal interface material
TSV	Through-silicon via

Nomenclature (units of measure)

A	Area (m^2)
Ar	Aspect ratio
C	Constant coefficient
COP	Coefficient of performance
c_p	Specific heat ($J \cdot kg^{-1} \cdot K^{-1}$)
D	Diameter (m)
g	Gravitational acceleration (m/s^2)
h	Heat transfer coefficient ($W \cdot m^{-2} \cdot K^{-1}$)
H	Height (m)
I	Electrical current (A)
$I_0(\cdot), K_0(\cdot)$	Modified zero-order Bessel functions of the first and the second types, respectively
$k, k_\parallel, k_\perp$	Thermal conductivity ($W \cdot m^{-1} \cdot K^{-1}$)
K_m	TEC module thermal conductance (W/K)
l	Length (m)
n	Constant index
N	Number; number of TEC pellets
Nu	Nusselt number
P	Pitch (m)
Pr	Prandtl number
P_w	Perimeter (m)
q	Heat transfer rate (W)
q''	Heat flux (W/m^2)
q_c, q_h	Heat load at cold and hot sides, respectively (W)
q_{te}	Electrically driven TEC power (W)
R	Electrical resistance (V/A); thermal resistance (K/W)
r	Radius (m)
Ra	Rayleigh number
R_{bi}	Heat sink base to inlet thermal resistance (K/W or °C/W)
Re	Reynolds number
R_{hs}	Heat sink thermal resistance (K/W)
R_{jc}	Junction to case or cold side thermal resistance (K/W)
R_m	TEC module electrical resistance (Ω)
R_{tim}	Thermal resistance at the thermal interface (K/W)
R''_{tim}	Specific thermal resistance at the thermal interface ($K \cdot cm^2/W$)
s	Seebeck coefficient (V/K)
S_m	TEC module Seebeck coefficient (V/K)
T	Temperature (K or °C)
t	Thickness (m)
T_{h0}	TEC hot-side temperature in supplier's datasheet (K or °C)
u	Velocity (m/s)
U	Velocity vector (m/s)
V	Voltage (V); electric field (V)
v	Volume (m^3)
VF	The volumetric flow rate (m^3/s, L/min)

w	Width (m)	
x	x coordinate (m)	
Z	Figure of merit (1/K)	
ZT	$Z \cdot T$, figure of merit	
Δp	Pressure drop (Pa)	

Greek letters

α	Coefficient of thermal expansion (1/K)
β	Volumetric fraction
ε	Emissivity; heat exchanger effectiveness;
η_f	Fin effectiveness
μ	Dynamic viscosity (N·s/m^2)
ρ	Density (kg/m^3)

Chapter 4: Stress and reliability analysis for interconnects

Abbreviation

AF	Acceleration factor
ALT	Accelerated life testing
ATC	Accelerated thermal cycling
CCGA	Ceramic column grid array
CDI	Cumulative damage index
CSP	Chip-scale packages
CTE	Coefficient of thermal expansion
DFR	Design for reliability
DMA	Dynamic mechanical analysis
DOF	Degree of freedom
ENIG	Electroless nickel immersion gold
EPC	Elastic-plastic-creep
FCOB	Flip-chip-on-board
FEA	Finite element analysis
FTTF	First-time-to-failure
IC	Integrated circuits
IMC	Intermetallic compound
MTTF	Mean time to failure
OSP	Organic solderability preservative
PBGA	Plastic ball grid array
PQFP	Plastic quad flat package
RoHS	Reduction of hazardous substances
SEM	Scanning electron microscopy
SHPB	Split Hopkinson pressure bar
SiP	System-in-package
TC	Thermal cycling
TS	Thermal shock
TSV	Through-silicon via
VQFN	Very-thin quad-flat-no-lead package
WEEE	Waste electrical and electronic equipment

Nomenclature (units of measure)

A	Area (m^2), fatigue coefficient in fatigue model
a	Strain rate sensitivity of hardening in Anand model, multiplier of stress in creep model, material constant
A_0	Initial area (m^2)
a_T	Shift factor
b	Fatigue strength exponent, width of beam cross-section
C	Preexponential factor (1/s) constant coefficient, complex constant in creep model
C_1, C_2	Material constants in shift function
D	Diameter (m), displacement (m)
E	Young's modulus (Pa, 1 MPa=10^6 Pa, 1 GPa=10^9 Pa)
F	Force (N)
f	Frequency (Hz); friction factor
g	Acceleration of gravity (9.8 m/s^2)
G	Shear modulus (Pa)
h	Heat transfer coefficient (W·m^{-2}·K^{-1}), thickness or height (m)
H	Height (m)
I	Electrical current (A), moment of inertia (kg·m^2)
K	Strength coefficient, bulk modulus (GPa) (1 GPa=10^9 Pa)
k_B	Boltzmann constant
K_I	Stress intensity factor (MPa·m$^{1/2}$)
l	Support span in bending test (m)
L	Length (m)
m	Mass (kg)
M	Moment (N·m)
N	Number; number of fatigue cycles
n	Strain hardening exponent, actual fatigue cycle in different environment
N_f	Fatigue cycle
p	Pressure (Pa)
Q, Q_{act}	Apparent activation energy (J/mol)
R	Thermal resistance (K/W or °C/W), electrical resistance (Ω)
T	Temperature (K or °C)
TR	Transmissibility
t	Thickness (m); time(s)
U	Voltage (V)
V	Voltage (V)
v	Volume (m^3)
W	Strain energy density (Pa)
x	Displacement; x-coordinate (m)
y	y-coordinate (m), distance from surface to middle plane (m)
z	z-coordinate (m)

Greek letters

α	Coefficient of thermal expansion or CTE (1/K)
γ	Shear strain
ε	Normal strain
$\dot{\varepsilon}_C$	Creep strain rate (1/s)

ε_{ea}	Elastic strain amplitude
ε_p	Plastic strain amplitude
ε_t	Thermal strain
ν	Poisson's ratio
ρ	Density (kg/m^3)
σ	Normal stress (Pa)
σ_a	Stress amplitude for nonzero mean stress case (Pa)
σ_{ar}	Stress amplitude for zero mean stress case (Pa)
σ'_f	Fatigue strength coefficient (Pa)
σ_m	Mean stress (Pa)
τ	Shear stress (Pa)
ω	Natural frequency in radians per second

Chapter 5: Reliability and failure analysis of encapsulated packages

Abbreviation

AFM	Atomic force microscopy
CPI	Chip package interaction
C-SAM	C-mode scanning acoustic microscopy
MEC	Chemical microetch
MSL	Moisture sensitivity level
PBGA	Plastic ball grid array
ppm	Part per million
SIF	Stress intensity factor
TQFP	Thin quad flat pack
UBM	Under bump metallization
VPR	Vapor phase reflow

Nomenclature (units of measure)

A	Area (m^2)
C	Moisture concentration (kg/m^3)
c_p	Specific heat (J·kg^{-1}·K^{-1})
DM	Moisture diffusivity (m^2/s)
D_{pj}	Displacement in j direction (m)
E	Modulus of elasticity (Pa or MPa)
G	Shear modulus (Pa)
G_t	Strain energy release rate (J/m^2)
H	The thickness of the plastic (m)
J	Concentration flux (kg/m^2)
k	Thermal conductivity (W·m^{-1}·K^{-1})
L	Length (m)
m	Mass (kg)
n	Normal direction
p	Pressure (Pa)
P_{b_1}, P_{b_2}	Critical stress loads (Pa)

p_s	Saturated pressure (Pa)
Q_{act}, Q_s	Activation energy (J/mol)
R_u	Universal gas constant (J·mol^{-1}·K^{-1})
s	Solubility (mg·mm^{-3}·MPa^{-1})
T	Temperature (°C)
t	Time (s)
v	Specific volume (m^3/kg)
V	Volume (m^3)
x, y	Cartesian coordinates in x and y directions (m)

Greek letters

α	Coefficient of thermal expansion (1/K)
SC	Swelling coefficient
ε	Shear strain
δ	Length (m)
ν	Poisson's ratio
ρ	Density (kg/m^3)
σ	Normal stress (Pa)
τ	Shear stress (Pa)
ϕ	C/s, the normalized concentration
φ	Mode mixity

Chapter 6: Thermal and mechanical tests for packages and materials

Abbreviation

DMA	Dynamic mechanical analyzer
EMC	Epoxy molding compound
IMC	Intermetallic compound
SAC	Sn-Ag-Cu solders
TTSP	Time—temperature superposition principle
UTS	Ultimate tensile stress

Nomenclature (units of measure)

a_T	Shift factor
C_b, C_1, C_2, C_3, C_4	Material constants
C_{th}	Heat capacitance used in structure functions (W·s/K or J/K)
E	Elastic modulus (Pa)
E'	Storage modulus (Pa)
E''	Loss modulus (Pa)
E_0	Elastic modulus at $t=0$ (Pa)
E_∞	Elastic modulus at $t=\infty$ (Pa)
I	Electrical current (A)
KC	Proportionality coefficient of temperature rise over the voltage change (°C/V)
Q	Activation energy (J/mol)

q	Heat transfer rate (W)
q_c	Cooling power (W)
R_{hs}	Heat sink thermal resistance (K/W)
R_{ja}	Junction-to-ambient thermal resistance (K/W or °C/W)
R_{jb}	Junction-to-board thermal resistance (K/W or °C/W)
R_{jc}	Junction-to-case or cold-side thermal resistance (K/W)
R_m	TEC module electrical resistance (Ω)
R_{th}	TEC module Seebeck coefficient (V/K)
S_m	TEC module Seebeck coefficient (V/K)
T	Temperature (°C)
t	Time (s)
$T_{a,0}$	Initial ambient air temperature before heating power is applied (°C)
$T_{a,t}$	Final ambient air temperature when steady state is reached (°C)
T_g	Glass transition temperature (°C)
T_j	Junction temperature (°C)
ΔV	Voltage change in the sensing diode reading (V)

Greek letters

δ	Phase lag between stress and strain
ε_0	Maximum amplitude of strain
$\dot{\varepsilon}$	Strain rate (s^{-1})
$\dot{\varepsilon}_{cr}$	Creep strain rate (s^{-1})
σ	Normal stress (Pa)
σ_0	Maximum amplitude of stress (Pa)
σ_y	Yield stress (Pa)
ω	Frequency of strain oscillation (Hz)

Chapter 7: System-level modeling, analysis, and design

Abbreviation	
BEOL	Back end of line
BLT	Bond line thickness
CTE	Coefficient of thermal expansion
DA	Die attach
DRIE	Deep reactive ion etching
FEA	Finite element analysis
GPU	Graphics processing unit
HBM	High bandwidth memory
MJMC	Microjet microchannel structure
PCB	Printed circuit board
RDL	Redistribution layer
SEM	Scanning electron microscope
SLID	Solid–liquid interdiffusion
SMC	Silicon microcooler
TC	Temperature cycling
TCB	Thermocompression bonding
TCOB	Board-level temperature cycling

TFI	TSV-free interposer
TSI	Through-silicon interposer
TSV	Through-silicon via
TV	Test vehicle
UBM	Under bump metallization

Nomenclature (units of measure)

C_1, C_2, C_3, C_4	Material constants
c_p	Specific heat (J·kg^{-1}·K^{-1})
D	Diameter (m)
D_x, D_y, D_z	Displacements in x, y, z directions, respectively (m)
$EH_{h1}, EH_{h2},$ EH_{c1}, EH_{c2}	Enthalpies of the inlet and outlet fluids at the hot side and the inlet and outlet fluids at the cold side, respectively (W)
H	Jet to wall distance (m)
l	Length (m)
n	Number
N_{db0}	Nominal decibels (dB)
N_f	Fatigue life (cycle)
q	Heat transfer or heat exchange rate (W)
r	Distance (m)
R_{sa}	Heat sink to ambient thermal resistance (K/W or °C/W)
S1	Maximum principal stress (Pa)
SP, SP_0	Sound power (W)
S_{xz}	Shear stress (Pa)
T	Temperature (°C)
$T_{c1}, T_{c2}, T_{h1}, T_{h2}$	Temperatures of cold inlet, cold outlet, hot inlet, and hot outlet, respectively (°C)
T_s	Heat sink temperatures (°C)
VF	Volumetric flow rate (m^3/s, L/min)
w	Microchannel width (m)
W_{cr}	Creep strain energy density (Pa)
W_{in}	Inelastic strain energy density (Pa)
W_{pl}	Plastic strain energy density (Pa)
z	Z coordinate

Greek letters

$\dot{\varepsilon}$	Strain rate (s^{-1})
$\dot{\varepsilon}_{cr}$	Creep strain rate (s^{-1})
ρ	Density (kg/m^3)
σ	Stress (Pa)

Appendix 2 Conversion factors

Energy	1J(0.2388 cal)	$=9.4782 \times 10^{-4}$ Btu
Force	1N	$=0.22481$ lbf
Heat transfer rate	1W	$=3.4121$ Btu/h
Heat flux	1W/m^2	$=0.3170$ Btu$\cdot$h$^{-1}\cdot$ft^{-2}
Heat transfer coefficient	1W$\cdot$m$^{-2}\cdot$K^{-1}	$=0.17611$ Btu$\cdot$h$^{-1}\cdot$ft$^{-2}\cdot°$F^{-1}
Length	1m	$=39.37$ in
Pressure	1Pa	$=1.4504 \times 10^{-4}$ lbf/in^2
		$=1.4504 \times 10^{-4}$ Psi
		$=4.015 \times 10^{-3}$ in(water)
Temperature	K	$=(5/9)(°\text{F}+459.67)$
		$=°\text{C}+273.15$
Thermal conductivity	1W$\cdot$m$^{-1}\cdot$K^{-1}	$=0.57779$ Btu$\cdot$h$^{-1}\cdot$ft$^{-1}\cdot°$F^{-1}
Thermal resistance	1K/W	$=0.52753$ $°$F$\cdot$h$^{-1}\cdot$Btu^{-1}
Volume	1m^3	$=6.1023 \times 10^{4}$ in^3
		$=35.315$ ft^3
		$=264.17$ gallon(U.S.)
	1ppm	$=10^{-6}$

Index

Note: 'Page numbers followed by "*f*" indicate figures and "*t*" indicate tables.'

A
ABAQUS, 228
   finite element code, 277–278
ABCD transmission matrixes of differential TSVs, 47
Accelerated life testing (ALT), 149–150
Accelerated tests, 227–228
Accelerated thermal cycling tests (ATC tests), 150–151
Acceleration factor (AF), 150
Actual reliability test, 143
Adhesion
   strength enhancement at Cu, 284–287
   test methodology, 270–271
AF. *See* Acceleration factor (AF)
AFM. *See* Atomic force microscopy (AFM)
Air cooling for electronic devices, 101–105
AlphaGo, 1
ALT. *See* Accelerated life testing (ALT)
Anand constitutive model, 157
Annular TSVs, 47–48
ANSYS, 228
   code, 267
   FEA software, 138
   FLUENT solver, 332–333
Arrhenius equation, 248
ATC tests. *See* Accelerated thermal cycling tests (ATC tests)
Atomic force microscopy (AFM), 285–286
AuSn layer, 360
AuSn SLID system, 349
Avogadro constant (NA), 140–141

B
Back end of line (BEOL), 86, 370
Back grinding (BG), 370
Back-to-face integration, 86
Ball-grid array (BGA), 69, 370–371, 385, 385f

BC. *See* Boundary condition (BC)
BCB. *See* Benzocyclobutene (BCB)
Beam theory, 156–157
Bend loading, 169
Benzocyclobutene (BCB), 74–75
BEOL. *See* Back end of line (BEOL)
Best-fit Weibull life distribution model, 150–151
Beyond-Moore technologies, 2
BG. *See* Back grinding (BG)
BGA. *See* Ball-grid array (BGA)
Bilinear elastic-plastic model, 223
Blackbody, 63
Block g-level vibration tests, 193
BLT. *See* Bond line thickness (BLT)
Board-level drop test, 209–210
   impact tests, 227
Board-level solder joint reliability analysis, 385–387
Board-level temperature cycling (TCOB), 370
Boltzmann constant, 140–141
Bond line thickness (BLT), 68, 350
Bounce diagram, 21, 22f
Boundary condition (BC), 250–251, 352
   during vapor phase reflow, 251
Boussinesq's approximation, 95
Brazing process, 336
Bump layout design optimization, 236
Buoyancy effect, 95
Burn-in rack system modeling with multiple trays
   model development for rack-level test system, 343–344
   model results, 345–348
Burn-in test of electronic packaging, 299–306
   image of TEC and X-ray image of internal elements, 303f

Burn-in test of electronic packaging (*Continued*)
  system testing at room temperature and hot temperature, 301f
  TEC details list for system, 303t
  thermal design configuration for system, 302f

C
C4 solder joint thermal cycling reliability, 379–384
Capacitance–voltage characteristic of signal/ground TSVs, 35, 35f
Capacitive couplings, 21
Carbon nanotubes (CNTs), 28, 70–71
  CNT/Cu-CNT TSVs, 36–41
  filling ratio, 38–39
  TSVs, 88
Cartesian coordinate system, 89
CCGA. *See* Ceramic column grid array (CCGA)
CDI approach. *See* Cumulative damage index approach (CDI approach)
Central processing unit (CPU), 325
Ceramic column grid array (CCGA), 194–195
CFD. *See* Computational fluid dynamics (CFD)
Characteristic impedance, 19, 34
  CPW, 20
  of lossless TL, 18
Charpy impact test and analysis, 217
Charpy test, 210, 215–217
Chip bonded with heat spreader, 357–360
Chip cooling, silicon-based microcooler for, 332–335
Chip package interaction (CPI), 245, 246f
Chip-level cooling, 325–327
Chip-on-wafer bonding process, 370–371
Chip-scale packages (CSP), 169
Circuit modeling, 7
Closed-form thermal conductivity model for TSV chips, 77–79
CMOS inverter. *See* Complementary metal–oxide–semiconductor inverter (CMOS inverter)
CNTs. *See* Carbon nanotubes (CNTs)
Coaxial through-silicon vias, 41–43, 41f
Coefficient of performance (COP), 115

Coefficient of thermal expansion (CTE), 137, 263
Coffin-Manson low cycle fatigue model, 144
Coffin-Manson strain-based model, 143
Cold plate clamping, 297
Compact third-order Butterworth filter, 53
Complementary metal–oxide–semiconductor inverter (CMOS inverter), 13–14, 15f
Compliance matrix of material, 136
Computational fluid dynamics (CFD), 89, 332–333, 344
Computational modeling and simulation, 228
COMSOL solver, 332–333
Constant g-level vibration tests, 193
Constant resistor, 212
Continuity equation, 89
Controlled pulse drop at board-level, 209–210
Convection heat transfer, 63
Conventional 3D FEA model, 178–180
Conventional fin-tube heat exchanger, 331
Conventional planar package, 74
Cooler system after assembly, 361
Cooling system modeling and design, 330–342. *See also* Thermoelectric cooling (TECs)
  heat exchanger HEX-A modeling, 335–338
  modeling and design of heat exchanger HEX-B, 338–342
  silicon-based microcooler for chip cooling, 332–335
COP. *See* Coefficient of performance (COP)
Coplanar waveguide (CPW), 20
  effective permittivity and characteristic impedance of, 20
Copper (Cu), 152–153
CPI. *See* Chip package interaction (CPI)
CPU. *See* Central processing unit (CPU)
CPW. *See* Coplanar waveguide (CPW)
Creep, 131
  activation energy, 317–318
  constitutive models, 317–318
  model, 140–141
  strain energy-based fatigue life models, 155

strain fatigue model, 145–146
test and analysis, 312–318
Creep-fatigue properties
creep model, 140–141
EPC model, 141–142
solder fatigue life models, 142–149
Critical film thicknesses, 62, 62t
Critical loads, 273, 274t
Crosstalk effect, 43–44
CSP. See Chip-scale packages (CSP)
CTE. See Coefficient of thermal expansion (CTE)
Cu-Ni-Sn ternary IMC, 152–153
Cumulative damage index approach (CDI approach), 144
fatigue analysis, 198–200
Cumulative damage theory. See Palmgren-Miner rule
Cyclic bend test, 169
fatigue model development for, 189–191
finite element analysis for, 176–189
of specimen, 171–172
three-point bend tests, 169
Cyclic stress–strain, 151

D

DA material. See Die attach material (DA material)
Daisy chain, 212
resistance, 194–195
Darpa-funded project, 100
Darveaux's model, 148
Decoupling capacitors, 26
Deep reactive ion etching technique (DRIE technique), 335
Degree of freedom (DOF), 178–180
Deionized water, 333
Delamination growth, criterion for, 275–277
Depletion capacitance, 31–32
Design-for-reliability (DFR), 10, 225–226
case study, 227–239
die strength characterization, 231–232
finite element analysis and optimization, 232–236
problem statement, 229–231
experimental validation, 236–238
using finite element method, 227–228
methodology for drop test, 225–226

multidisciplinary approach in, 11f
optimization of package design, 228–229
Device scaling, 3
DFR. See Design-for-reliability (DFR)
Diamond-like coating (DLC), 101
Die attach material (DA material), 349, 352
Die stacking process, 230–231, 232f
Die strength characterization, 231–232
Dielectric layer, 28
Dielectric loss, 23
Differential signaling, 44–48
DIP. See Dual in-line package (DIP)
Displacement-controlled isothermal mechanical fatigue test, 143, 155
Distance to neutral point effect (DNP effect), 385
DLC. See Diamond-like coating (DLC)
DMA. See Dynamic mechanical analysis/analyzer (DMA)
DNP effect. See Distance to neutral point effect (DNP effect)
DOF. See Degree of freedom (DOF)
Dorn's equation, 140–141
Double power law model, 140–141, 317
Double-ring cold plate, 297
DRIE technique. See Deep reactive ion etching technique (DRIE technique)
Drop impact test and analysis, 150, 209–227
Charpy impact test and analysis, 217
design-for-reliability methodology for drop test, 225–226
drop reliability test and analysis, 212–217
finite element analysis for drop impact test, 217–225
test vehicle and set up, 211–212
Drop life model, 224
Dual in-line package (DIP), 3–4
Dye penetration method, 10
Dynamic hardening, 210
Dynamic mechanical analysis/analyzer (DMA), 139–140, 297, 318
Dynamic resistance, 212

E

Eddy current loss, 23
Effective complex conductivity, 38–39

Effective conductivity of Cu-CNT, 39
Effective density and heat capacity of interposer, 80
Effective inductance, 53
Effective permittivity
 CPW, 20
 of microstrip line, 18–19
 of slotline, 19–20
Effective solder joint model, 156–157
Effective thermal conductivity, 76–77
Elastic model of solder, 220–222
Elastic modulus (E), 131–132, 166–167, 210, 308–312
Elastic-plastic interfacial fracture, 271
Elastic-plastic model, 168
Elastic-plastic-creep model (EPC model), 141–142
 constitutive model, 366
Elastomeric TIM, 69
Electrical analysis for advanced packaging, 13
Electrical modeling and design
 fundamental theory, 13–28
  electrical analysis for advanced packaging, 13
  EMC, 27–28, 27f
  interference, 27–28
  power allocation, 23–27
  signal distribution, 13–23
 modeling, characterization, and design of through-silicon, 28–53
Electrical performance and design methodology, 6–7
Electroless nickel immersion gold (ENIG), 152–153
Electromagnetic compatibility (EMC), 27–28, 27f
Electromagnetic interference (EMI), 13
Electronic equipment, 192
Electronic packaging, 1, 9, 13
 burn-in test of, 299–306
Electronic products, 9
Electronics systems, heat generation in, 73–74
EMC. See Electromagnetic compatibility (EMC); Epoxy molding compound (EMC)
EMI. See Electromagnetic interference (EMI)

Encapsulated packages
 fracture mechanics analysis in IC package, 270–282
 heat transfer and moisture diffusion in plastics IC packages, 245–262
 reliability enhancement in PBGA package, 282–289
 thermal-and moisture-induced stress analysis, 263–270
 typical IC packaging failure modes, 245
End-notched flexural (ENF), 272
Endothermic reaction, 73
Energy equation, 90
Energy-based fatigue models, 147–149, 192
Energy-based life models, 155
ENF. See End-notched flexural (ENF)
Engineering stress–strain curve, 133
ENIG. See Electroless nickel immersion gold (ENIG)
EPC model. See Elastic-plastic-creep model (EPC model)
Epoxy molding compound (EMC), 320, 321f–322f, 371
Equivalent circuit model of TSVs, 7, 29, 29f
Equivalent heat transfer coefficient, 108
Equivalent series inductance (ESL), 26
Equivalent series resistance (ESR), 26
ESL. See Equivalent series inductance (ESL)
ESR. See Equivalent series resistance (ESR)
Exponential model, 317

F

FA. See Failure analysis (FA)
Fabrication process, 335
Face-to-face integration, 86
Failure analysis (FA), 10
 thermal cycling test and, 152–153
Failure mechanisms for electronic solders, 149–150
Fan-out wafer-level packaging (FOWLP), 3
Fatigue, 131
Fatigue cycle, 206
Fatigue failure, 194–195
 criterion, 143
Fatigue model development for cyclic bend test, 189–191
FCOB. See Flip-chip-on-board (FCOB)
FE. See Finite element (FE)

FEA. *See* Finite element analysis (FEA)
FEA modeling and simulation, 154
FEA models, 369
Fick's second law of diffusion, 248
Field-programmable gate array (FPGA), 4
Fin efficiency, 72
Fin equation, 72
Finite element (FE), 247
  simulation, 143
Finite element analysis (FEA), 10, 138, 252, 253f, 264, 267–270, 349, 370
  for cyclic bend test, 176–189
    submodeling method for bend test, 178–181
  design-for-reliability using, 227–228
  for drop impact test, 217–225
    elastic model of solder, 220–222
    finite element model, 219–220
    life prediction for drop test, 224–225
    material property of solder, 217–219, 221t
    plastic model of solder, 223–224
    simulation results, 220–225
  of interfacial fracture, 277–278
  model, 155–157
  parametric study using FEA simulation, 162–166
  properties of packaging materials, 253t
  modeling
    for four-point cyclic bend test, 184–189
    and simulation, 352–353
    for three-point cyclic bend test, 181–184
  and optimization, 232–236
    analysis for stacking middle die, 233–235
    analysis for stacking top die, 235–236
    of bump layout design, 236
    finite element model, 232–233
  simulation, 10–11
    results and discussion, 157–162
  for vibration test, 200–208
    modal analysis of flip-chip-on-board assembly, 200–202
    quasi-static analysis for vibration fatigue, 202–207
Finned surfaces, 71–72
First-time-to-failure (FTTF), 197–198
Flip-chip BGA, 74
Flip-chip-on-board (FCOB), 148

assemblies, 193
  modal analysis, 200–202
  vibration test fatigue data, 198t
  Weibull plots of test results for, 197f
Floating substrate effect, 34–36
FloTHERM software, 336
Flow rule, 142
Flow stress, 148
Forced convection, 62–63
Four-point bend test, 169
  correlation between three-point bend and, 172–173
Four-point cyclic bend test
  at 25 and 125°C, 176
  for correlation, 174–175
  FE analysis modeling
    correlation between three-point bend and four-point bend test, 184–185
    and result discussion, 185–189
FOWLP. *See* Fan-out wafer-level packaging (FOWLP)
FPGA. *See* Field-programmable gate array (FPGA)
Fracture mechanics
  analysis in IC package, 270–282
    interfacial fracture analysis, 273–278
    interfacial fracture toughness, 281–282
    measurement, 272–273
    results, 278–281
  approach, 149
Free convection, 62–63
Free fall board-level, 209–210
Free fall product level, 209–210
Frequency-modified Coffin-Manson model, 145
Frequency-scanning test, 195–196
FTTF. *See* First-time-to-failure (FTTF)

G
Garofalo hyperbolic sine model, 317
Garofalo steady-state creep behavior, 140–141
Generalized Maxwell model, 139–140, 139f
Glass transition temperature, 320
Goodman criterion, 143–144
Governing equations, 89–92
GPUs. *See* Graphic processing units (GPUs)
Graphene, 70–71

Graphic processing units (GPUs), 2, 370–371
Graphite sheet, 69
"Green" IC packaging technologies, 282
Grid cells, 90

H

Hardening rule, 142
Harmonic motion, 192
Harmonic response analysis, 207–208
HBM chips. *See* High Bandwidth Memory chips (HBM chips)
Heat
　conduction, 59–60
　　micro-and nanoscale effect on, 60–62
　　through radial systems, 67
　convection, 62–63
　exchanger HEX-A modeling, 335–338
　exchanger HEX-B modeling, 338–342
　flux, 59–60
　generation in electronics systems, 73–74
Heat transfer
　correlations, 63, 64t–65t
　from extended fin surfaces, 71–73
　in plastics IC packages, 245–262
　　boundary condition during vapor phase reflow, 251
　　constitutive relations for water vapor, 249–250
　　FE analysis, 252
　　governing equations for, 248
　　initial and boundary conditions, 250–251
　　popcorn cracking of SOJ packages, 247f
　　solution procedure, 251–252
High aspect ratio CNT TSV, 75
High Bandwidth Memory chips (HBM chips), 370–371
High heat flux cooling technologies, 327–328
High-frequency impedance, 26
Hooke's law, 132–135
Hybrid MJMC silicon microcooler, 332, 333f
Hydraulic leak test, 367–369
Hygrostrain, 264
Hygrostress development, 263–264
Hygrothermal stress analysis, case study for, 264–270
　design and material properties, 265
　materials used for 100TQFP test vehicle, 266t
　finite element analysis, 267–270
　observed failure modes of TQFP package, 264
Hyperbolic sine model, 317–318

I

IC. *See* Integrated circuit (IC)
ICECOOL project, 100
IMC. *See* Intermetallic compound (IMC)
Impedance, 6
　and admittance of simplified TL model, 34
　of capacitor, 26
　matching, 21, 21f
　of power distribution network, 24, 25f
In situ measurement, 10
In-line TSV array, 79
In-plane thermal conductivity, 84
Indium (In), 349
Indium-based solder, 349
Inductance, 25, 25f
Inductive couplings, 21
Infrared (IR)
　camera, 341–342, 362–364
　voltage drop, 24
Initial stress-free condition, 165–166
Insulation layer, 79
Integrated circuit (IC), 1
　challenges and solutions, 4–5
　chips, 200–202
　devices and applications, 1–2
　　applications to driving electronic products, 3f
　　semiconductor technology road map, 2f
　package, 245–247, 246f
　　CPI, 246f
　　failure modes, 245
　　fracture mechanics analysis in, 270–282
　packaging evolution, 1–5
　performance and design methodology, 6–11
　semiconductor packaging trend, 3f
Integrated passive devices (IPDs), 50–53
Interconnects, 13–14, 14f
　charging circuit, 15f
　discontinuity, 20–21, 20f

package, 19f
stress and reliability analysis
  design-for-reliability, 227–239
  mechanical properties of materials, 131–149
  reliability test and analysis methods, 149–227
Interface failure, 215–217
Interface toughness, 276
Interfacial fracture
  analysis, 273–278
    criterion for delamination growth, 275–277
    finite element analysis, 277–278
  mechanism, 271
  toughness measurement, 272–273
Interference, 27–28
Intermetallic compound (IMC), 152–153, 210, 307, 349
  brittle failure, 215–217
  layer effect on life prediction, 162–166
Internal impedance, 32
Interstrata bond layer, 87–88
Inverter circuit and switching models, 13–14, 14f
IPDs. *See* Integrated passive devices (IPDs)
Isothermal cyclic bend test and analysis
  correlation between three-point bend and four-point bend test, 172–173
  four-point cyclic bend test for correlation, 174–176
  test vehicle and experimental procedure, 170–172
  three-point cyclic bend test, 173–174

J
J-Std-20C standard, 283
JEDEC. *See* Joint Electron Device Engineering Council (JEDEC)
JESD51–14 standard, 293
Jet-scrubbing technologies, 285
Joint Electron Device Engineering Council (JEDEC), 293
Joule heating, 73
Junction temperature, 7
Junction-to-ambient thermal resistance (R_{ja}), 294–297
Junction-to-case thermal resistance (R_{jc}), 294–296, 296f

K
Kinematic hardening, 142
Kirchhoff's voltage law, 15–17

L
Laminated substrate, 83–84
Land-grid array (LGA), 69
Lansmont Model 65/81 drop impact tester, 211, 211f
Law creep behavior breakdown, 314–316
Lead-free solder, 210
LGA. *See* Land-grid array (LGA)
$LiCoO_2$-based 18650 battery, 73–74
Life prediction, 155–168
  for drop test, 224–225
  IMC layer effect, 162–166
  model for solder joints, 143
Liquid cooling
  for electronic devices, 106–112
    analytical model for finned heat sink, 106–109
    analytical results and comparison with experimental measurements, 109–111
    anatomy of thermal resistance elements, 112
  of 3D package, 99–100
Liquid-based TIM, 69
Lithium-ion batteries, 73
Load capacitance, 14
Logic errors, 7
Loop inductance, 26
  of pair of TSVs, 32
Lossless TL, 18

M
Magnetic loss tangent, 23
Material mechanical test and characterization, 306–321
  material test and characterization for polymers, 318–321
  material test and characterization for solder alloys, 306–318
    creep test and analysis, 312–318
    tensile test and analysis, 307–312
Material property model, 154–155
Maxwell model, 139
Maxwell–Eucken's equation, 76, 80
Maxwell's equations, 15–16
  curl equation, 23

MC method. *See* Monte Carlo method (MC method)
Mean free path (MFP), 36–38, 62, 62t
Mean time to failure (MTTF), 143, 197–198
MEC technologies. *See* Microetch technologies (MEC technologies)
Mechanical bending test and analysis, 169–192
　fatigue model development for cyclic bend test, 189–191
　finite element analysis for cyclic bend test, 176–189
　isothermal cyclic bend test and analysis, 170–176
Mechanical failures, 69
Mechanical properties of materials, 131–149
　creep-fatigue properties, 140–149
　stress–strain relationship, 131–137
　viscoelastic properties, 139–140
　viscoplastic Anand model of solder, 137–138
Mentor FloTHERM software, 90, 297–298
Metal–oxide–semiconductor (MOS), 13–14
　effect, 28–32
Metal–oxide–semiconductor field-effect transistors (MOSFETs), 13–14
MFP. *See* Mean free path (MFP)
Micro-scale effect on heat conduction, 60–62
Microbumps/underfill layers, 95
Microchannels, 106, 109–110
Microcooler system, 348–369
　chip bonded with heat spreader, 357–360
　cooler system after assembly, 361
　FEA modeling and simulation, 352–353
　reliability test and hydraulic leak test, 367–369
　shear test and simulation of model, 353–357
　silicon microcooler system, 350–352, 369
　thermal cycling simulation, 366–367
　thermomechanical coupling simulation, 361–366
Microcooler TV, 341–342
Microetch technologies (MEC technologies), 283, 285–286
Microjet array impinging cooling method, 328
Micropatterns, 336
Microstrip line, 18–19
Miner's cumulative fatigue damage ratio, 198
Miner's linear superposition principal, 147
MLG. *See* Multilayer graphene (MLG)
Modal analysis of flip-chip-on-board assembly, 200–202
Module parameters, 116–117
Moisture diffusion in plastics IC packages, 245–262
　boundary condition during vapor phase reflow, 251
　constitutive relations for water vapor, 249–250
　FE analysis, 252
　governing equations for, 248
　initial and boundary conditions, 250–251
　moisture distributions during preconditioning, 252–255
　popcorn cracking of SOJ packages, 247f
　solution procedure, 251–252
　temperature and moisture distribution during VPR, 256–257
　vapor pressure buildup in delaminated gap during VPR, 258–262
Moisture preconditioning, 247
Moisture sensitivity levels (MSL), 282
　reflow soldering temperature impact on nongreen PBGA, 283–284
Moisture-induced stress analysis, 263–270
Momentum equation, 90
Monte Carlo method (MC method), 166
Moore's law, 1, 228–229
More than Moore technologies (MtM technologies), 2
Morrow energy-based fatigue model, 147–148
Morrow's energy-based model, 143
MOS. *See* Metal–oxide–semiconductor (MOS)
MOSFETs. *See* Metal–oxide–semiconductor field-effect transistors (MOSFETs)
MSL. *See* Moisture sensitivity levels (MSL)

MtM technologies. *See* More than Moore technologies (MtM technologies)
MTTF. *See* Mean time to failure (MTTF)
Multilayer graphene (MLG), 70–71
Multiple chips, 6–7
Multiwalled carbon nanotube (MWCNT), 36–38, 75
Mutual capacitance, 21
Mutual inductance, 21
   between adjacent lines, 25
   between power lines, 26
MWCNT. *See* Multiwalled carbon nanotube (MWCNT)

N
n-channel metal–oxide–semiconductor (NMOS), 13–14
NA. *See* Avogadro constant (NA)
Nanoscale effect on heat conduction, 60–62
NASTRAN, 228
Natural convection, 62–63
Necking phenomena, 132–133
Newton's law of cooling, 63
Nickel (Ni), 307
NMOS. *See* n-channel metal–oxide–semiconductor (NMOS)
Non-TSV Interposer, 370
Nongreen PBGA packages, 283
   reflow soldering temperature impact on MSL of, 283–284
Nonlinear strain hardening, 132–133, 142
Nonuniformity factors, 163–164
Normal stress, 134–135
Norton's power law model, 316
   constitutive model, 140–141
Numerical modeling, 89–96
   governing equations, 89–92
   package thermal performance, 95–96
   and simulation, 229
   of 2.5D package, 92–95

O
Octant model, 168
Ohmic heating, 73
Ohm's law, 24
One-dimensional thermal resistance network, 114

One-level global-local modeling approach, 181–182
One-level submodeling method for bend test, 178–181, 180f
Organic solderability preservative (OSP), 170
Organic substrate, 384
OSP. *See* Organic solderability preservative (OSP)
Oxide capacitance, 36
Oxide charge, 32

P
p-channel metal–oxide–semiconductor (PMOS), 13–14
Package assembly process, 283
Package thermal
   performance, 95–96
   tests, 293
      burn-in test of electronic packaging, 299–306
      steady-state thermal test, 293–298
      thermal performance, 294f
      transient thermal test, 298–299
Package warpage, 165–166. *See also* Wafer warpage
   analysis and optimization, 378
   assembly-induced, 370
   diamond heat spreader thickness effect, 357–360, 360f
Package-level cooling, 327–329
Package-level FEA model, 178–180
Package-level TC reliability, 379–380
Package-level thermal analysis and design, 74–89
   analytical solutions for analysis of through-silicon vias, 74–83
      porous media model, 76–77
      submodels of through-silicon vias, 77–83
   submodels for substrate and redistribution layer, 83–85
   thermal model for 3D integrated circuit package, 86–89
Package-level thermal enhancement, 97–101
   liquid cooling of 3D package, 99–100
   with new materials, 100–101

Package-level thermal enhancement (*Continued*)
  thermal performance under natural convection, 97–98
Package-on-package (POP), 86
Packaged microelectronic system reliability, 149–150
Palmgren-Miner rule, 144
Parametric study using FEA simulation, 162–166
Paris law, 149
Partial element equivalent circuit method (PEEC method), 42
PBGA. *See* Plastic ball grid array (PBGA)
PCBs. *See* Printed circuit boards (PCBs)
PCT. *See* Pressure cooker test (PCT)
PDN. *See* Power distribution network (PDN)
PECVD process. *See* Plasma-enhanced chemical vapor deposition process (PECVD process)
PEEC method. *See* Partial element equivalent circuit method (PEEC method)
Performance metrics, 293
Permittivity, 23
Pin-grid array (PGA), 69
Plane strain, 136
Plane stress, 136
Plasma-enhanced chemical vapor deposition process (PECVD process), 335
Plastic ball grid array (PBGA), 148, 151–152, 152f
  assemblies, 193
    vibration test fatigue data, 198t
    Weibull plots of test results for, 197f
  packages, 282
    characterization and improvement of moisture sensitivity, 283–288
    reliability assessment, 288–289
    reliability enhancement, 282–289
Plastic IC packages, 263
  heat transfer and moisture diffusion in, 245–262
Plastic model of solder, 223–224
Plastic quad flat package (PQFP), 211
Plastic strain fatigue model, 144–145
Plastic-encapsulated IC package, 263
Plasticity theory, 142
Plated through hole (PTH), 83–84

PMOS. *See* p-channel metal–oxide–semiconductor (PMOS)
Poisson effect, 133–134
Polymer
  material test and characterization for, 318–321
  materials, 131
POP. *See* Package-on-package (POP)
"Popcorn" cracking phenomenon, 245–247, 247f
Porous media model, 76–77
Portable electronics devices, 209–210
Power
  allocation, 23–27
  integrity, 6, 6f
  law breakdown creep behavior, 140–141
  noise margin, 24
  supply noise, 24
  supply system, 23–24, 23f
Power distribution network (PDN), 6, 13
  impedance, 7, 48–50
PQFP. *See* Plastic quad flat package (PQFP)
Preconditioning, moisture distributions during, 252–255
Pressure cooker test (PCT), 286–287
Pressure drop, 109
Printed circuit boards (PCBs), 325, 326f
Prony series, 321
Propagation constant, 18, 34
  of lossless TL, 18
Propagation velocity, 18
  of microstrip line, 18–19
PTH. *See* Plated through hole (PTH)

Q

Quality factor, 53
Quarter global FEA model, 156–157
Quarter model, 155
Quasi-static analysis, 193
  for vibration fatigue, 202–207
    determination of quasi-static load, 202–203
    determination of stress amplitude, 203–205
    harmonic response analysis, 207–208
    vibration fatigue life prediction, 205–207

Index

R

Rack-level test system, model development for, 343–344
Radial systems, heat conduction through, 67
Rate-dependent plastic model, 223
Rate-independent plasticity theory, 142
Receiver circuits, 13–14
Redistribution layer (RDL), 76, 370–371
Reduction of Hazardous Substances (RoHS), 151–152
Reference temperature, 165–166
Reflection coefficient, 21
Reflow soldering temperature impact on MSL of nongreen PBGA, 283–284
Reliability
   enhancement in PBGA package, 282–289
   issues, 9–11
   of semiconductor devices and products, 9–10
Reliability tests, 10, 227–228, 367–369
   and analysis methods, 149–227
      mechanical bending test and analysis, 169–192
      thermal cycling test and analysis, 150–168
   and analysis methods
      drop impact test and analysis, 209–227
      vibration test and analysis, 192–209
RLCG parameters, 32–34
RoHS. *See* Reduction of Hazardous Substances (RoHS)

S

S-parameter
   matrixes of differential TSVs, 47
   of TSV pair, 34
SAC solder. *See* Sn-Ag-Cu solder (SAC solder)
SAM. *See* Scanning acoustic microscope (SAM)
Saturated vapor pressure, 261–262
Scanning acoustic microscope (SAM), 10, 264, 265f
Scanning electron microscope (SEM), 149, 152–153
Schubert's model, 164–165
SDD. *See* Symbolically defined device (SDD)
SEM. *See* Scanning electron microscope (SEM)
Semiconductor packages, 3
Shear stress, 134–135, 134f
Shear test
   simulation, 369
   and simulation of model, 353–357
SHPB tester. *See* Split Hopkinson pressure bar tester (SHPB tester)
SIF. *See* Stress intensity factor (SIF)
Signal
   distribution, 13–23
   integrity, 6, 6f
   voltage, 18
Silicon capacitance and conductance, 33–34
Silicon capacitance between signal/ground TSVs, 35, 36f
Silicon microchannel cooler (SMC), 349
Silicon microcooler system, 350–352, 369
Silicon-based microcooler for chip cooling, 332–335
Simultaneous switching noise (SSN), 13
Single-walled CNT (SWCNT), 36–38
Sinusoidal stress, 139
SiP technology. *See* System-in-package technology (SiP technology)
Skin effect, 23, 28
SLID system. *See* Solid-liquid interdiffusion system (SLID system)
Slotline, effective permittivity of, 19–20
SLT. *See* System-level testing (SLT)
Small-outline J-leaded packages (SOJ packages), 247, 247f
SMC. *See* Silicon microchannel cooler (SMC)
Sn-Ag-Cu solder (SAC solder), 307
   lead-free solder, 140–141
SOJ packages. *See* Small-outline J-leaded packages (SOJ packages)
Solder, viscoplastic Anand model of, 137–138
Solder alloys, material test and characterization for, 306–318
   creep test and analysis, 312–318
   tensile test and analysis, 307–312
Solder constitutive models, 168
Solder creep behavior, 154
Solder fatigue life models, 142–149
   strain-based fatigue models, 144–147

Solder fatigue life models (*Continued*)
  stress-based fatigue models, 143–144
Solder joint, 10
  failures, 210, 217–218
    criterion, 152–153
Solder life prediction model, 148
Solder mask materials, mechanical
    properties of, 284–287
Solid-liquid interdiffusion system (SLID
    system), 349
SOLIDWORKS software, 344, 345f
Solomon's low cycle fatigue model, 145
Specific contact resistance, 79
Specific thermal resistance, 67–68
Split Hopkinson pressure bar tester (SHPB
    tester), 217–218
Spreading thermal resistance, 108–109
SS. *See* Stainless steel (SS)
SSN. *See* Simultaneous switching noise
    (SSN)
Stainless steel (SS), 336
Steady-state creep, 140–141
Steady-state thermal test, 293–298
  four-wire test of package and board
    electrical resistance, 295f
  thermal test of 2. 5D package, 295f
Stiffness, 132
Strain energy density, 190
Strain hardening phenomenon, 132–133
Strain-based fatigue models, 144–147
  creep strain fatigue model, 145–146
  energy-based fatigue models,
    147–149
  fracture mechanics approach, 149
  plastic strain fatigue model, 144–145
  total strain fatigue model, 146–147
Strain-hardening law, 141–142
Stress
  amplitude, 203–205
  analysis, 10
  factors, 9–10
  issues, 9–11
  stress-based drop life model, 224
  stress-based fatigue models, 143–144
Stress intensity factor (SIF), 270–271
Stress–strain
  behavior of solder joints, 202
  relationship, 131–137
    3D stress components, 135f

comparison between engineering and
    true stress–strain curves, 134f
  curve for ductile material, 133f
  typical shear test illustration, 134f
Structure-material-assembly-reliability-
    thermal codesign methodology
    (SMART codesign methodology),
    370, 372f
Submodeling method for bend test,
    178–181
Substrate copper layer, 94
Substrate loss, 28
Substrate vias, 91
Substrate-integrated waveguide, 51
Supply voltage, 24, 24f
Surface mounted components, 151–152
SWCNT. *See* Single-walled CNT (SWCNT)
Sweep vibration reliability test, 196–198
Syed's model, 164–165
Symbolically defined device (SDD), 36, 36f
System-in-package technology (SiP
    technology), 228–229
System-level testing (SLT), 69
System-level thermal modeling, 325. *See
    also* Thermal modeling, analysis,
    and design
  burn-in rack system modeling with multiple
    trays, 343–348
  chip-level cooling, 325–327
  codesign modeling and analysis for
    advanced packages, 370–388
  board-level solder joint reliability
    analysis, 385–387
  C4 solder joint thermal cycling
    reliability, 379–384
  methodology, 372–373
  package warpage analysis and
    optimization, 378
  test vehicle and experimental procedure,
    370–371
  thermal performance evaluation,
    387–388
  wafer warpage analysis and optimization,
    377–378
  wafer warpage simulation and validation,
    373–377, 376f
  and design of electronic system, 326f
  mechanical modeling and design for
    microcooler system, 348–369

modeling and design of cooling system with cooler and heat exchanger, 330–342
package-level cooling, 327–329
at tray, rack, and room levels, 329–330
System-on-single chip, 86

T

TC. *See* Thermal cycling (TC)
TC test. *See* Thermal cycling test (TC test)
TCB process. *See* Thermocompression bonding process (TCB process)
TCOB. *See* Board-level temperature cycling (TCOB)
TDI test. *See* Transient dual interface test (TDI test)
TDP. *See* Thermal design power (TDP)
TECs. *See* Thermoelectric cooling (TECs)
Telegrapher equations, 17
Temperature dependence of thermal conductivity of selected materials, 60, 62f
Temperature-sensitive sensor, 293
Tensile test and analysis, 307–312
Test vehicle (TV), 341–342
TFI. *See* TSV-Free Interposer (TFI)
Thermal adhesive, 69
Thermal conductivity, 60, 61f
Thermal cycling (TC), 10
  fatigue life analysis
    FEA modeling and life prediction, 155–168
    material property model, 154–155
  loading, 150
  loads, 10
  reliability test model, 352
  simulation, 366–367
Thermal cycling test (TC test), 143
  and analysis, 150–168
    thermal cycling fatigue life analysis, 154–168
    thermal cycling test and failure analysis, 152–153
    thermal environments for electronic products, 151t
Thermal design power (TDP), 7, 8f
Thermal gel material, 68–69
Thermal grease, 68–69

Thermal interface materials (TIMs), 59, 68, 297
Thermal interfacial resistance, 67–71, 68f
Thermal mechanical stress—induced package failures, 245
Thermal metrics of die stack, 87–88
Thermal model for 3D integrated circuit package, 86–89
Thermal modeling, analysis, and design. *See also* System-level thermal modeling
  air cooling for electronic devices with vapor chamber configurations, 101–105
  liquid cooling for electronic devices, 106–112
  numerical modeling, 89–96
  package-level thermal analysis and design, 74–89
  package-level thermal enhancement, 97–101
  principles, 59–74
    heat conduction, 59–60
    heat conduction through radial systems, 67
    heat convection, 62–63
    heat generation in electronics systems, 73–74
    heat transfer from extended fin surfaces, 71–73
    micro-and nanoscale effect on heat conduction, 60–62
    thermal interfacial resistance, 67–71
    thermal radiation, 63
    thermal resistance, 63–67
  TECs, 112–126
Thermal performance
  and design methodology, 7–9
  evaluation, 387–388
  under natural convection, 97–98
Thermal radiation, 63
  effect, 95
Thermal resistance, 7, 63–68, 118
  elements, 112
  network, 107
Thermal runaway, 7, 8f
Thermal shock (TS), 10, 149–150
Thermal strain, 137
Thermal stress, 137
Thermal-induced stress analysis, 263–270

Thermal-induced stress analysis (*Continued*)
  case study for hygrothermal stress analysis, 264–270
  development of thermal stress, 263
  hygrostress development, 263–264
Thermal-related package tests, 293
Thermocompression bonding process (TCB process), 349–352
Thermoelectric cooling (TECs), 91, 112–126, 117t, 119f. *See also* Cooling system modeling and design
  analytical solution
    at module level, 115–118
    at pellet level, 113–115
  optimization, 118–123
    comparison with previous studies, 119–121
    thermal resistances, 124–126
    through T_j minimization, 121–123
Thermomechanical coupling simulation, 352–353, 361–366
3-die stacked package, 230, 230f
3-die stacking process, 236, 237f
Three-dimension (3D)
  half model, 382
  IC packaging, 4
  integration, 228–229
  microelectronics system, 86
  package, 97–98
    under direct cooling condition, 99f
    liquid cooling, 99–100
    quarter model, 168
    simulation model, 387–388
    thermal model for 3D integrated circuit package, 86–89
Three-point bend test, 169, 231, 231f, 272, 272f
  correlation between four-point bend test and, 172–173
  loading curve, 232f
Three-point cyclic bend test, 173–174
  correlation between four-point bend test and, 184–185
  FE model modeling for, 181–184
Through-plane thermal conductivity, 85, 95
Through-silicon interposer (TSI), 370
Through-silicon vias (TSVs), 28, 30f, 74
  analytical solutions for analysis, 74–83
  circuit modeling, 29–34

  metal–oxide–semiconductor effect, 29–32
  RLCG parameters, 32–34
equivalent circuit model, 7
etch, 91
fabricated copper, 78f
floating substrate effect, 34–36
minimum TSV capacitance, 36
modeling, 28–36
MOS capacitance *vs.* applied voltage, 31f
optimization, 36–43
  CNT/Cu-CNT TSVs, 36–41
  coaxial through-silicon vias, 41–43, 41f
pair of, 29f
signal/power integrity, 43–50
structure, 28, 78f
submodels, 77–83
technology, 3, 6–7, 370
TSV-based solenoid inductors, 51, 51f–52f
TSV/IPD interposer, 50–53
TIM1. *See* Type 1 thermal interface material (TIM1)
Time constant, 14
Time delay, 15
Time-independent plastic strain, 137–138
Time-temperature superposition principle (TTSP), 320
TIMs. *See* Thermal interface materials (TIMs)
TLs. *See* Transmission lines (TLs)
Total inelastic strain, 141–142
Total strain fatigue model, 146–147
Transient dual interface test (TDI test), 293
Transient thermal test, 298–299
Transient voltage waveform, 21, 22f
Transmission coefficient, 21
Transmission lines (TLs), 13
  model, 17f
  theory, 16–17
Trapezoidal microvia, 84
Tray-level model, 325, 344
TS. *See* Thermal shock (TS)
TSI. *See* Through-silicon interposer (TSI); TSV interposer (TSI)
TSV interposer (TSI), 4
TSV-Free Interposer (TFI), 370
TSVs. *See* Through-silicon vias (TSVs)
TTSP. *See* Time-temperature superposition principle (TTSP)

TV. *See* Test vehicle (TV)
2.5D
  integration, 228–229
  numerical modeling of package, 92–95
  packaging, 4
Two-dimension (2D)
  axisymmetric FEA model, 232–233
  FE simulations, 247
Two-level submodeling method for bend test, 178–180, 179f
Type 1 thermal interface material (TIM1), 68

U

UBM. *See* Under bump metallization (UBM)
Ultimate tensile stress (UTS), 308
Under bump metallization (UBM), 245, 370–371

V

Vapor chamber heat sink (VCHS), 101–102
Vapor phase reflow (VPR), 247, 251f
  boundary condition during, 251
  temperature and moisture distribution during, 256–257
  vapor pressure buildup in delaminated gap during, 258–262
VCHS. *See* Vapor chamber heat sink (VCHS)
Vertically integrated 3D circuits, 86
Very-thin quad-flat-no-lead package (VQFN package), 169–170, 173–174
Vibration
  fatigue
    failure of solder joints, 193
    life prediction, 205–207
  loading, 10, 150
  test and analysis, 192–209
    cumulative damage index fatigue analysis, 198–200
    finite element analysis for vibration test, 200–208
    frequency-scanning test, 195–196
    sweep vibration reliability test, 196–198
    test vehicle and setup, 193–195

Virtuoso, 94
Viscoelastic properties, 139–140
Viscoelasticity, 139
Viscoplastic Anand model of solder, 137–138, 180
Voltage
  drop, 24
  of load capacitance, 14
  voltage-controlled TSV capacitance, 36
Voltage regulating modules (VRMs), 325
Volume
  fraction, 80
  ratio of copper and silicon, 77
  volume-averaged method, 366–367
  volume-averaged strain energy density, 192
  volume-averaging technique, 162–164
  volume-weight method, 203
  volumetric heat removal techniques, 86
Von Mises stresses, 203
Von Mises yield criterion, 142
VPR. *See* Vapor phase reflow (VPR)
VQFN package. *See* Very-thin quad-flat-no-lead package (VQFN package)
VRMs. *See* Voltage regulating modules (VRMs)

W

Wafer process simulation, 372–373
Wafer warpage. *See also* Package warpage
  analysis and optimization, 377–378
  simulation and validation, 373–377, 376f
Wafer-level packages (WLPs), 3–4
Waste electrical and electronic equipment (WEEE), 151–152
Water vapor, constitutive relations for, 249–250
Wavelength, 15–16
WEEE. *See* Waste electrical and electronic equipment (WEEE)
Weibull plots of cycle, 176
Williams–Landel–Ferry Equation (WLF Equation), 321
  shift function, 139–140
WLF Equation.
  See Williams–Landel–Ferry Equation (WLF Equation)

WLPs. *See* Wafer-level packages (WLPs)
Work hardening. *See* Nonlinear strain
  hardening

Y
Yield criterion, 142

Yield stress, 308
Young's modulus. *See* Elastic modulus

Z
Z factor, 118